Studies in Mechanobiology, Tissue Engineering and Biomaterials

Volume 3

Series Editor
Amit Gefen, Ramat Aviv, Israel

Further volumes of this series can be found on our homepage:
http://www.springer.com/series/8415

Lynne E. Bilston
Editor

Neural Tissue Biomechanics

Springer

Lynne E. Bilston
Neuroscience Research Australia
Barker Street
Randwick, NSW 2031
Australia
e-mail: L.Bilston@neura.edu.au

ISSN 1868-2006 e-ISSN 1868-2014
ISBN 978-3-642-13889-8 e-ISBN 978-3-642-13890-4
DOI 10.1007/978-3-642-13890-4
Springer Heidelberg Dordrecht London New York

Cover design: WMXDesign GmbH, Heidelberg

Printed on acid-free paper

Springer is part of Springer Science+Business Media (www.springer.com)

Preface

Biomechanics of the nervous system have been studied for decades, and interest in the role of mechanical forces in the growth, development, physiological functioning, and injury of the nervous system has exploded in the last few years.

This book aims to provide a comprehensive overview of the current scientific understanding of the biomechanical behaviour of the nervous system, from the macroscopic (whole organism) level, down to the molecular level where transduction of mechanical signals occurs. This includes both the intrinsic biomechanical properties of the neural tissues and associated structures, and also a picture of how these tissues respond physiologically and/or pathologically to mechanical forces.

In modern biomechanics, a broad range of techniques are used to study the nervous system, ranging from traditional engineering techniques, such as mechanical testing, mathematical and computational modelling, to emerging imaging techniques at the interface between medical science and physics, such as magnetic resonance elastography. Biological techniques also range from advanced microscopy, such as atomic force microscopy, to cellular patch clamp biophysics, and electrophysiology. Biomechanics is a very multidisciplinary research area, and applications of many of these techniques from multiple domains to specific neural biomechanics problems are described in the chapters of this book.

Biomechanical studies of the nervous system encompass both fundamental research, such as that into the mechanisms of mechanotransduction, and also applied research, aimed at solving a particular clinical problem, as well as many topics in between. This book covers a range of applications along this continuum, aiming to expose the reader to both the breadth and depth of neural biomechanics.

I would like to thank all of the authors, all busy researchers and leaders in their fields, as well as the peer reviewers who contributed their time and expertise to the book.

Sydney, Australia Lynne E. Bilston

Contents

An Introduction to the Biomechanics of the Nervous System

Lynne E. Bilston

Abstract The nervous system is divided into the central nervous system, which includes the brain and spinal cord, and the peripheral nervous system which includes the nerves and associated structures that carry signals to and from the periphery. This chapter describes the organisation and anatomy of the nervous system, and provides an overview of the role played by mechanical factors in nervous system function and dysfunction.

1 Organisation of the Nervous System

The nervous system is usually classified into the central nervous system, composed of the brain and spinal cord, and the peripheral nervous system, which includes the peripheral nerves and ganglia that lie outside the brain and spinal cord. In addition, the central nervous system is protected by the meninges (pia mater, arachnoid mater, dura mater), and the peripheral nerves are protected by neural sheaths. At the distal ends of the efferent peripheral nerves are the muscles and other tissues, and at the ends of the afferent nerves are a broad range of receptors that sense the status of the peripheral tissues and report these to the brain. Some of these receptors are sensitive to chemical changes (chemoreceptors) and some to mechanical stimuli (mechanoreceptors).

L. E. Bilston (✉)
Neuroscience Research Australia and Prince of Wales Clinical School,
University of NSW, Sydney, Australia
e-mail: L.Bilston@neura.edu.au

Stud Mechanobiol Tissue Eng Biomater (2011) 3: 1–9
DOI: 10.1007/8415_2010_61

Published Online: 28 January 2011

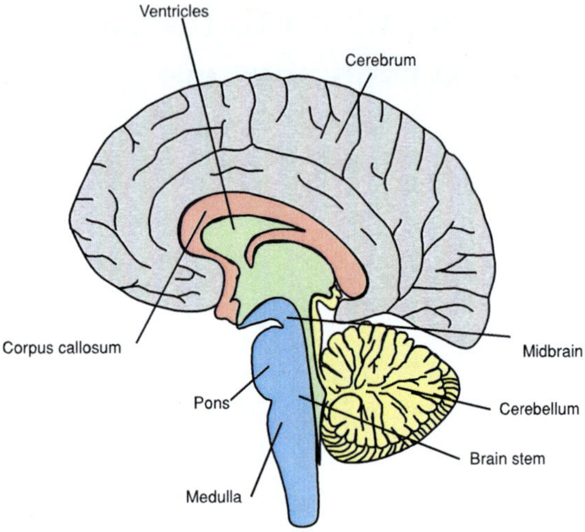

Fig. 1 The human brain in the mid-sagittal plane, showing the major structures. (© L. Bilston, used by permission)

1.1 The Brain

Brain tissue is made up of nerve cells (neurons), and supporting cells (astrocytes, glia), and is richly supplied with blood. Brain tissue can be either white or grey matter, with grey matter consisting of unmyelinated nerve fibres (axons) and cell bodies, and white matter consisting of myelinated axons. The white colour comes from the myelinated sheath created by oligodendrocytes surrounding these axons. The brain is divided into the cerebral hemispheres, the brainstem (midbrain, pons, and medulla), and the cerebellum. The left and right hemispheres are separate, joined through a dense white matter bundle called the corpus callosum. Deep in the centre of the brain, are fluid-filled spaces known as the ventricles. Cerebrospinal fluid is manufactured within the ventricles and flows out to surround the brain and spinal cord, providing some cushioning and buoyancy for the brain within the skull. The arrangement of these structures is outlined in Fig. 1.

1.2 The Spinal Cord

The spinal cord is a long structure, with two pairs of nerve roots exiting at each vertebral level. At the distal end, the spinal cord tapers into the cauda equina, a long bundle of nerve roots that exit through the lumbar and sacral vertebrae. Like the brain, the spinal cord also has both white and grey matter, with the white matter tending to be around the periphery, while the grey matter forms an h-shaped core within the spinal cord (Fig. 2).

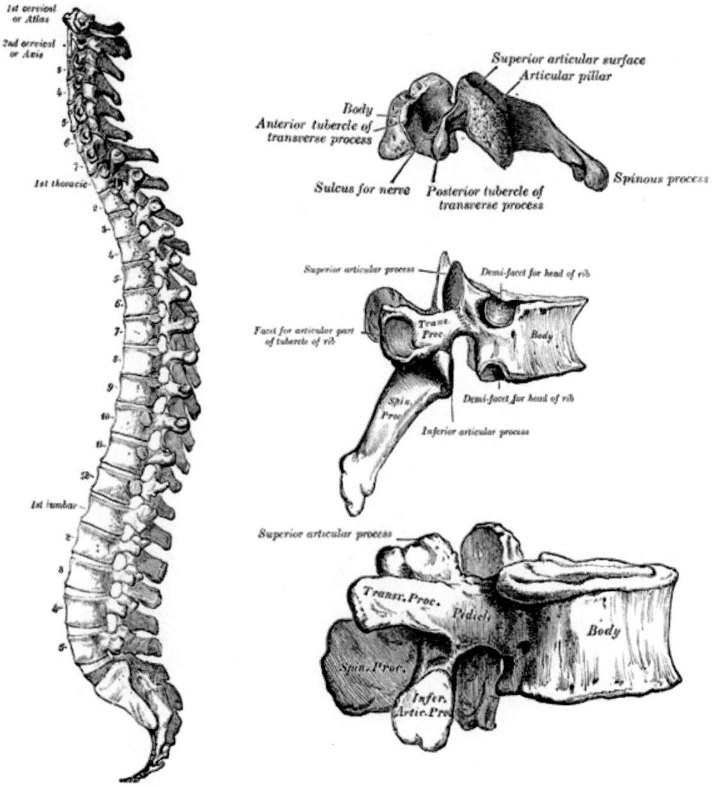

Fig. 2 The spinal column (*left*), and typical cervical (*top right*), thoracic (*middle right*) and lumbar vertebrae (*bottom, right*) (from Gray's Anatomy, 1918)

1.3 The Meninges

Both the brain and the spinal cord are enclosed in three layers of fibrous tissue, with the fine delicate pia mater being adhered to the surface of the brain and spinal cord. This is enclosed by the sub-arachnoid space, which is filled with cerebrospinal fluid and a fine network of connective tissue, and surrounded by the arachnoid mater. Immediately over the top of the arachnoid mater lies the dura mater—a tough protective membrane of collagenous tissue. In the skull, the dura mater is adhered to the skull and passes down into the gap between the hemispheres of the brain to join the tough fibrous falx cerebri, and at the base of the brain it also joins the tentorium cerebelli, that separates the cerebrum from the cerebellum. In the spinal region the dura is free to move within the vertebral canal (Fig. 3).

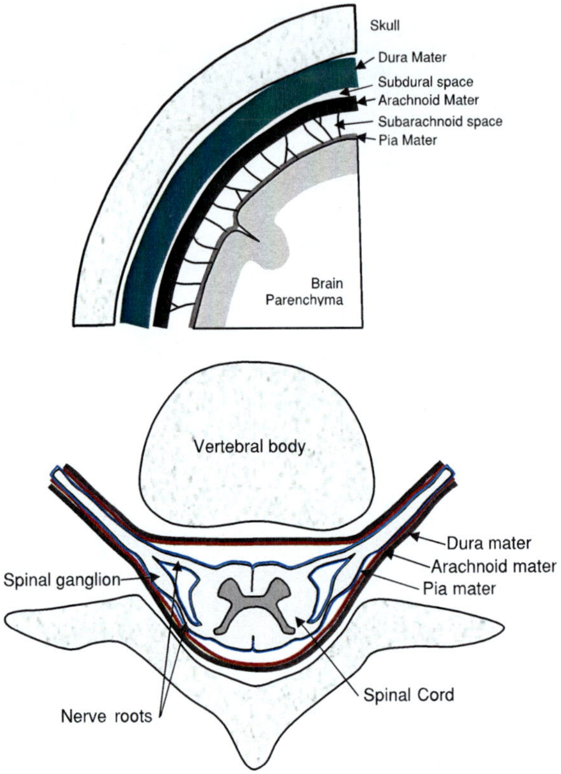

Fig. 3 Schematic of the meninges in the cranium (*top*) and in the spinal column

1.4 The Peripheral Nervous System

The peripheral nervous system consists of all the nerves that connect the spinal cord and brain to the target organs and tissues. The peripheral nervous system is divided into the autonomic nervous system, that regulates various 'automatic' body functions such as cardiac and respiratory function, temperature and digestion, to name but a few, and the somatic nervous system, that senses stimuli and controls the voluntarily activated muscles (Fig. 4).

1.5 Receptors

While peripheral nerves themselves can transduce mechanical stimuli, particularly at their distal ends, they are commonly conduits for signals arising at sensory receptors to the CNS. While many receptors detect chemical changes (e.g. chemoreceptors that detect oxygen or CO_2 concentrations in the blood for feedback

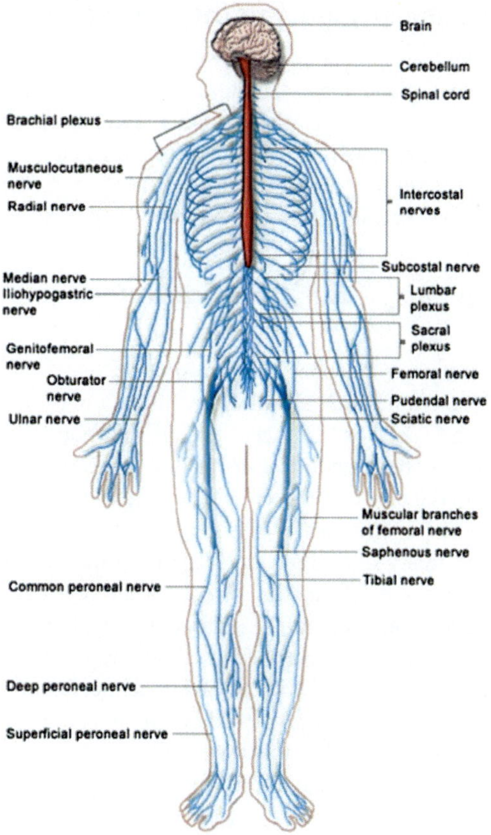

Fig. 4 The peripheral nervous system (from http://en.wikipedia.org/wiki/File:Nervous_system_diagram.png)

into the cardiac and respiratory system control), many detect mechanical signals. They are known as mechanoreceptors. These include the receptors for our sense of touch, proprioception, and pain. The Meissner corpuscles, located superficially in skin, detect light touch. Pacinian corpuscles, located deeper in the skin and elsewhere in the body, detect deep pressure and vibration. Merkel cells also detect light pressure, and are located in the dermis of the skin. The Ruffini corpuscle, located in the dermis, detects skin stretch. Muscle spindles, located in muscle belly tissue, detect muscle stretch. Golgi tendon organs, located at the muscle–tendon interface, senses muscle force. Baroreceptors are located in blood vessels, and detect blood pressure. The hair cells of the cochlea detect vibration of the hairs in the fluid of the cochlea, thus creating the sense of sound. These are discussed in more detail in the mechanotransduction chapter.

2 Mechanical Inputs to the Nervous System

Both normal (physiological) inputs and injuries mechanical inputs act on the body in a typical humans lifetime. Physiological mechanical inputs are an essential part of normal function, allowing the body to self-regulate many autonomic functions, as well as providing sensory input. In most cases, injurious loads to the body do not result in neural injury, because the nervous system is protected from much of this loading.

2.1 Physiological Mechanical Inputs

As described in the previous section, mechanical inputs to the body are part of normal function. Such physiological inputs may be essential for normal function, such as the need for loading of muscle and bone tissue to prevent atrophy leading to muscle wasting and diseases such as osteoporosis.

Within the nervous system, physiological inputs include mechanical stimuli that are detected by the sensory system and perceived as touch, pressure or stretch. There are many other mechanical signals that are not consciously perceived, but are part of essential functions, such as the regulation of blood pressure based on baroreceptors located in the aortic arch and carotid sinus. There are also low pressure baroreceptors located in the venous system and heart wall that help to regulate blood volume. Gastrointestinal mechanoreceptors are involved in regulation of digestion, from the detection of stomach wall stretch after food consumption to motility in the lower GI tract. A range of skin, joint, muscle and vestibular responses to the local mechanical environment are integrated to give us a sense of where our body is in space, known as proprioception.

In addition, the nervous system needs to be able to function in the face of normal levels of mechanical loading, such as movement of the limbs and articulation of the joints. Both the central and peripheral nervous systems have protective mechanisms to allow for this. The brain's motion inside the skull is controlled by the meninges, the falx cerebri and tentorium cerebelli. It also floats in a bath of cerebrospinal fluid to protect it from minor impacts. The spinal cord can move up and down within the vertebral canal to a certain degree, together with the surrounding meninges. Its motion is constrained by the denticulate ligaments, which prevent excessive motion and thus limit local stretch of the nerve roots.

The peripheral nerves are protected by the dense connective tissue epineurium, which surrounds a nerve bundle. This provides some mechanical support and strength to protect the nerve tissue from external mechanical loads, while allowing the nerve to slide as the body moves.

2.2 Injurious Mechanical Inputs

When the load imparted to the nervous system exceeds its tolerance to mechanical forces, then injury can occur. The extent of injury to neural tissue depends on both the amount of load (or deformation) and the rate at which the tissue is loaded. Neural dysfunction can be either temporary (e.g. temporary loss of consciousness) or permanent (e.g. quadriplegia). The capacity of the nervous system to regenerate depends on where the injury occurs, with the peripheral nervous system having greater capacity to heal and regenerate than the central nervous system.

In trauma, neural injury occurs at high loading rates, and often involves large amounts of deformation. Brain injury, spinal injury and peripheral nerve injuries are discussed further in later chapters.

Chronic loading of neural tissue can also result in neural damage. This is often due to compression of the neural tissue. Examples include compression of brain tissue by tumours, or enlarged ventricles in hydrocephalus, compression of spinal nerve roots in degenerative spinal disorders, and loading of the median nerve in the wrist in carpal tunnel syndrome. In some of these cases, it is likely that the injury mechanism is a mixture of mechanical compression and ischemia due to a compromised blood supply to the tissue. Biomechanics of structural brain disorders are discussed in a later chapter.

3 Importance of Mechanics in the Nervous System

Finally, in this chapter, we discuss why studying biomechanics of the nervous system is important for understanding normal function and injury prevention and treatment.

3.1 Biomechanics of Neural Development

Biomechanics play a central role in the nervous system, from the early phases of growth in utero. Brain development, including the development of the gyri and sulci (folds and recesses) in the surface of the brain [1–3], have been hypothesised to be influenced by mechanical loading in the developing tissue. Mechanical loading may influence neural stem cell differentiation into specific cell types [4]. CNS axons have also been shown to grow more rapidly when exposed to elongational loads [5]. These mechanical factors in neural development are not well understood, and there is considerable interest in understanding these factors, since abnormal brain development, including brain morphogenesis is thought to underlie several serious neurological and psychiatric conditions. In addition, understanding how mechanical forces can influence neural cell growth might provide important information on how to repair damaged neural tissue.

3.2 Nervous System Diseases with a Mechanical Component

There are a broad range of neurological disorders that include a mechanical component as part of the pathophysiology. In the brain, the so-called structural neurological disorders, including hydrocephalus (where the ventricles of the brain enlarge and the brain tissue becomes compressed), pseudotumour cerebri (also known as intracranial hypertension), and slit ventricle syndrome (where the ventricles collapse after insertion of shunt to divert excess CSF from the ventricles), are all thought to involve mechanical processes, possibly in conjunction with other physiological changes. Brain tumours compress the surrounding tissue, and/or result in elevated intracranial pressure, in addition to infiltrating healthy tissue. Subdural haematoma, which is a large collection of blood in the subdural space that often accompanies head injury, causes damage in part by compressing the brain.

Spinal disorders with a strong mechanical component include syringomyelia, where fluid accumulates in the central canal or parenchyma of the spinal cord, compressing the spinal cord tissue from the inside. Spinal tumours, stenosis of the spinal canal, and degenerative spinal disease can also compress spinal cord tissue causing neurological dysfunction.

Peripheral nerves are also susceptible to mechanically related disorders. These include nerve root compression, by a herniated spinal disk for example, and carpal tunnel syndrome, where the median nerve is compressed at the wrist. The so-called "Saturday night palsy" is arm weakness caused by compression of the radial nerve near the brachial plexus, often due to falling asleep with the arm compressed by a chair back or similar firm object. There are several anatomical locations where a peripheral nerve may be entrapped, giving rise to neurological symptoms due to local compression, stretch or tethering of the nerve.

3.3 Neural Injury

Traumatic injury is perhaps one of the most obvious of mechanically mediated disorders of the nervous system. Brain injury has been a major health problem in most developed countries for decades, and was one of the issues that motivated some of the earliest studies of the mechanics of the brain. Brain injury is one of the most common causes of serious traumatic injury and death, and occurs most commonly as a result of road trauma. Spinal cord injury is much less common, but also devastating and a significant public health concern along with brain injury, due to the high ongoing costs of care for the neurologically injured patient.

Peripheral nerve injuries are also significant, and can occur in both high energy trauma (e.g. road crashes) and lower-energy events such as sports. Injuries to the brachial plexus in adults typically occur during trauma to the shoulder, but they also occur during childbirth in some cases due to loading of the shoulder joints in the birth process.

Understanding of this broad range of mechanically mediated neurological disorders requires interdisciplinary research into biomechanics, tissue viscoelasticity, and neurophysiology. This book aims to summarise the current state-of the-art across a broad range of these areas.

References

1. Geng, G., Johnston, L.A., Yan, E.: Biomechanisms for modelling cerebral cortical folding. Med. Imag. Anal. **13**, 920–930 (2009)
2. Richman, D., Stewart, R., Hutchinson, J., et al.: Mechanical model of brain convolutional development. Science **189**, 18–21 (1975)
3. Xu, G., Knutsen, A., Dikranian, K., et al.: Axons pull on the brain, but tension does not drive cortical folding. J. Biomech. Eng. **132**, 071013 (2010)
4. Tang, W.: Biomechanics at the micro scale. In: Proceedings of IEEE, p. 5 (2005)
5. Smith, D., Wolf, J., Meaney, D.: A new strategy to produce sustained growth of central nervous system axons: continuous mechanical tension. Tissue Eng. **7**, 131–139 (2001)

Brain Tissue Mechanical Properties

Lynne E. Bilston

Abstract Brain tissue is a complex multiphase material. Its mechanical behaviour arises from the inherent viscoelastic nature of the neural and supporting cellular components and their interaction with the vasculature and interstitial fluid. Brain behaves as a very soft, highly nonlinear viscoelastic solid. It has been mechanically characterised in shear, tension and compression, but despite decades of research into its behaviour, there is still considerable debate about its precise mechanical properties. This chapter aims to present the most reliable mechanical data for brain tissue, and to explain how this behaviour is affect by age, disease and a range of experimental conditions under which mechanical measurements are made.

1 Introduction

Brain tissue is a structurally and mechanically complex material. It is a fluid-bathed multiphase material, made up of several different cell types, and is densely vascularized. Mechanically, it is very soft, and its mechanical response is strongly dependent on the magnitude, type and rate of loading applied to it. More specifically, it is a highly nonlinear, strain-rate sensitive viscoelastic material. In this chapter, the mechanical behaviour of brain tissue is reviewed, and some of the challenges of measuring brain tissue mechanical properties are outlined. More detailed discussion of the methodological issues in brain testing is published elsewhere [5, 10, 25].

L. E. Bilston (✉)
Neuroscience Research Australia and Prince of Wales Clinical School,
University of NSW, Sydney, Australia
e-mail: l.bilston@neura.edu.au

Stud Mechanobiol Tissue Eng Biomater (2011) 3: 11–24
DOI: 10.1007/8415_2010_36

Published Online: 15 December 2010

2 History

Although there is one study from the 1950s [20], brain tissue properties began to be of major interest in the 1960s onward, largely motivated by the need to understand mechanisms underlying traumatic brain injury, which was (and remains) a major public health problem. In the 1960s, Dodgson studied the creep properties of brain tissue [15], followed by the thesis of Koeneman [28] who conducted creep and oscillatory studies on animal brain tissue. There were then several studies attempting to characterise the large deformation viscoelastic properties of brain, whose results are difficult to interpret due to a range of methodological issues [17–19, 21, 30, 41, 46].

Even among these early studies there are large variations in the reported mechanical properties of brain tissue, a problem that has continued up to the current day. Many of the studies reported in the literature suffer from methodological flaws that limit the validity of the data. Common errors include the extraction of "linear viscoelastic" parameters from experiments conducted at large deformations (see below for further explanation), the use of tissue many hours post-mortem, when substantial tissue property changes have occurred, and dehydration of samples, or use of osmotically inappropriate bathing solutions, so that tissue swelling or dehydration occurs, altering the measured tissue response. This chapter does not attempt to thoroughly review all of these older studies, focussing instead on those studies that are most robustly conducted, making the data more reliable.

3 In vitro Behaviour

Many studies of brain tissue mechanical behaviour have used excised tissue samples, or in vitro tissue. These are, necessarily, dead tissue, in which the effects of perfusion pressure and active metabolism are not present. Such tissues may also have undergone changes after death of the donor (whether human or animal) including microbiological and/or enzymatic degradation. These are well-known to affect tissue mechanical properties [5, 10, 25], and the most accurate data is obtained on very fresh samples, tested within a few hours of death. Even then, the properties are likely to be somewhat different from live tissue (e.g. [47]).

Another key issue is the loading regime used to test the tissue. In the wider rheology field, complex materials are typically characterised first in the linear viscoelastic regime, where the extracted mechanical properties, such as the shear modulus or elastic modulus, are independent of the magnitude of the strains applied. In practice, this occurs below a specific strain, called the linear viscoelastic strain limit. The linear viscoelastic strain limit is estimated by a series of tests in which the strain is slowly increased until the estimates of material functions begin to change. From test data at strains below this limit, the linear viscoelastic material functions can be derived. These include the relaxation modulus,

G(t), the storage and loss moduli as functions of frequency, $G'(\omega)$, $G''(\omega)$ respectively, and the creep modulus, J(t). These functions are commonly used as building blocks for constitutive models that describe more complex tissue behaviour at larger strains, and it is thus important to obtain accurate estimates of these functions.

For brain tissue, the linear viscoelastic strain limit is very small, of the order of 0.1–0.3% [6, 35]. This means that for most mechanical processes of practical interest, such as brain injury or neurosurgery, brain behaves as a nonlinear viscoelastic material. It also means that much of the brain mechanical properties data in the literature, which uses linear viscoelastic methods to extract material functions from test conducted outside of the linear viscoelastic regime, is flawed. Fortunately, several more recent studies have used appropriate methods to obtain linear viscoelastic properties for brain tissue.

4 Linear Viscoelastic Properties

The linear viscoelastic strain limit for brain tissue has been estimated in shear tests, both from oscillatory shear tests, in which the strain is slowly increased while holding the oscillation frequency constant [6, 35], and from relaxation tests, where tests have been conducted at increasing strains to determine when the shape of the relaxation curve changes [6]. The best conducted studies of fresh tissue samples estimate the linear viscoelastic strain limit to be less than 0.5% strain [6, 35]. Similar tests have not been conducted in tension or compression, and thus the linear viscoelastic properties in tension and compression are not well characterized, nor is it known whether the linear viscoelastic limit in tension and compression is different to shear. One indentation study in compression [19] suggested that for very small sinusoidal indentations (25 μm), brain behaved linearly, but behaved nonlinearly for larger indentations (300 μm). The precise compressive strains are not known in this study, and indentation is not a purely compressive loading method, as shear strains are also created around the indentation site.

4.1 Shear

The linear viscoelastic storage and loss moduli for brain tissue increase with frequency, as shown in Fig. 1. That is, it increases in stiffness with increasing loading rate. The shear moduli are of the order of a few kilopascals, indicating that brain is indeed a very soft solid, at least at the loading rates measured.

The linear viscoelastic relaxation modulus for brain tissue shows power-law type relaxation (see Fig. 2).

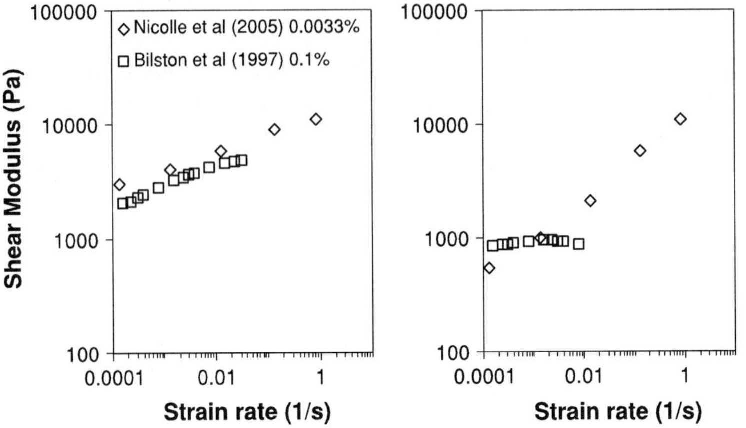

Fig. 1 Shear storage and loss moduli for brain tissue in the linear viscoelastic regime

5 Nonlinear Viscoelastic Properties

Once the strain exceeds the linear viscoelastic strain limit, brain tissue behaves in a complex nonlinear manner. It retains the strain-rate sensitivity seen in the linear viscoelastic regime, but the apparent stiffness depends on both the strain and the loading type.

5.1 Shear

In shear, once shear strains exceed the linear viscoelastic limit, brain shows shear thinning behaviour. That is, it appears softer at larger strains. This is shown in

Fig. 2 Linear viscoelastic shear relaxation modulus for brain tissue [7]

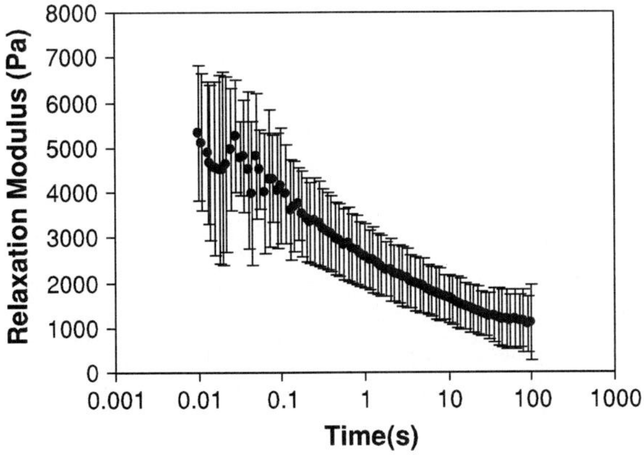

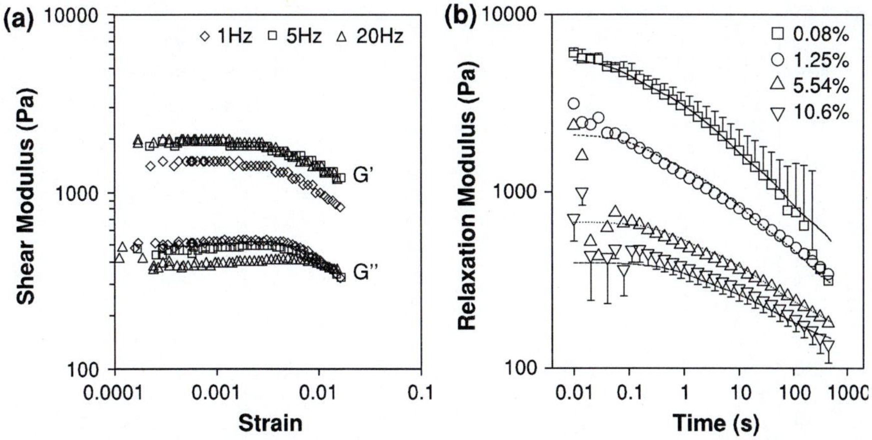

Fig. 3 Shear thinning behaviour of brain tissue at strains in the nonlinear regime. **a** Storage and loss moduli. **b** Relaxation modulus. Data from Bilston et al. [7]

Fig. 3, where for increasing strains, the apparent storage and loss moduli, and the relaxation modulus appear to decrease with increasing strain.

In large strain oscillatory shear tests, the stress response to a sinusoidal input strain is non-sinusoidal, and for correct analysis of this type of test, this needs to be taken into account (e.g. [26, 48]), although this is often not done in the literature.

Relaxation tests have been often conducted on brain tissue [3, 7, 8, 40, 42], in order to circumvent this difficulty, and much of the data in the literature is summarised in Fig. 4.

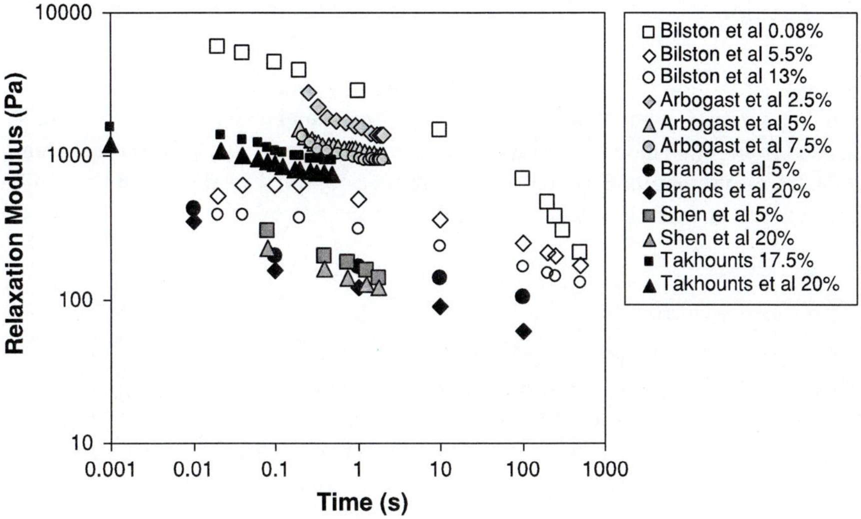

Fig. 4 Relaxation behaviour of brain tissue at large strains

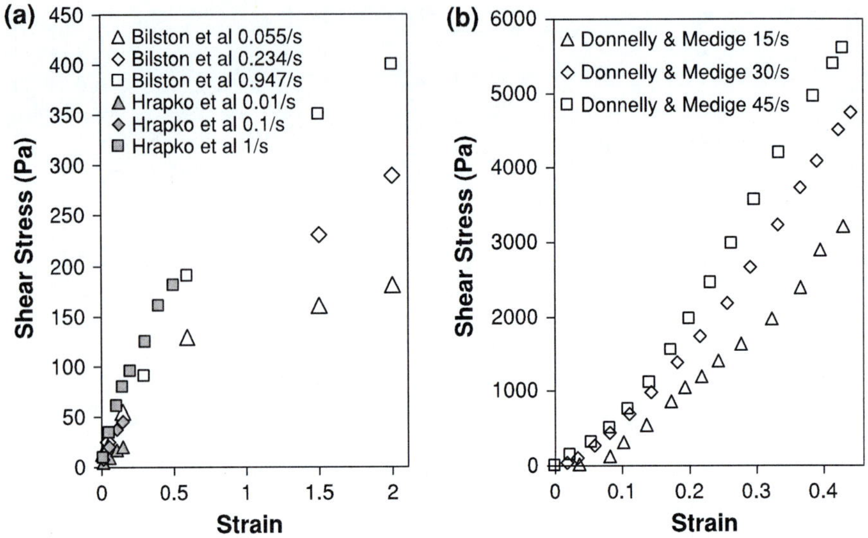

Fig. 5 Constant strain rate response of brain tissue. **a** loading rates of up to 1/s **b** High loading rates

Constant strain rate tests have also shed light on the behaviour of brain tissue at large strains. These confirm strong strain-rate sensitivity, and suggest that in shear, brain tissue can sustain quite large strains without macroscopic structural failure (Fig. 5).

5.2 Tension

Relatively few studies have tested brain tissue in tension, due to the difficulty of gripping samples without damaging them. The available data [33, 45] is summarised in Fig. 6. These studies note that strain-rate sensitivity is observed, along with strain-softening that is consistent with that observed in shear. Failure limits seem to be somewhat lower in tension than shear, however.

5.3 Compression

There have been several studies of brain tissue in compression at large strains, over a wide range of strain rates [9, 18, 32, 43]. These include constant loading rate compression tests (Fig. 7) and relaxation tests (Fig. 8). These studies have confirmed the strong strain-rate sensitivity of brain tissue, with brain being substantially stiffer at higher loading rates. Onset of yield or failure from the available data is in the range of 30–50% in compression.

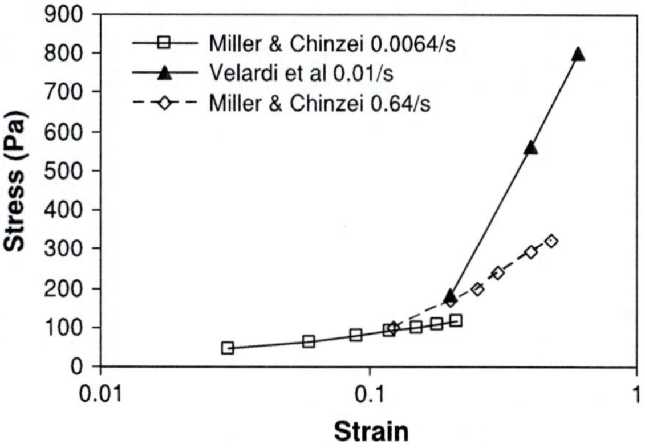

Fig. 6 Tensile properties of brain tissue at large strains

Indentation tests have also been conducted, and these are usually analyzed using a finite element model to obtain material parameters for a particular assumed constitutive model. An example of this is Miller [31].

Relaxation tests in compression have all been conducted well above the linear viscoelastic limit [9, 18, 32, 43], but they show a consistent relaxation modulus shape, indicating that the reduced relaxation modulus (where G(t) is normalised by its peak value) is not affected by strain or strain rate. Whether this is true at very low strains is not known, and given that in shear the small strain relaxation is different to the large strain relaxation, this should be kept in mind when utilising this data. Only relatively short relaxation tests have been conducted in compression (up to 5 s) and none of the data appears to have reached a plateau within this time, which is consistent with the shear relaxation which continued over hundreds of seconds.

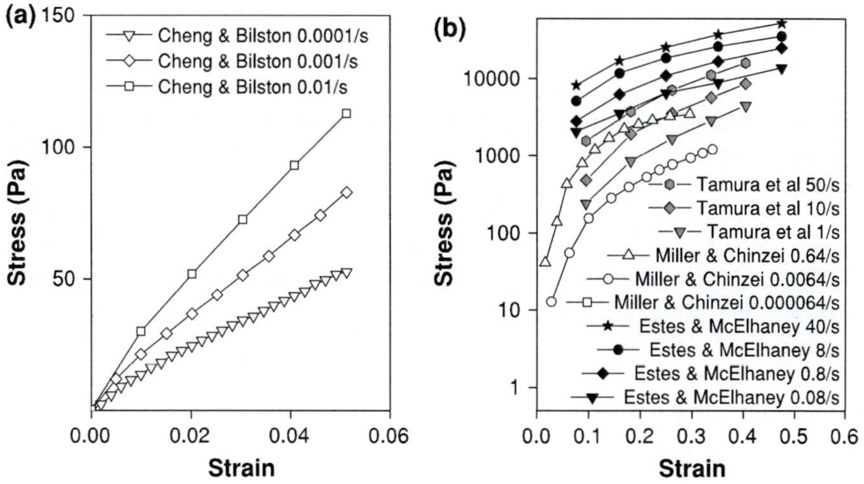

Fig. 7 Compressive properties of brain tissue. **a** small strains. **b** large strains

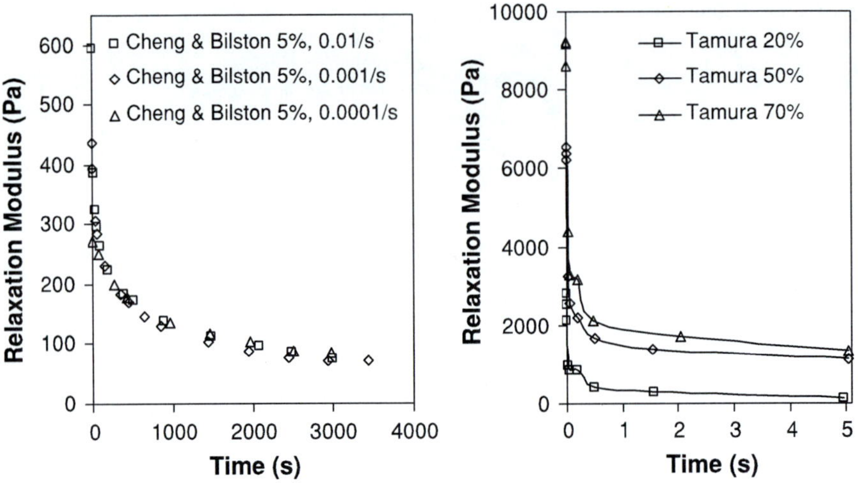

Fig. 8 Relaxation properties of brain tissue in compression. Note that the high strain and strain rate tests of Tamura et al. [43] were conducted at 50/s

Note that very few of the studies conducted at high loading rates have accounted for sample inertia, which may be significant at high loading rates.

5.4 Other Properties

For some mechanical processes, the bulk behaviour of brain tissue is not sufficient to describe the condition of interest. For example, if one is interested in hydrocephalus or brain edema, then it is important to understand how fluid moves within the brain parenchyma itself, as well as how the bulk tissue responds to mechanical loading. There have been a few studies that have attempted to characterize the permeability of brain tissue by fitting multiphase models to the response of brain tissue in compression. These include Chinzei and Miller [11], who concluded that a simple poroelastic model could not capture the strain rate sensitivity of brain tissue, and Cheng and Bilston [9], who used a more complex poroviscoelastic model to include this viscoelastic response. They obtained an estimate of brain permeability of $4.08 \times 10^{-12} m^4/Ns$, with an long-term elastic modulus of 350 Pa. This permeability is very low, suggesting that there is substantial resistance to fluid flow through brain parenchyma.

6 In vivo Properties

A small, but increasing, number of studies have been conducted in vivo, that is, on living subjects. These have been conducted on both animals and humans. While some studies show that there are significant changes at the time of death, due to loss of perfusion pressure [19, 30, 47], this has not been universally observed [24]. Longer term changes in tissue properties, attributed to tissue degradation have been observed in animal studies with the largest changes occurring beyond 6 h of death [22, 24]. These tend to increase the measured stiffness of the tissue.

Recently, a novel technique for estimating the linear viscoelastic properties of brain tissue, magnetic resonance elastography (MRE), has been developed. This technique uses a wave propagation method to estimate either the elastic or viscoelastic properties of brain tissue. Very small strains are involved, since displacements are only a few microns, and thus linear viscoelastic properties are able to be extracted in vivo. There are a variety of different methods of analysing the raw data, which comes in the form of time series of displacement maps of the shear waves propagating through the brain in three orthogonal coordinate directions, but most involve solving the equations for wave propagation in a linear elastic or linear viscoelastic media. An example of brain storage and loss moduli maps at a vibration frequency of 80 Hz are shown in Fig. 9. Loading rate effects can be investigated by use of different vibration frequencies within limits imposed by high frequency attenuation of the waves at the upper end, and ability to reliably estimate waves at the lower end (Fig. 10). Similar to rheometric test results, Klatt et al. [27] have shown using MRE shows that brain shear modulus increases its shear modulus with loading frequency.

This technique not only has the potential to allow measurement of human brain properties in vivo, it can also be used to estimate brain property changes with age and disease. There has been one study that showed the brain appears to soften with increasing age in adults [38], and a pilot study has demonstrated that tumours can be differentiated from healthy brain tissue on the basis of their stiffness using MRE [51]. Multiple sclerosis patients have also been shown to have lower brain shear moduli than age matched healthy volunteers [50].

MRE has also been used in animal studies, including one that looked at the effects of interstitial tissue pressure and time post-mortem on tissue behaviour [47], and rat and mouse brain properties [4, 29].

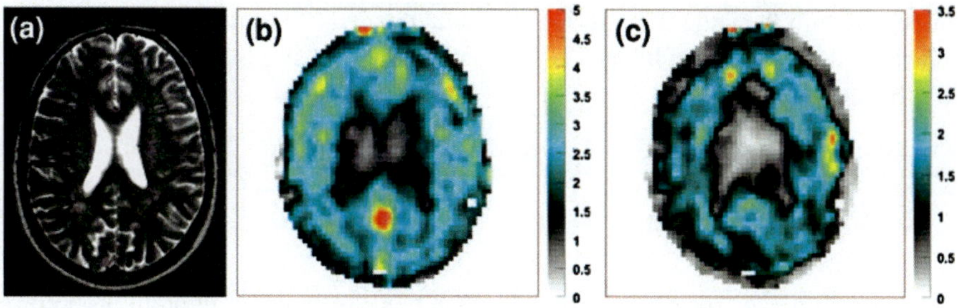

Fig. 9 Magnetic resonance elastography of human brain tissue. **a** Anatomy, **b** Storage modulus **c** Loss modulus, in kPa. Figure courtesy of M. Green

Fig. 10 Magnetic resonance elastography measurements of storage and loss moduli for brain in vivo

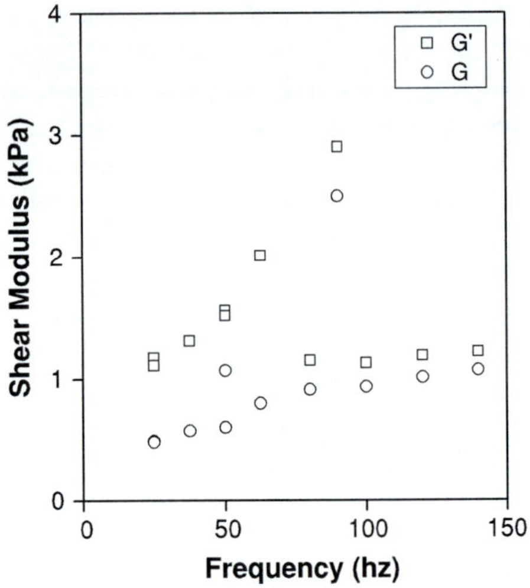

Intraoperative measurements of brain properties in vivo have also been made, both in humans and animals. In humans, Schiavone et al. [39] used an aspiration method during neurosurgical procedures, to estimate brain properties using an hyperelastic model. In animals, indentation tests have been performed, including in macaques [19], and pigs [24, 34]. The complex loading conditions require the use of a finite element simulation together with an assumed constitutive model to estimate brain tissue properties.

7 Age

The effect of age on brain tissue properties has only been occasionally studied, and the results are conflicting. One study suggests that the shear moduli increase with age during development [44] while other studies suggest the opposite, at least at large strains [23, 37]. Further studies are required to determine this over a wider range of ages and loading types, and to match the observed changes in mechanical properties with the underlying tissue structural changes. In older adults, Sack et al. showed a decrease in the linear viscoelastic modulus of in vivo brain tissue using MR Elastography [38].

8 Regional Variations

There is still debate over the degree to which there is variation in brain properties with location in the brain and also whether brain tissue is mechanically anisotropic to a significant degree. Diffusion tensor imaging clearly shows that at least the major white matter tracts in the brain are anisotropic with respect to the diffusion of water. That is, water molecules can diffuse more easily along the fibre tracts than radially within them. Only a few studies have examined mechanical anisotropy, with some concluding that white matter is largely homogeneous and isotropic [1, 35, 41] while others suggest that some regions are anisotropic, with the fibre direction approximately twice as stiff as the radial direction (corpus callosum [37], brainstem [2, 37]).

Regional properties have also only been studied occasionally, and there is also some inconsistency in the reported results. The brainstem is thought to be stiffer than the cerebral cortex [1], and small differences have been reported between the corona radiata, corpus callosum and cerebral grey matter [37]. Another study, however, found little difference between the midbrain and corpus callosum [16].

9 Disease

Several diseases are known to affect the mechanical properties of tissues, and surgeons have long known that abnormal tissue can often be distinguished from normal tissue due to differences in texture or stiffness. However, few studies have characterised such mechanical changes in neural tissues. New technologies such as MR elastography may allow mechanical measurements to be made in vivo, and this might usefully provide information to help with diagnosis and monitoring of treatments.

Cancerous tissue in accessible parts of the body (e.g. the breast) is often initially detected due to changes in stiffness compared to the surrounding healthy tissue. There have been very few studies of neural tissue cancer mechanics. Some studies have focussed on the mechanical effects of tumours on surrounding tissues [12], mechanical effects during surgery [49], or mechanical effects on tumours themselves (e.g. [14]). One study, using MR elastography, has concluded that tumours in the brain are substantially stiffer than surrounding tissues, although the absolute values obtained for both normal and cancerous tissue are suspect [51]. This latter finding needs to be verified in other studies.

Degenerative or demyelinating diseases have also been shown to alter brain mechanical properties. Wuerfel et al. have shown that in multiple sclerosis, there is a small (13%) but significant decrease in the shear modulus of the whole brain [50], but this was not related to more usual measures of disease severity. In an animal model of Alzheimer's disease, it was seen that the anisotropic and viscoelastic shear modulus of corpus callosum were decreased [29].

Hydrocephalus is a condition where fluid accumulates in the ventricles of the brain and they become enlarged. While there have been no direct measurements of changes in brain tissue mechanical properties, several researchers have suggested that changes in brain stiffness may influence the development of hydrocephalus (e.g. [36]), or may occur as a result of compression of brain tissue by fluid pressure in the ventricles. This has largely been assessed through clinical measures of brain ‘compliance’ [13], but not through direct mechanical measurements to date.

10 Conclusions

Brain tissue mechanical behaviour has a long and complex history, with many methodologically flawed studies appearing in the literature. More recent studies using rigorous rheological methods and in vivo biomechanical imaging technologies have given us greater understanding of the complex nonlinear viscoelastic behaviour of this very soft, incompressible solid. These techniques have great promise for better understanding how brain tissue mechanical behaviour changes as humans develop and age, and also how both mechanical loading and disease processes might affect the brain mechanically.

References

1. Arbogast, K.B., Margulies, S.S.: Regional differences in mechanical properties of the porcine central nervous system. In: Proceedings of the 41st Stapp Car Crash Conference, SAE (1997)
2. Arbogast, K.B., Margulies, S.S.: Material characterization of the brainstem from oscillatory shear tests. J. Biomech. **31**(9), 801–807 (1998)
3. Arbogast, K.B., Meaney, D.F., et al.: Biomechanical characterization of the constitutive relationship for the brainstem. In: Proceedings of the 39th Stapp Car Crash Conference, Coronado, CA, SAE (1995)
4. Atay, S.M., Kroenke, C.D., et al.: Measurement of the dynamic shear modulus of mouse brain tissue in vivo by magnetic resonance elastography. J. Biomech. Eng. **130**(2), 021013 (2008)
5. Bilston, L.E., Clarke, E.C., et al.: Brain tissue mechanical properties—making sense of 5 decades of test data. The Pathomechanics of Tissue Injury and Disease, and the Mechanophysiology of Healing. Gefen, A., Kerala, Research Signpost, pp. 1–18 (2008)
6. Bilston, L.E., Liu, Z., et al.: Linear viscoelastic properties of bovine brain tissue in shear. Biorheology **34**(6), 377–385 (1997)
7. Bilston, L.E., Liu, Z., et al.: Large strain behaviour of brain tissue in shear: some experimental data and differential constitutive model. Biorheology **38**(4), 335–345 (2001)
8. Brands, D.W.A., Bovendeerd, P. H. M., et al.: The large shear strain dynamic behaviour of in vitro porcine brain tissue and a silicone gel model material. Stapp Car Crash Conference, SAE (2000)
9. Cheng, S., Bilston, L.E.: Unconfined compression of white matter. J. Biomech. **40**(1), 117–124 (2007)
10. Cheng, S., Clarke, E.C., et al.: Rheological properties of the tissues of the central nervous system: a review. Med. Eng. Phys. **30**(10), 1318–1337 (2008)
11. Chinzei, K., Miller, K.: Compression of swine brain tissue: experiment in vitro. J. Mech. Eng. Lab. **50**(4), 106–115 (1996)
12. Clatz, O., Bondiau, P.-Y., et al.: In silico tumor growth: application to glioblastomas. Medical Image Computing and Computer-Assisted Intervention—MICCAI 2004. In: Barillot, C., Haynor, D.R., Hellier, P. vol. **3217** pp. 337–345. Springer, Berlin (2004)

13. Czosnyka, M., Czosnyka, Z.H., et al.: Age dependence of cerebrospinal pressure-volume compensation in patients with hydrocephalus. J. Neurosurg. **94**(3), 482–486 (2001)
14. Deisboeck, T., Guiot, C.: Surgical impact on brain tumor invasion: a physical perspective. Ann. Surg. Innov. Res. **2**(1), 1 (2008)
15. Dodgson, M.C.H.: Colloidal structure of brain. Biorheology **1**(1), 21–30 (1962)
16. Donnelly, B.R., Medige, J.: Shear properties of human brain tissue. J. Biomech. Eng.**119**(4), 423–432 (1997)
17. Engin, A.E., Wang, H.C.: A mathematical model to determine viscoelastic behavior of in vivo primate brain. J. Biomech. **3**(3), 283–296 (1970)
18. Estes, M.S., McElhaney, J.H.: Response of brain tissue to compressive loading. ASME Paper 70-BHF-13 (1970)
19. Fallenstein, G.T., Hulce, V.D., et al.: Dynamic mechanical properties of human brain tissue. J. Biomech. **2**(3), 217–226 (1969)
20. Franke, E.K.: The response of the human skull to mechanical vibrations. Wright-Patterson Air Force Base, Ohio. WADC Technical Report No. 54–24 (1954)
21. Galford, J.E., McElhaney, J.H.: A viscoelastic study of scalp, brain, and dura. J. Biomech. **3**, 211–221 (1970)
22. Garo, A., Hrapko, M., et al.: Towards a reliable characterisation of the mechanical behaviour of brain tissue: The effects of post-mortem time and sample preparation. Biorheology **44**(1), 51–59 (2007)
23. Gefen, A., Gefen, N., et al.: Age-dependent changes in material properties of the brain and braincase of the rat. J. Neurotrauma **20**(11), 1163–1177 (2003)
24. Gefen, A., Margulies, S.S.: Are in vivo and in situ brain tissues mechanically similar? J. Biomech. **37**(9), 1339–1352 (2004)
25. Hrapko, M., Dommelen, J.A.W.v., et al.: The influence of test conditions on characterization of the mechanical properties of brain tissue. J. Biomech. Eng. **130**(3), 031003 (2008)
26. Hyun, K., Kim, S.H., et al.: Large amplitude oscillatory shear as a way to classify the complex fluids. J. Non-Newtonian Fluid Mech. **107**(1–3), 51–65 (2002)
27. Klatt, D., Hamhaber, U., et al.: Noninvasive assessment of the rheological behavior of human organs using multifrequency MR elastography: a study of brain and liver viscoelasticity. Phys. Med. Biol. **52**(24):7281–7289 (2007)
28. Koeneman, J.-B.: Viscoelastic properties of brain tissue. M.Sc. thesis, Case Institute of Technology (1966)
29. Larrat, B., Chan, Q.C., et al.: Anisotropic viscoelastic properties of the corpus callosum—application of high-resolution 3D MR-elastography to an Alzheimer mouse model. Ultrasonics symposium IEEE (2007)
30. Metz, H., McElhaney, J., et al.: A comparison of the elasticity of live, dead, and fixed brain tissue. J. Biomech. **3**, 453–458 (1970)
31. Miller, K.: Biomechanics of soft tissues. Med. Sci. Monitor **6**(1), 158–167 (2000)
32. Miller, K., Chinzei, K.: Constitutive modelling of brain tissue: experiment and theory. J. Biomech. **30**(11–12), 1115–1121 (1997)
33. Miller, K., Chinzei, K.: Mechanical properties of brain tissue in tension. J. Biomech. **35**(4), 483–490 (2002)
34. Miller, K., Chinzei, K., et al.: Mechanical properties of brain tissue in-vivo: experiment and computer simulation. J. Biomech. **33**(11), 1369–1376 (2000)
35. Nicolle, S., Lounis, M., et al.: Shear linear behavior of brain tissue over a large frequency range. Biorheology **42**(3), 209–223 (2005)
36. Pena, A., Harris, N.G., et al.: Communicating hydrocephalus: the biomechanics of progressive ventricular enlargement revisited. Acta Neurochirurgica **81**, 59–63 (2002) (Suppl.)
37. Prange, M.T., Margulies, S.S.: Regional, directional, and age-dependent properties of the brain undergoing large deformation. J. Biomech. Eng. **124**(2), 244–252 (2002)
38. Sack, I., Beierbach, B., et al.: The impact of aging and gender on brain viscoelasticity. Neuroimage **46**(3), 652–657 (2009)

39. Schiavone, P., Chassat, F., et al.: In vivo measurement of human brain elasticity using a light aspiration device. Med. Image Anal. **13**(4), 673–678 (2009)
40. Shen, F., Tay, T.E., et al.: Modified Bilston nonlinear viscoelastic model for finite element head injury studies. J. Biomech. Eng. **128**(5), 797–801 (2006)
41. Shuck, L.Z., Advani, S.H.: Rheological response of human brain tissue in shear. J. Basic Eng. **94**, 905–911 (1972)
42. Takhounts, E., Crandall, J.R., et al.: On the importance of nonlinearity of brain tissue under large deformations. Stapp Car Crash J. **47**, 79–92 (2003)
43. Tamura, A., Hayashi, S., et al.: Mechanical characterization of brain tissue in high-rate compression. J. Biomech. Sci. Eng. **2**(3), 115–126 (2007)
44. Thibault, K.L., Margulies, S.S.: Age-dependent material properties of the porcine cerebrum: effect on pediatric inertial head injury criteria. J. Biomech. **31**(12), 1119–1126 (1998)
45. Velardi, F., Fraternali, F., et al.: Anisotropic constitutive equations and experimental tensile behavior of brain tissue. Biomech. Model. Mechanobiol. **5**(1), 53–61 (2006)
46. Wang, H.C., Wineman, A.S.: A mathematical model for the determination of viscoelastic behavior of brain in vivo. II. Relaxation response. J. Biomech. **5**(6), 571–580 (1972)
47. Weaver, J.B., Perrinez, P.R., et al.: The effects of interstitial tissue pressure on the measured shear modulus in vivo. Medical Imaging: Physiology, Function, and Structure from Medical Images, In: Proceedings of SPIE. Manduca, A., Hu, X.P., SPIE. vol. **6511**. pp. 1A-1–1A-11 (2007)
48. Wilhelm, M., Maring, D., et al.: Fourier-transform rheology. Rheologica Acta **37**(4), 399–405 (1998)
49. Wittek, A., Miller, K., et al.: Patient-specific model of brain deformation: application to medical image registration. J. Biomech. **40**(4), 919–929 (2007)
50. Wuerfel, J., Paul, F., et al.: MR-elastography reveals degradation of tissue integrity in multiple sclerosis. NeuroImage **49**(3), 2520–2525 (2010)
51. Xu, L., Lin, Y., et al.: Magnetic resonance elastography of brain tumors: preliminary results. Acta Radiologica **48**(3), 327–330 (2007)

Spinal Cord Mechanical Properties

Elizabeth C. Clarke

Abstract Knowledge of the mechanical properties of the spinal cord is useful for understanding spinal cord injury mechanisms and thresholds and developing realistic spinal cord models (for example, computer models for spinal cord injury or surgical simulations). The response of the spinal cord to mechanical loading has been studied under tension and compressive indentation, and its' behaviour has been shown to be non-linear viscoelastic. This chapter discusses testing modes that have been used to study spinal cord mechanical behaviour, presents a summary of reported mechanical properties of the spinal cord, including some mechanical testing data and constitutive model parameters, and discusses the effects of various specimen and experimental conditions on the mechanical behaviour of the spinal cord. Areas for future research and refinement of testing protocols are also presented.

1 Introduction

Mechanical properties of the spinal cord describe the physical response of this tissue to mechanical loading. Knowledge of these properties is important for several applications covering diverse experimental and clinical disciplines. For example, mechanical properties are needed when developing realistic and accurate computational and mechanical models involving the spinal cord (e.g. finite element models of the spinal cord to study injury, and models for spine/spinal cord surgery simulations). Also, understanding the response of the spinal cord to various mechanical loading conditions can provide insight into spinal cord injury

E. C. Clarke (✉)
Prince of Wales Medical Research Institute, Barker St, Randwick NSW, 2031, Australia
e-mail: e.clarke@powmri.edu.au

Stud Mechanobiol Tissue Eng Biomater (2011) 3: 25–40

DOI: 10.1007/8415_2010_15

Published Online: 15 June 2010

mechanisms and thresholds, for both traumatic and non-traumatic causes. Furthermore, knowledge of spinal cord mechanical properties is required in the field of tissue engineering in order to match the properties of tissue engineered spinal cord constructs to the real spinal cord in its natural environment.

Although the applications of this knowledge are diverse and important, there have been relatively few studies of spinal cord mechanical properties, and these few studies have had a wide range of experimental variables leading to variability in the mechanical properties reported for the spinal cord. There are also gaps in our knowledge to enable useful application of spinal cord properties. For example spinal cord mechanical properties have been investigated at strain rates in the order of 0.001–0.3 s^{-1}, but to enable clinical application of test data, mechanical properties are required at much higher and much lower strain rates, in the order of >5.0 s^{-1} (e.g. order of magnitude of spinal cord injury, SCI) and <5E–7 s^{-1} (e.g. order of magnitude of tissue growth for tissue engineering purposes).

This chapter will begin with a discussion of the test methods used to study spinal cord mechanical properties, an explanation of the meaning and implications of terms used to describe spinal cord properties and the impacts of some experimental variables on the reported properties of the spinal cord. This will be followed by a summary of spinal cord mechanical properties and an interpretation of these. Finally the gaps in our knowledge of spinal cord properties will be discussed and some future directions recommended. Note that this chapter only deals with the spinal cord and not para-spinal tissues.

1.1 Background to Soft Tissue Mechanical Testing

There are a number of standard methods used for the investigation of soft tissue mechanical properties. Some examples are tensile testing, confined and unconfined compression testing, indentation testing (localised compression) and shear/rheological testing (Fig. 1). These tests are most often performed in vitro. As the name suggests, tensile testing involves applying tension to (i.e. stretching) the specimen. Confined compression testing applies compression to the specimen within an enclosed chamber, while unconfined compression testing applies compression along one axis of the specimen leaving it free to deform along the other axes. These latter two typically apply the compression over the whole of one face of the specimen whereas indentation testing applies a compressive load to a localised area of the specimen through a probe (the indenter; usually spherical). Soft tissues can also be tested under shear loading, which can be applied in a linear or rotational manner.

A number of different types of tests can be performed using the loading methods shown in Fig. 1. The most straightforward, and therefore the most common, is to investigate the relationship between applied strain and measured stress by simply loading the specimen to a predefined limit or until the specimen fails. These tests can provide information about the tissue modulus ('stiffness') and failure stress and/or strain. Creep and relaxation tests are also commonly

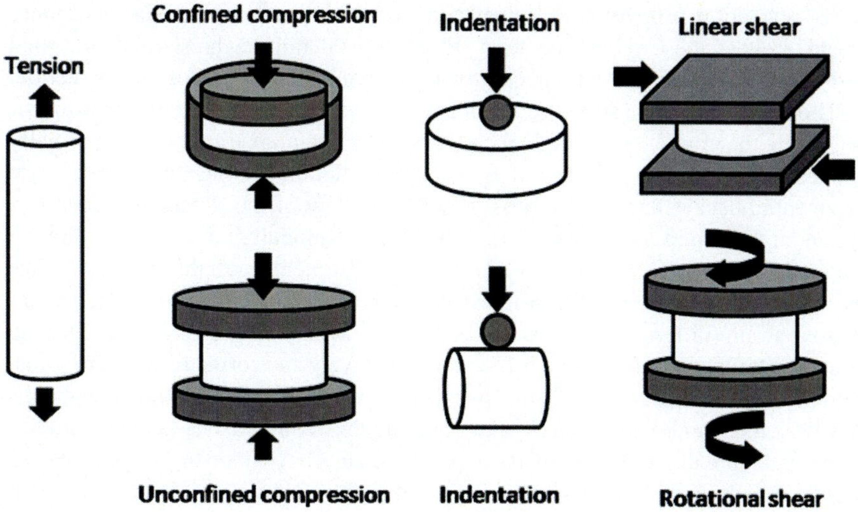

Fig. 1 Typical test methods used to investigate mechanical properties of soft tissues

performed. Relaxation testing involves the application of strain to a predetermined limit and then observing the reduction in stress over time. Similarly creep testing involves the application of loading to a predetermined stress and then observing the increase in strain over time.

2 Mechanical Testing of Spinal Cord

2.1 Testing Modes and Applications of Test Data

The spinal cord is essentially a long cylinder and this shape has made it a convenient specimen for tensile testing. Thus the majority of studies of spinal cord mechanical properties have been performed under tension, with the average cross sectional dimensions being used to calculate stress from the measured force (e.g. [3, 8, 11, 15]). Tensile testing of spinal cord has been performed under loading at constant strain rate (to investigate its stress–strain relationship and estimate the modulus of the spinal cord) (e.g. [29]) and under periods of constant applied stress (to investigate the stress–relaxation behaviour and relaxation modulus of the spinal cord) (e.g. [30]). These tests are often performed within the same study in order to devise constitutive models for the spinal cord (e.g. [3, 12, 16]). Only the earliest report of spinal cord tensile properties performed loading until failure [29], and all subsequent studies have performed testing below yield. Two possible reasons for this are that the threshold for physiological damage of the spinal cord appears lower than the threshold for structural damage [1] (making the structural failure properties of spinal cord of lower practical value), and because the stress–relaxation properties of spinal cord are of interest, and testing to failure would prevent these relaxation tests. Physiological failure thresholds for spinal cord tissue have been investigated and will be mentioned in the following section.

The stress–strain and stress–relaxation responses of the spinal cord are non-linear [8]. While the gradient of the spinal cord tensile stress–relaxation curve is normally defined as a function of time (the relaxation modulus, typically a three-term exponential decay function), surprisingly the gradient of the stress–strain curve is commonly reported as a constant (e.g. the tangent modulus) at a given strain (or range of strains) (e.g. [21, 26, 27, 29]). There are problems with this practice. Since the stress–strain response is non-linear the gradient is dependent upon the instantaneous strain (or range of strains used), so this practice of reporting modulus at an arbitrary strain level can make it difficult to objectively and correctly compare spinal cord moduli between studies. Fitting a constitutive model to stress–strain data points and reporting model parameters may make moduli comparisons between studies easier and more objective. A small number of studies have done this for spinal cord tensile properties and these will be discussed in the following section. However it should be noted that most constitutive relationships are phenomenological in nature and, while useful for comparison, provide little in terms of illuminating the physics underpinning the material behaviour.

Although tensile testing of spinal cord is the most common method, indentation compression studies have been performed, and such data would arguably be more directly applicable than tensile testing data alone. This is because spinal cord compression is believed to be one of the prominent mechanisms involved in both traumatic and non-traumatic spinal cord injury. Also, the majority of laboratory animal studies of spinal cord injury (e.g. 'contusion' models) use high speed indentation (around 350–1,000 mm/s) to produce the injury [22]. Although those laboratory studies use indentation methods and record load and displacement data, they are generally used to investigate the injury pathology or new therapies for spinal cord injury so do not normally report mechanical property data. Perhaps one of the reasons for this is that it is not straight-forward to convert force and displacement data from localised indentation loading into stress and strain data. One approach to overcome this is to use a reverse modeling approach (e.g. [13, 25]). A finite element model of the experiment is developed (e.g. the specimen and indenter) and a material property model (constitutive model) for the tissue (e.g. a hyperviscoelastic or poroelastic material model) (for descriptions see [7]) is used as the material property input to the finite element model. The parameters of the material property model are changed until the finite element model-predicted response matches the actual (experimental) response of the tissue. The material model parameters can be optimized manually but more sophisticated routines have been developed to help refine the process (e.g. [23]). One of the challenges with this approach is that finite element models may have difficulty resolving physical failure (or large deformations) thus the experimental tests must be performed below the failure limits of the specimen (or at relatively low deformations) in order to closely match the experimental and model outputs. Such data would be extremely useful to the biomechanics community because the 'standard' mechanical testing apparatuses are often limited to operation at strain rates orders of magnitude lower than injurious loading conditions, and it is well known that spinal cord mechanical properties are very sensitive to loading rate [3].

While spinal cord mechanical properties have been reported separately for the various loading modes, combining the results from these studies would be useful, but this has not been attempted. Simple comparisons between stress and strain data of spinal cord under different loading modes (e.g. tension vs. compression) may provide insight into mechanisms behind different injury causes (e.g. compression/contusion vs. distraction mechanisms of SCI). Furthermore, combining test data from various loading modes to produce more comprehensive constitutive models for spinal cord mechanical behaviour would be useful when developing computer models with complex loading patterns (e.g. finite element models for spinal cord injury and spinal surgery simulations). One of the challenges in combining such information is understanding and compensating for the effects of variability in testing conditions. Changes in experimental variables such as species, testing environment, degree of dissection and preconditioning protocols are all likely to affect the measured properties (see [8]), therefore more research is required into the effects of experimental variables on spinal cord mechanical properties, and devising methods to compensate recorded data for such changes to allow the data to be combined.

There are several other mechanical testing methods used to study soft tissue mechanical properties, such as rheological testing and magnetic resonance elastography (which exploits mechanical wave propagation through soft tissues and motion sensitive MRI sequences to determine in vivo tissue properties) (e.g. [16]), however these have not yet been used to study spinal cord mechanical properties. Such methods hold promise in attaining a more complete picture of spinal cord mechanical properties, particularly in vivo human spinal cord properties.

2.2 Methodological Considerations

2.2.1 Testing Environment

The choice of testing environment (i.e. in vivo vs. in vitro testing) is perhaps one of the most important for any study of biological tissues. The obvious advantage of in vivo testing is that the tissue remains as close as possible to its natural state, in that temperature, hydration and structural integrity are maintained. However, in vivo testing is more challenging and, in the case of human subjects, cannot be performed with traditional mechanical testing methods as these are too invasive. Emerging non-invasive test methods like Magnetic Resonance Elastography will make in vivo human investigations possible. In contrast, the advantage of in vitro testing of the spinal cord is simplicity over in vivo testing, for example there is no need to maintain animals under anaesthesia or devise apparatuses to clamp the spine or spinal cord. However, again the disadvantages of in vitro testing of the spinal cord are loss of hydration and temperature regulation, thus in vitro testing of spinal cord usually involves artificial temperature and hydration regulation such as continuous exposure to warm saline or enclosure in an environmental chamber during testing if available.

One of the other disadvantages of in vitro testing is the delay between death and testing. The mechanical behaviour of spinal cord has been demonstrated to change with post-mortem time delay, for example the modulus appears to increase after death [26].

2.2.2 Boundary Conditions

Another methodological consideration during mechanical testing of spinal cord is the choice of interface between the specimen and testing apparatus. Each loading mode has different requirements.

For tensile testing of spinal cord, it is important to prevent slip between the specimen and the clamps. One approach for in vitro testing has been the use of cyanoacrylate adhesive ('superglue') to fix the spinal cord ends to custom clamps that are then gripped by standard mechanical test-bench clamps [3, 11, 15]. A more elegant but also more challenging approach (freeze-clamping) has been recently developed for in vitro testing; the spinal cord ends are placed within metal canisters that are filled with dry ice and the cooled metal then contracts around the spinal cord as a means of radial clamping [26]. A radial clamping method has also been used for in vivo tensile testing of spinal cord [17–19]. That approach used clamps (similar to 'hose-clips') that fastened around the spinal cord applying light radial pressure. Radial clamping methods have the advantage that they can be used for in vivo testing of spinal cord, but come with the disadvantage that only the perimeter of the spinal cord is coupled to the testing apparatus and the boundary conditions at the interior portion of the spinal cord are uncertain.

Normally, axial compression testing of soft tissues aims to limit the friction between the testing plates and the specimen (usually through the use of water based lubricants) to allow the specimen to expand laterally when compressed axially. However axial compression testing of spinal cord has not been performed thus this approach has not been tested for spinal cord. In theory, indentation testing would benefit from lubricants to limit friction between the indenter and the specimen, however this is not normally done (e.g. [4, 20]. Typically spinal cord indentation testing is performed with an unlubricated plastic or metal spherical indenter. The effects of indenter friction on the spinal cord response to compressive indentation are not known, but it is possible that a high friction interface may reduce deformation and/or increase the reaction force at the indenter interface.

2.2.3 Preconditioning

Preconditioning is a common procedure that is performed prior to mechanical testing of biological soft tissues. It involves cyclically loading and unloading the specimen in order to produce a consistent pre-testing state and smaller inter-sample variability [2, 14]. Preconditioning has been performed in the most recent studies of spinal cord mechanical properties. However methods vary between laboratories and the effects of changing preconditioning protocols (i.e. number of cycles, maximum strain and strain rate) on the measured mechanical properties are not fully understood. In the recent studies where preconditioning was performed on spinal cord, the number of cycles ranged from 10 to 15, however preconditioning studies on other soft tissues has suggested that 6–7 preconditioning cycles are sufficient. Presumably, preconditioning with a higher number of cycles is more conservative and the only negative effect is an increase in testing duration. Also, in studies where all specimens are tested under the same loading conditions (i.e. common strains and strain rates) it would be logical for the specimens to be preconditioned at the same common strains and strain rates as the actual test. However in studies where the actual tests are performed over a range of strains and strain rates, the most appropriate method of preconditioning is not clear. Recently it has been confirmed that preconditioning strain significantly affects the measured spinal cord mechanical properties, with a higher preconditioning strain having a softening affect on the spinal cord [8]. That same study recommends preconditioning all specimens within a study to the same (maximum) strain. Similarly, the effects of preconditioning strain rate on spinal cord mechanical properties is also expected to affect the measured tissue properties thus it would be prudent to use a common preconditioning strain rate for all tests within a study. It is important to keep these effects in mind when preparing for any mechanical testing study, but also when comparing previously reported spinal cord mechanical properties from different studies.

2.2.4 Tissue Source

It is also clear that certain host specimen characteristics (e.g. age and species) affect the mechanical properties of spinal cord. A study of neonatal and adult rat spinal cord has demonstrated that the age of the host animal at the time of mechanical testing affects the mechanical behaviour of the spinal cord [11]. That study showed that the younger spinal cord was significantly softer than the adult spinal cord. This finding is important for several reasons. First, in vitro studies of human tissues tend to come from a high mean age group of tissue donors, thus the reported properties of the human spinal cord could have a relatively high modulus. The change in spinal cord properties with age is also important to keep in mind when comparing mechanical properties between different studies, and often this is made more difficult by the use of spinal cords from different animal species because animals of similar ages are not necessarily at an equivalent developmental stage. The biological processes that contribute to these changes in mechanical properties with age have not been investigated, but possibly it is caused by changing tissue compositions (e.g. ratio of gray to white matter) or that certain tissue elements experience changes in mechanical properties with development and ageing.

3 Mechanical Properties of Spinal Cord

3.1 Intact Spinal Cord

The mechanical properties of the intact spinal cord have been investigated using tension and indentation testing. Both of these loading modes have been used to investigate the stress–strain and stress–relaxation properties of spinal cord, however indentation testing is associated with difficulties in converting measured force to stress. Therefore, although 'stress' responses are sometimes reported for indentation testing of spinal cord, this term in this testing scenario does not have a straightforward meaning. Similarly, although some studies report mechanical properties of the spinal cord under 'compression', they are, strictly speaking, 'indentation' studies because the compression is applied to the spinal cord over a localised area of the specimen, not the entire specimen.

3.1.1 Tensile Properties of Intact Spinal Cord

Stress–Strain Relationship

A summary of tensile stress–strain data for intact spinal cords are presented in Fig. 2, and a summary of experimental protocols for studies of spinal cord tensile mechanical properties are presented in Table 1.

One of the earliest reports of spinal cord mechanical properties was made by Tunturi in 1978 [29], involving quasi-static loading (5 g increments) of canine spinal cord to failure. It is the only report of failure properties of the spinal cord under tension, and it also provided a unique comparison of spinal cord tensile properties with selective dissection of adjoining tissues (e.g. denticulate ligaments and dura). The report did not provide enough information to convert reported load

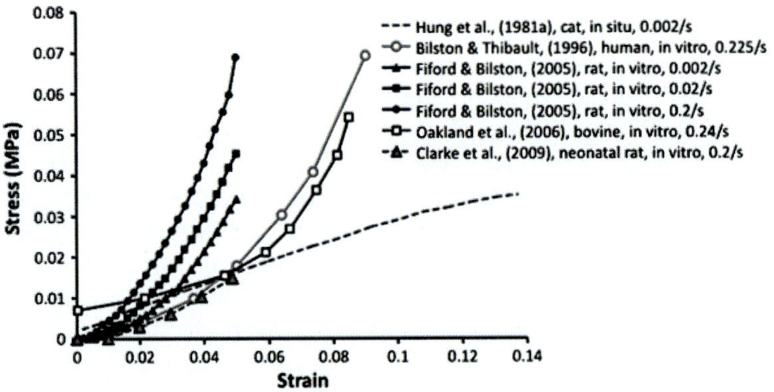

Fig. 2 Tensile stress–strain data summarised from several studies of spinal cord properties. The reference source and strain rate of the test are indicated

Table 1 Summary of testing variables for studies of spinal cord tensile stress–strain behaviour

References	Species	Env.	N	Precond.	Max. strain (%)	Strain rate	Modulus
Tunturi [29]	Dog	In vivo	1	None	19 (fail)	Quasi-static (5 g incr)	0.012–0.017 MPa
Hung et al. [18]	Cat	In vivo	3	None	1, 16	0.002 s^{-1}	0.252, 0.295 MPa
Hung et al. [19]	Cat	In vivo	4	None	8–12	0.0008 s^{-1}	0.4 MPa
Hung and Chang [17]	Puppy (3–5 kg)	In vivo	3	None	1.70	0.003 s^{-1}	0.265 MPa
Bilston and Thibault [3]	Human (30–84 years)	In vitro	3	None	~10	0.048, 0.120, 0.225 s^{-1}	1.02, 1.17, 1.37 MPa
Fiford and Bilston [15]	Rat (~230 g)	In vitro	6	10 cycles	5	0.002, 0.02, 0.2 s^{-1}	See Table 2
Oakland et al. [26]	Cow	In vitro	1	15 cycles	~8.5	0.24 s^{-1}	1.19 MPa
Clarke et al. [11]	Rat (14 days)	In vitro	8	10 cycles	5	0.002, 0.02, 0.2 s^{-1}	See Table 2

Note that N represents the number of specimens for which data was provided in the original article then reproduced in Fig. 2, not the total number of specimens used in the study into stress and appeared to use only a single specimen, thus data points are not shown in Fig. 2, however it did demonstrate that rupture of the spinal cord occurred at approximately 18% strain, beginning with a longitudinal split between the white matter columns and separation by shearing until failure occurred.

Subsequent studies of spinal cord mechanical properties have focused on investigating the effects of other variables on its mechanical behaviour, for example, the effects of strain rate, age and post-mortem delay.

A series of studies of feline and canine spinal cord mechanical properties under tension were conducted by Hung and Chang and colleagues in the early 1980s [5, 17–19]. Data from one of their studies is shown in Fig. 2, but data from their other studies is very similar. Their data show relatively linear stress–strain relationships (see Fig. 2), with reported 'pseudo moduli' in the range of 0.25–0.3 MPa, but no apparent effect of strain rate on modulus in the slow range of 0.0008–0.003 s^{-1}. In contrast, more recent studies of spinal cord mechanical properties have demonstrated non-linear ('J-shaped') stress–strain responses and increasing modulus with increasing strain rate (e.g. [3] (human); [11] (neonatal rat); [15] (rat)). An example of the effect of strain rate on tensile stress–strain data are shown for the study by Fiford and Bilston in Fig. 2. Similar patterns were reported in the other studies [3, 11], however these are not reproduced in Fig. 2 for clarity. Those three studies reported constitutive models and optimised model parameters to fit their data points. The constitutive models (the elastic and relaxation functions) described by Fiford and Bilston [15] and Clarke et al. [11] appear in Eqs. 1 and 2 respectively, and optimised parameters for these two studies are shown in Table 2.

$$\sigma^e(\varepsilon) = A(e^{B\varepsilon} - 1) \tag{1}$$

$$G(\varepsilon, t) = G_0 + (1 + \beta\dot{\varepsilon})\left[G_1 e^{-v_1 t} + G_2 e^{-v_2 t} + G_3 e^{-v_3 t}\right] \tag{2}$$

where A, B, Gi, vi and β are material parameters, ε and $\dot{\varepsilon}$ are the strain and strain rate respectively, and the strain history as a function of time (t) is given by:

$$\varepsilon(t) = \begin{cases} \dot{\varepsilon}_0 t, & \text{for } 0 < t < t_0 \\ \dot{\varepsilon}_0 t_0, & t > t_0 \end{cases} \tag{3}$$

Another recent study by Oakland et al. [26] also demonstrated non-linear stress–strain relationships for spinal cord under tension. Data from that study is also presented in Fig. 2. Note that in Fig. 2, several of the studies have conducted tensile stress–strain studies at a strain rate near 0.2 s^{-1}. Some of these (i.e. [3, 11, 26]),

Table 2 Constitutive model parameters for Eqs. 1, 2 for two studies of spinal cord properties

References	A MPa	B	G_0	G_1	v_1	G_2	v_2	G_3	v_3	β
Fiford and Bilston [15]	0.0288	21.22	0.7913	0.0190	0.7103	0.0890	0.0216	0.1001	0.0027	1.3418
Clarke et al. [11]	0.0081	16.75	0.5123	0.0142	1,000	0.2894	0.0121	0.1841	0.2736	0

The reference sources are indicated in the table

show remarkable similarity given that these were conducted in different host species, different laboratories and using different equipment and protocols (see Table 1). However, it is clear by comparing the overall data-set that often experimental variables have an effect on the mechanical behaviour of the spinal cord. Although some of those investigators report moduli values and some report constitutive models with optimised model parameters, it is clear from the variation among the recorded data points that values may only apply to that particular dataset under those particular testing conditions. For applications of this data that require a modulus value for the spinal cord, consult Tables 1 and 2 for the most appropriate estimate (i.e. the closest conditions to the application of interest).

It may appear surprising that the failure properties of spinal cord have not been investigated since Tunturi's study in 1978, but perhaps this is because functional changes of nerve cells have been demonstrated to occur at lower strains than structural changes [1], and so structural failure properties are not appropriate as an estimate for injury threshold of the spinal cord.

Stress–Relaxation Relationships

One of the earliest studies of spinal cord stress–relaxation behaviour was conducted by Tunturi in 1980 [30], involving in vitro dog spinal cord. Tunturi demonstrated rapid relaxation of the spinal cord, reaching approximately 60% of the original stress within approximately 10–20 s. There was insufficient information to enable conversion of force or displacement data into stress and strain data, and the maximum applied strain is not clear. However Tunturi did present unique histological images and descriptions of the failure of spinal cord white matter.

Subsequent studies of spinal cord stress–relaxation behaviour following tension have focused on the effects of strain rate and maximum applied strain, in order to fit comprehensive constitutive models to the test data. Some data from these studies is shown in Fig. 3, and these studies are summarized in Table 3. Note that the different studies investigated relaxation under varying time durations and sampling rates. Only a very short period of relaxation was reported in the study by Bilston and Thibault [3], but data were collected at a high sampling rate, whereas

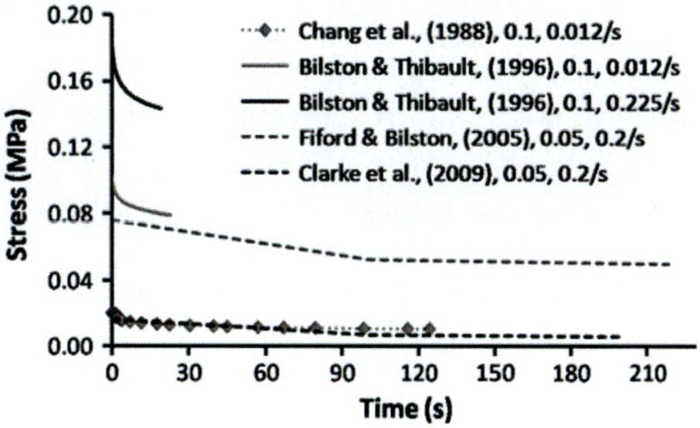

Fig. 3 Tensile stress–relaxation of spinal cord summarised from several studies. The reference source, applied strain and strain rate of the test are indicated

Table 3 Summary of testing variables for studies of spinal cord tensile stress–relaxation behaviour

References	Species	Env	Max. strain	Strain rate	Equilibrium time
Tunturi [30]	Dog	In vitro	Unspecified	Quasi-static (60 g applied)	<15 min
Chang et al. [6]	Cat	In vivo	0.5–11%	0.003–0.012 s^{-1}	~2 min
Bilston and Thibault [3]	Human (30–84 years)	In vitro	~10%	0.048–0.225 s^{-1}	<25 s
Fiford and Bilston [15]	Rat	In vitro	2–5%	0.002–0.2 s^{-1}	~8–15 min
Clarke et al. [11]	Neonatal rat	In vitro	2–5%	0.002–0.2 s^{-1}	~8–15 min

The reference sources are indicated in the table

for the studies by Fiford and Bilston [15] and Clarke et al. [11], relaxation was reported for over several minutes but sampled at a slower rate.

In 1988 Chang et al. [6] were first to investigate tensile stress–relaxation properties in cat spinal cords over a range of applied strains and strain rates. Notably, that study was also the only investigation of spinal cord relaxation behaviour conducted in vivo. Their data hinted that changes in applied strain do not noticeably change the percentage of stress relaxation but changes in applied strain rate do change the percentage of stress–relaxation, where higher strain rates (~0.012 vs. ~0.003 s^{-1}) caused a larger proportion of the stress to relax (~40 vs. ~20%). Within each of the subsequent studies investigating spinal cord stress–relaxation behaviour, data support the theory that increasing the applied strain rate increases the proportion of stress–relaxation (e.g. the data from Bilston and Thibault tested at 0.225 and 0.012 s^{-1} shown in Fig. 3) [3, 11, 15], however when comparing the proportions of stress relaxation between the various studies the pattern is not consistent. For example, Bilston and Thibault [3] reported ~20% stress–relaxation for a strain rate of 0.225 s^{-1} (in vitro human spinal cords), however Chang et al. [6] reported a larger proportion of stress–relaxation (~40%) for spinal cords tested at a lower strain rate (~0.012 s^{-1}) (in vivo cat spinal cords). Perhaps this difference could be accounted for in the differences in time to equilibrium (25 s for Bilston and Thibault vs. 2 min for Chang et al.), but it is also likely that experimental variables (such as species and testing environment in the example above) have an effect on the stress–relaxation behaviour of the spinal cord. Therefore if spinal cord stress–relaxation data are to be used, the data should be sourced from the study with test conditions that are closest to the application of interest.

3.1.2 Indentation Properties of Intact Spinal Cord

Stress–Strain and Stress–Relaxation Relationships

A large number of studies have utilised high speed compressive indentation of the spinal cord in order to produce experimental spinal cord injury in anaesthetised animals ('contusion models'), however those studies do not typically report force–deformation data or mechanical properties of the spinal cord during the impact, therefore they will not be discussed here. In fact, very few studies have investigated compressive indentation properties of the spinal cord for the purposes of investigating spinal cord 'mechanical properties', and those that have done so have not reported 'stress–strain' data. This is likely because it is not straightforward to convert force and displacement data from localised indentation loading into stress and strain data. The calculation of classical mechanical properties relies on the loading being applied to an entire specimen, thus using the cross sectional dimensions of the spinal cord and/or indenter to convert the displacement of the spinal cord under the indenter into 'strain', or convert the reaction force measured at the indenter into 'stress', are technically incorrect. One approach to estimate the mechanical properties of the spinal cord under compressive indentation would be to use a reverse computational modelling approach, where mechanical property data inputs are manipulated until a match between the measured and model-estimated deformations of the spinal cord is found. Numerous studies of other soft tissues have employed reverse modeling approaches to develop material models (e.g. [13, 25], and relatively sophisticated routines now exist to refine the parameter optimization process (e.g. [23]).

The mechanical response of the spinal cord under compressive indentation has been investigated in a single study involving in vivo quasi-static loading of cat spinal cord [20]. That study reported a characteristic non-linear ('J-shaped') stress–strain curve, similar to studies of spinal cord under tension. In contrast to the tensile studies of spinal cord, they reported no apparent change in mechanical behaviour from increasing loading rate. Although that finding may indeed be correct, it is more likely that the loading rates used in that study (in the order of $<0.0084\ s^{-1}$) were simply too low to observe any changes in mechanical behaviour with loading rate. A similar study has investigated relaxation behaviour of the spinal cord under compressive indentation [4]. That study involved in vivo compressive indentation of dog spinal cord, although the loading rate was not clear. Again, stress–strain data and mechanical properties were not reported in that study, however they did report a very high degree of relaxation ($\sim$80%) after several hours of indentation, suggesting that equilibrium relaxation stresses reported at earlier timeframes may be inaccurate.

3.2 Constituent Parts of the Spinal Cord

Although this chapter discusses mechanical properties of the spinal cord, a number of studies have also investigated mechanical properties of selected spinal cord tissues and these will be briefly discussed. Ichihara et al. [21] compared the tensile behaviour of spinal cord gray and white matter using traditional mechanical testing methods. They used a biopsy needle to core longitudinal gray and white matter specimens from bovine spinal cord and tested them in vitro at a moderate strain rate (0.05 s^{-1}) to failure. They reported a higher modulus and lower failure strain for gray matter, and supported their findings by demonstrating a match between finite element model predicted deformations of whole spinal cord with actual deformations measured using magnetic resonance imaging. A different study by Ozawa et al. [27] compared in vitro rabbit spinal cord gray and white matter moduli using a non-conventional technique (pipette aspiration pressure). They found no significant differences between gray and white matter moduli. Perhaps the reason for this contention is that in the latter study by Ozawa et al. the deformation of the specimen area at the pipette tip was not large enough to see the divergence in moduli of gray and white matter that is seen in the data reported by Ichihara et al.

A study by Maikos et al. compared mechanical behaviour of rat brain and spinal cord dura in vitro under tension [24]. They applied a slow strain rate (0.0014 s^{-1}) to failure to fit and compare constitutive model parameters, and they also applied a high strain rate (19.4 s^{-1}) to 10% strain for 10 s to compare stress–relaxation constants for brain and spinal cord dura. While that study is important for presenting spinal cord dura mechanical testing data and is particularly useful because of the constitutive model they presented for spinal cord dura, it is a shame that a single study has not compared the dynamic mechanical behaviour of the spinal cord with and without dura attached since the first quasi-static study by Tunturi in 1978, as this may give an indication of the level of 'protection' afforded by the dura to the spinal cord.

Finally, a study by Schreiber et al. [28] has investigated the effects of myelin and glia on the tensile properties of the spinal cord. In their study, a group of chicken embryos were injected with myelin suppression agents and the spinal cords were later tested in vitro under tension and compared with results from an untreated group. Schreiber et al. reported significantly lower modulus and tensile strength of spinal cords in the demyelinised group, indicating that myelin has a significant effect on the mechanical behaviour of the spinal cord.

4 Future Work

It is clear that there is much work to be done in order to gain a full picture of the mechanical properties of the spinal cord. First, only tensile testing of the spinal cord has been used in a sufficient number of studies to make comparisons amongst studies and comment on the effects of experimental variables. Other testing methods have either not been used or have not been extended to investigate mechanical properties of the spinal cord (as opposed to simply logging the mechanical responses to loading). Investigating spinal cord mechanical properties, particularly injury and/or failure thresholds, under different loading modes is important because recent studies have discovered a link between spinal cord injury mechanisms and the ensuing spinal cord injury pathology and severity [9, 10]. Similarly, the mechanical properties of spinal cord require further investigation at much faster (e.g. >5 s) and much slower (e.g. <5E–7 s^{-1}) strain rates in order to better understand mechanisms and thresholds of traumatic and non-traumatic (e.g. tumor) causes of SCI, and to provide useful data for tissue engineering/regeneration purposes. Furthermore, the anisotropy of the spinal cord has not been investigated thoroughly but recent studies suggest that direction of spinal cord loading plays a role in spinal cord injury pathology and severity [12]. Unfortunately though, as various laboratories investigate some of these specific research questions, more experimental variability is introduced and comparisons between different studies are difficult. For this reason, a thorough investigation of the effects of separate experimental variables on measured mechanical properties is required in order to allow future comparisons between studies conducted under different conditions. There also needs to be a greater understanding of the relationships between spinal cord structure, function and mechanical properties. The spinal cord is a complex, viscoelastic structure, and the mechanical properties of component tissues, combined with their complex arrangement are not understood. Further investigation into spinal cord component tissue properties and the effects of structural arrangement of these on mechanical properties of the cord are required. This may provide insight into the viscoelastic nature of the spinal cord, including mechanisms for stress–relaxation and effects of preconditioning. Finally, with the advent and refinement of non-invasive techniques to study tissue properties in vivo, it is hoped that these methods can be used to investigate human spinal cord mechanical properties in its natural environment.

References

1. Bain, A.C., Raghupathi, R., Meaney, D.F.: Dynamic stretch correlates to both morphological abnormalities and electrophysiological impairment in a model of traumatic axonal injury. J. Neurotrauma **18**, 499–511 (2001)
2. Bilston, L.E., Liu, Z., Phan-Thien, N.: Large strain behaviour of brain tissue in shear: some experimental data and differential constitutive model. Biorheology **38**, 335–345 (2001)
3. Bilston, L.E., Thibault, L.E.: The mechanical properties of the human cervical spinal cord in vitro. Ann. Biomed. Eng. **24**, 67–74 (1996)
4. Carlson, G.D., Warden, K.E., Barbeau, J.M. et al.: Viscoelastic relaxation and regional blood flow response to spinal cord compression and decompression. Spine **22**, 1285–1291 (1997)
5. Chang, G.L., Hung, T.K., Bleyaert, A., Jannetta, P.J.: Stress–strain measurement of the spinal cord of puppies and their neurological evaluation. J. Trauma **21**, 807–810 (1981)
6. Chang, G.L., Hung, T.K., Feng, W.W.: An in vivo measurement and analysis of viscoelastic properties of the spinal cord of cats. J. Biomech. Eng. **110**, 115–122 (1988)
7. Cheng, S., Bilston, L.E.: Unconfined compression of white matter. J. Biomech. **40**, 117–124 (2007)
8. Cheng, S., Clarke, E.C., Bilston, L.E.: The effects of preconditioning strain on measured tissue properties. J. Biomech. **42**, 1360–13602 (2009)
9. Choo, A.M., Liu, J., Dvorak, M., Tetzlaff, W., Oxland, T.R.: Secondary pathology following contusion, dislocation, and distraction spinal cord injuries. Exp. Neurol. **212**, 490–506 (2008)

10. Choo, A.M., Liu, J., Lam, C.K., Dvorak, M., Tetzlaff, W., Oxland, T.R.: Contusion, dislocation, and distraction: primary hemorrhage and membrane permeability in distinct mechanisms of spinal cord injury. J. Neurosurg. **6**, 255–266 (2007)
11. Clarke, E.C., Cheng, S., Bilston, L.E.: The mechanical properties of neonatal rat spinal cord in vitro, and comparisons with adult. J. Biomech. **42**, 1397–1402 (2009)
12. Clarke, E.C., Choo, A.M., Liu, J., et al.: Anterior fracture-dislocation is more severe than lateral: a biomechanical and neuropathological comparison in rat thoracolumbar spine. J. Neurotrauma **25**, 371–383 (2008)
13. Cox, M.A., Gawlitta, D., Driessen, N.J., Oomens, C.W., Baaijens, F.P.: The non-linear mechanical properties of soft engineered biological tissues determined by finite spherical indentation. Comp. Methods Biomech. Biomed. Eng. **11**, 585–592 (2008)
14. Fallenstein, G.T., Hulce, V.D., Melvin, J.W.: Dynamic mechanical properties of human brain tissue. J. Biomech. **2**, 217–226 (1969)
15. Fiford, R.J., Bilston, L.E.: The mechanical properties of rat spinal cord in vitro. J. Biomech. **38**, 1509–1515 (2005)
16. Green, M.A., Bilston, L.E., Sinkus, R.: In vivo brain viscoelastic properties measured by magnetic resonance elastography. NMR Biomed. **21**, 755–764 (2008)
17. Hung, T.K., Chang, G.L.: Biomechanical and neurological response of the spinal cord of a puppy to uniaxial tension. J Biomech. Eng. **103**, 43–47 (1981)
18. Hung, T.K., Chang, G.L., Chang, J.L., Albin, M.S.: Stress–strain relationship and neurological sequelae of uniaxial elongation of the spinal cord of cats. Surg. Neurol. **15**, 471– 476 (1981a)
19. Hung, T.K., Chang, G.L., Lin, H.S., Walter, F.R., Bunegin, L.: Stress–strain relationship of the spinal cord of anesthetized cats. J. Biomech. **14**, 269–276 (1981b)
20. Hung, T.K., Lin, H.S., Bunegin, L., Albin, M.S.: Mechanical and neurological response of cat spinal cord under static loading. Surg. Neurol. **17**, 213–217 (1982)
21. Ichihara, K., Taguchi, T., Shimada, Y., Sakuramoto, I., Kawano, S., Kawai, S.: Gray matter of the bovine cervical spinal cord is mechanically more rigid and fragile than the white matter. J. Neurotrauma. **18**, 361–367 (2001)
22. Kwon, B.K., Oxland, T.R., Tetzlaff, W.: Animal models used in spinal cord regeneration research. Spine **27**, 1504–1510 (2002)
23. Lei, F., Szeri, A.Z.: Inverse analysis of constitutive models: biological soft tissues. J. Biomech. **40**, 936–940 (2007)
24. Maikos, J.T., Elias, R.A., Shreiber, D.I.: Mechanical properties of dura mater from the rat brain and spinal cord. J. Neurotrauma **25**, 38–51 (2008)
25. Moerman, K.M., Holt, C.A., Evans, S.L., Simms, C.K.: Digital image correlation and finite element modelling as a method to determine mechanical properties of human soft tissue in vivo. J. Biomech. **42**, 1150–1153 (2009)
26. Oakland, R.J., Hall, R.M., Wilcox, R.K., Barton, D.C.: The biomechanical response of spinal cord tissue to uniaxial loading. Proc Inst Mech Eng [H]. **220**, 489–492 (2006)
27. Ozawa, H., Matsumoto, T., Ohashi, T., Sato, M., Kokubun, S.: Comparison of spinal cord gray matter and white matter softness: measurement by pipette aspiration method. J. Neurosurg. **95**, 221–224 (2001)
28. Shreiber, D.I., Hao, H., Elias, R.A.: Probing the influence of myelin and glia on the tensile properties of the spinal cord. Biomech. Model. Mechanobiol. **8**, 311–321 (2009)
29. Tunturi, A.R.: Elasticity of the spinal cord, pia, and denticulate ligament in the dog. J Neurosurg. **48**, 975–979 (1978)
30. Tunturi, A.R.: Viscoelasticity of dog spinal cord. Physiol. Chem. Phys. **12**, 373–378 (1980)

Constitutive Modelling of Brain Tissue for Prediction of Traumatic Brain Injury

J. A. W. van Dommelen, M. Hrapko and G. W. M. Peters

Abstract To develop protective measures for crash situations, an accurate assessment of injury risk is required. By using a Finite Element model of the head, the mechanical behaviour of the brain can be predicted for any acceleration and improved injury criteria can be developed and implemented into safety standards. Many head models are based on a detailed geometrical description of the anatomical components. However, for reliable predictions of injury, also an accurate constitutive model for brain tissue is required that is applicable for large deformations and complex loading conditions that occur during an impact to the head. This chapter deals with constitutive modelling of brain tissue. Different approaches towards modelling of the mechanical response of biological tissues are discussed. A short overview of the large strain behaviour of brain tissue and constitutive models that have been developed for this material is given. A non-linear viscoelastic model for brain tissue is then discussed in more detail. The model is based on a multi-mode Maxwell model and consists of a non-linear elastic mode in combination with a number of viscoelastic modes. For this model, also a numerical implementation scheme is given. The influences of constitutive non-linearities of brain tissue in numerical head model simulations are shown by comparing the performance of the model of Hrapko et al. with a simplified version, based on neo-Hookean elastic behaviour, and a third non-linear constitutive model from literature.

J. A. W. van Dommelen (✉), M. Hrapko and G. W. M. Peters
Materials Technology Institute, Eindhoven University of Technology,
P.O. Box 513, 5600 MB Eindhoven, The Netherlands
e-mail: J.A.W.v.Dommelen@tue.nl

M. Hrapko
e-mail: Matej.Hrapko@gmail.com

G. W. M. Peters
e-mail: G.W.M.Peters@tue.nl

Stud Mechanobiol Tissue Eng Biomater (2011) 3: 41–67

DOI: 10.1007/8415_2010_16

Published Online: 3 November 2011

1 Introduction

More than one third of all injuries are traumatic brain injury (TBI), which also represents one of the major causes of death resulting from traffic accidents [24]. Despite the major advances in prevention and treatment, head injury remains a major health and social problem. TBI can be caused when the head is suddenly struck by an object with or without the object penetrating the skull and the brain or by large accelerations or decelerations of the head. These injuries can be divided into primary injuries, which occur at the moment of impact, and secondary injuries, developing at a later stage. The majority of brain injuries are caused by diffuse axonal injury (DAI), which is characterised by microscopic damage of axons. DAI is typical for situations without any direct impact on the head, as it can be the result of rapid acceleration or deceleration. DAI is thought to be the most common and important pathology in mild, moderate, and severe traumatic brain injury [22]. It may develop over a period of hours or even days after an accident.

To develop protective measures, an accurate assessment of the risk of injury is required. In the early 1960s, the currently used Head Injury Criterion was developed [47] based on the Wayne State Tolerance Curve [16]. However, it is based on translational head acceleration only and it does not allow for a distinction between different injury mechanisms. By using a detailed Finite Element (FE) model of the head [e.g., 3, 7, 9, 17, 23, 42, 48], the behaviour of the brain can be predicted for any acceleration and improved injury criteria can be developed and implemented into safety standards. These FE models often contain a detailed geometrical description of the anatomical components but for reliable prediction of TBI, also an accurate description of the mechanical behaviour of the brain tissue is required.

This chapter deals with the constitutive modelling of biological materials and of brain tissue in particular. Various approaches for modelling the elastic and viscoelastic response of biological materials are given. The typical large strain response of brain tissue is shortly discussed, after which an overview of different models for the mechanical behaviour of brain tissue that have been proposed in literature is presented. One of these models is then discussed in more detail and also some aspects of a numerical implementation of this model in a finite element context are given. Finally, the influence of constitutive non-linearities on 3D head model predictions are shown.

2 Constitutive Modelling of Biological Materials

Simulations with finite element models of the human body require constitutive descriptions for the various materials that constitute the body. A constitutive model deals with the relation between the deformation (history) applied to a material and the stress that results from this deformation, which can formally be written as:

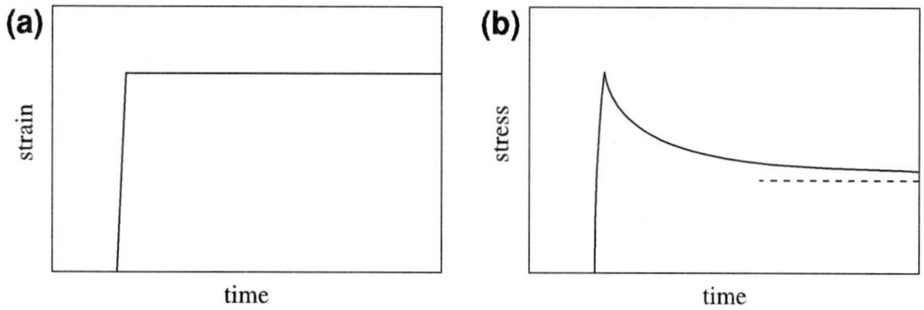

Fig. 1 Viscoelastic response to an approximate step in the strain

$$\boldsymbol{\sigma}(t) = \boldsymbol{\sigma}(\boldsymbol{F}, \tau | \forall \tau \leq t), \tag{1}$$

where $\boldsymbol{\sigma}$ represents the Cauchy stress tensor as a function of time t and $\boldsymbol{F}$ is the deformation gradient tensor.

The mechanical behaviour of biological tissue is the result of the properties of the individual microstructural components that constitute the material and the interplay between these components. Some attempts have been made to develop constitutive models for biological materials based on microstructural approaches. However, in impact simulations, mostly phenomenological material models are used. In general, properties of biological tissues are viscoelastic, i.e. their response is rate-dependent and they show stress relaxation at a constant strain level, as is illustrated in Fig. 1. Moreover, they are often non-linear, and anisotropic due to the specific microstructure (e.g., consisting of an arrangement of fibres and surrounding matrix material).

2.1 Elastic Models

For soft tissues, often hyperelastic models are used, which represent true elastic behaviour. Hyperelastic constitutive laws can be derived from a stored energy function W (also called strain energy function). This energy function is given in terms of the deformation gradient tensor $\boldsymbol{F}$, often through the invariants I_i of the right or left Cauchy–Green strain tensor ($\boldsymbol{C} = \boldsymbol{F}^{\mathrm{T}} \cdot \boldsymbol{F}$ or $\boldsymbol{B} = \boldsymbol{F} \cdot \boldsymbol{F}^{\mathrm{T}}$, respectively). For incompressible materials ($I_3 = \det(\boldsymbol{C}) = 1$), this can be written as:

$$W = W(I_1, I_2), \tag{2}$$

with

$$I_1 = \mathrm{tr}(\boldsymbol{C}) \quad \text{and} \quad I_2 = \frac{1}{2}\left[\mathrm{tr}^2(\boldsymbol{C}) - \mathrm{tr}(\boldsymbol{C}^2)\right]. \tag{3}$$

The constitutive relation for a material can then be derived according to:

$$\boldsymbol{S} = 2\frac{\partial W}{\partial \boldsymbol{C}} \quad \text{and} \quad \boldsymbol{\sigma} = \frac{1}{J}\boldsymbol{F} \cdot \boldsymbol{S} \cdot \boldsymbol{F}^{\mathrm{T}}, \tag{4}$$

where $\boldsymbol{S}$ is the second Piola–Kirchhoff stress tensor and $J = \det(\boldsymbol{F}) = \sqrt{I_3}$. The stored energy function must be chosen such that the obtained constitutive relation matches experimental data. Thereby, the stored energy function must satisfy some conditions: in the absence of a pre-stress, in the reference configuration, both the stored energy function and the Cauchy stress tensor must equal zero; for infinitesimal strains, the constitutive equation must represent a linear elastic material; and infinitely large strains must lead to infinitely large values of W. Since

$$\frac{\partial I_1}{\partial \boldsymbol{C}} = \boldsymbol{I} \quad \text{and} \quad \frac{\partial I_2}{\partial \boldsymbol{C}} = \mathrm{tr}(\boldsymbol{C})\boldsymbol{I} - \boldsymbol{C}, \tag{5}$$

for a stored energy function that is written as Eq. 2, the corresponding constitutive relation is given by:

$$\boldsymbol{S} = 2\frac{\partial W}{\partial I_1}\boldsymbol{I} + 2\frac{\partial W}{\partial I_2}(I_1\boldsymbol{I} - \boldsymbol{C}). \tag{6}$$

By using the Cayley–Hamilton theorem, the Cauchy stress tensor can be written as:

$$\boldsymbol{\sigma} = 2\frac{\partial W}{\partial I_1}\boldsymbol{B} - 2\frac{\partial W}{\partial I_2}\boldsymbol{B}^{-1}, \tag{7}$$

with $\boldsymbol{B} = \boldsymbol{F} \cdot \boldsymbol{F}^{\mathrm{T}}$ the left Cauchy–Green strain tensor (or Finger tensor). Note that for such an incompressible material, an undetermined hydrostatic term should be added to satisfy natural boundary conditions and that therefore a hydrostatic term that directly results from Eq. 6 has been left out.

A common choice for a stored energy function is:

$$W = \frac{C_1}{2}(I_1 - 3) + \frac{C_2}{2}(I_2 - 3), \tag{8}$$

which represents the Mooney–Rivlin model for an incompressible material. The corresponding constitutive relation is given by:

$$\boldsymbol{\sigma} = C_1\boldsymbol{B} - C_2\boldsymbol{B}^{-1} - p\boldsymbol{I}. \tag{9}$$

For $C_2 = 0$, this model reduces to a simple neo-Hookean model. The Mooney–Rivlin model represents a non-linear hyperelastic model. The non-linearities are however limited to certain deformation modes. For simple shear it can be shown that the Mooney–Rivlin model gives a linear response and that C_1 and C_2 have the same effect on this response. An important consequence is that the material parameters in this model cannot be completely characterised by shear experiments only. Other strain energy functions that have specifically been chosen to model the response of brain tissue will be discussed in further sections.

For soft tissues, a fiber-reinforced model may be used, where the strain energy function depends on the fiber stretch λ_{f}. The stress related to a strain energy term $W_{\mathrm{f}}(\lambda_{\mathrm{f}})$ is then found as:

$$S_{\mathrm{f}} = \frac{1}{\lambda_{\mathrm{f}}} \frac{\partial W_{\mathrm{f}}}{\partial \lambda_{\mathrm{f}}} \boldsymbol{n}_0 \boldsymbol{n}_0 \quad \text{with} \quad \lambda_{\mathrm{f}}^2 = I_4 = \boldsymbol{C} : \boldsymbol{n}_0 \boldsymbol{n}_0, \tag{10}$$

where $\boldsymbol{n}_0$ is the fiber direction in the initial configuration.

2.2 Viscoelastic Models

A viscoelastic material model can be represented by a mechanical analogue consisting of a certain arrangement of springs and dashpots. By placing a number of so-called Maxwell elements in parallel, a general viscoelastic framework can be obtained, see Fig. 2.

The behaviour of a Maxwell element can be described by a differential formulation or by an integral formulation. A differential approach could be favoured over an integral framework for the development of a non-linear viscoelastic model because of the possibility to derive it from physics and since it is a numerically more advantageous formulation. A general Maxwell-type differential constitutive equation can be written in the form [25]:

$$\boldsymbol{\sigma} + \lambda \overset{\nabla}{\boldsymbol{\sigma}} + \boldsymbol{f}(\boldsymbol{\sigma}, \boldsymbol{D}) = 2\eta_0 \boldsymbol{D}, \tag{11}$$

where now $\boldsymbol{\sigma}$ is the extra stress tensor (that differs from the Cauchy stress by a hydrostatic term), $\boldsymbol{D}$ is the deformation rate tensor and $\overset{\nabla}{\boldsymbol{A}} = \dot{\boldsymbol{A}} - (\nabla \boldsymbol{v})^{\mathrm{T}} \cdot \boldsymbol{a} - \boldsymbol{a} \cdot (\nabla \boldsymbol{v})$ represents the upper convected time derivative. In case $\boldsymbol{f}(\boldsymbol{\sigma}, \boldsymbol{D}) = \boldsymbol{0}$, this equation reduces to the upper convected Maxwell model and for small strain amplitudes, a linear viscoelastic model remains.

A linear viscoelastic model can also be written in an integral form, where the concepts of proportionality and superposition are considered. In this theory, the response to an arbitrary loading history is assumed to be given by a Boltzmann integral over an infinite number of small steps, which can be written as:

$$\boldsymbol{\sigma}(t) = \int_{-\infty}^{t} G(t-\tau) 2\boldsymbol{D} \mathrm{d}\tau \quad \text{with} \quad G(t) = G_\infty + \sum_{k=1}^{N} G_k \mathrm{e}^{-\frac{t}{\lambda_k}}, \tag{12}$$

with G_∞ the equilibrium shear modulus of the material. The so-called quasi linear viscoelastic (QLV) theory has been proposed by Fung [15]. This theory has

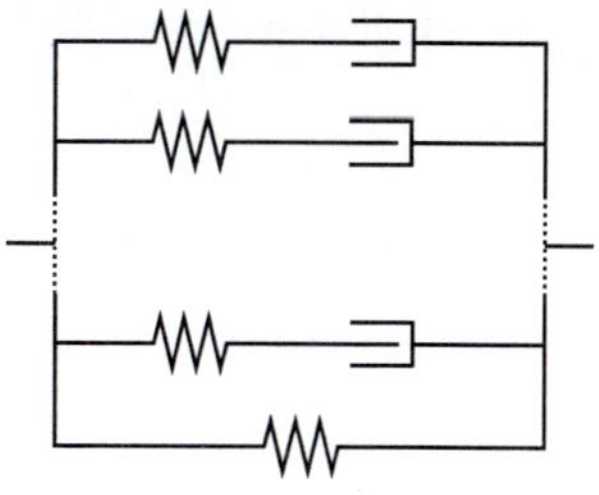

Fig. 2 Analogue of a viscoelastic model with multiple Maxwell elements and an elastic element

become widely used in injury biomechanics and has been applied for the constitutive or structural modelling of many soft biological tissues. The QLV theory is a generalisation of the linear viscoelastic theory and is also formulated in terms of a convolution integral:

$$\boldsymbol{\sigma}(t) = \int_{-\infty}^{t} M(t-\tau)\dot{\boldsymbol{\sigma}}_{\mathrm{e}}\mathrm{d}\tau \tag{13}$$

with

$$M(t) = M_\infty + \sum_{k=1}^{N} M_k \mathrm{e}^{-\frac{t}{\lambda_k}} \quad \text{and} \quad M(0) = M_\infty + \sum_{k=1}^{N} M_k = 1. \tag{14}$$

In this integral representation, the elastic response is separated from the relaxation function. The quasi linear viscoelastic theory assumes the time-dependent behaviour to be given by a Prony series. However a non-linear relation $\boldsymbol{\sigma}_{\mathrm{e}}(\boldsymbol{B})$ for the instantaneous elastic response to a deformation step can be used. In three dimensions, this instantaneous elastic response is commonly derived from a strain energy function.

3 Large Strain Response and Modelling of Brain Tissue

A constitutive model to be used in a numerical head model should accurately describe the mechanical response of brain tissue for the large deformations and complex loading histories and deformation modes that occur during an impact to the head. In this section, the typical response of brain tissue for large deformations is illustrated with some results of rheological shear experiments by Hrapko et al. [18]. Thereafter, an overview of various models for brain tissue that have been proposed in the literature is given. In the next section, one particular model will be presented in more detail.

3.1 Large Strain Behaviour of Brain Tissue

Large strain deformation of biological tissue may lead to damage which affects the mechanical behaviour of the tissue. The occurrence of mechanical damage[1] was

[1] It is important to make a clear distinction between functional and mechanical damage. Functional damage can be considered as injury, i.e. change or loss of functionality of the brain tissue, whereas mechanical damage only affects the mechanical properties of the tissue. At these strain levels, functional damage may still occur (as observed by for example Bain and Meaney [2] and Morrison et al. [33]) and at larger time scales also mechanical changes could develop.

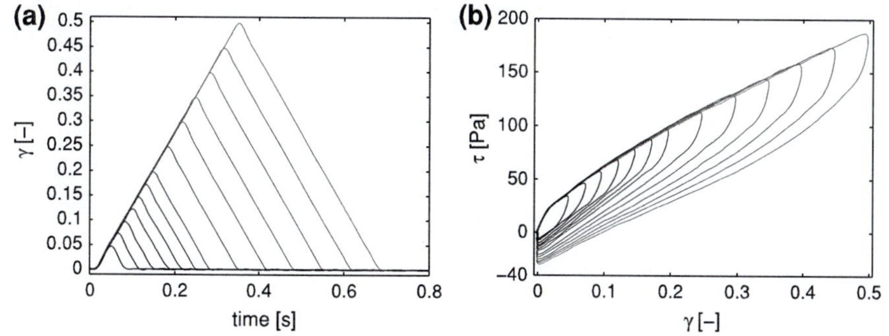

Fig. 3 Results of constant shear rate (1.5 s^{-1}) experiments with increasing strain amplitude. **a** Applied shear strain, **b** stress–strain response. Reprinted from Hrapko et al. [18], with permission from IOS Press

investigated by Hrapko et al. [18] with a series of constant shear rate experiments with increasing strain amplitude. Results from these experiments are presented in Fig. 3. In Fig. 3b, the 0.01 strain limit for linear behaviour is visible in the beginning of each loading part. There is no yield or failure visible for the strain range tested. By comparing the loading parts of different cycles, which were all applied at an identical strain rate, it can be observed that there is no significant immediate mechanical damage affecting the stress–strain behaviour due to previous shear deformations (up to strains of 0.45). These results show that mechanical measurements up to a strain of 0.45 are reproducible for short time scales. This observation is in agreement with results of Prange and Margulies [40] who reported no change in long term modulus and no structural changes in the tissue during stress relaxation experiments up to a strain of 0.5 and shear rates of 8.33 s^{-1}. Furthermore, no maximum in the stress–strain response was found, which is in agreement with the findings in other studies [1, 4, 13].

Experiments at different shear rates show the strain rate sensitivity of brain tissue, see Fig. 4. From these results, it can be observed that also in the non-linear range, the stress as a function of strain is strain rate dependent and that the response stiffens with increasing strain rate. In Fig. 5, results are shown from stress relaxation experiments. During the loading phase, the stress response weakens above the linear

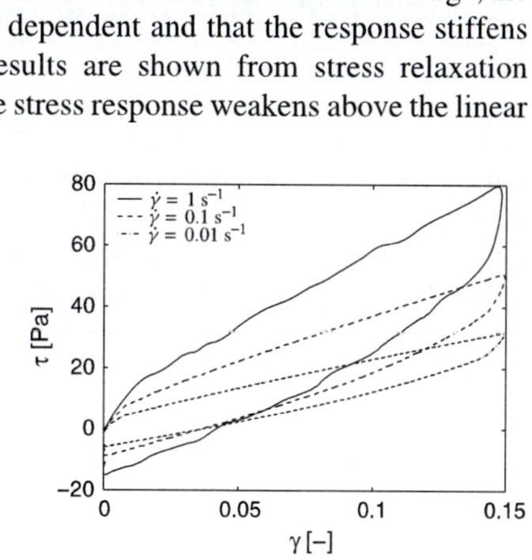

Fig. 4 stress–strain response obtained during constant shear rate experiments at different shear rates. Reprinted from Hrapko et al. [18], with permission from IOS Press

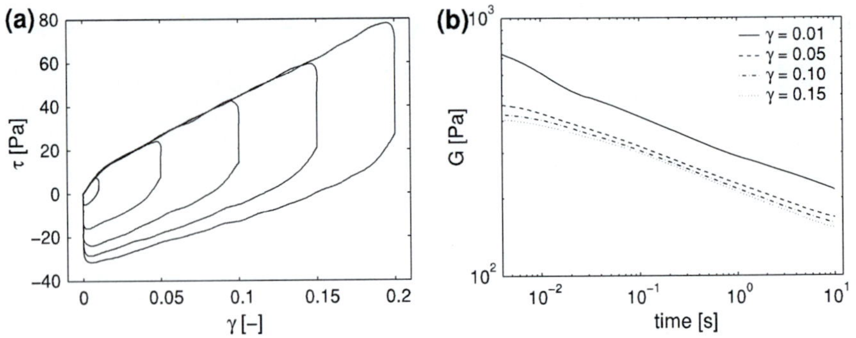

Fig. 5 Results of stress relaxation experiments in shear. **a** Stress–strain behaviour; **b** stress relaxation modulus. Reprinted from Hrapko et al. [18], with permission from IOS Press

viscoelastic strain limit. During stress relaxation, the relaxation modulus does not reach a plateau value within the time range allowed, see Fig. 5b. A decrease of the relaxation modulus with increasing strain was found in stress relaxation measurements by several authors [1, 4, 6, 18, 36, 38, 40]. In Hrapko et al. [18], this effect was decreasing with strain level for strains higher than the linear viscoelastic limit which differs from the results of Brands et al. [6] and Nicolle et al. [36]. The difference may be attributed to the non-homogeneous shear field in case of a conventional centred rotational shear setup. In Hrapko et al. [18] an approximately homogeneous shear field was obtained by placing the samples at the edge of the plate. Similar behaviour was observed in simple shear measurements on a translational shearing device [1, 40, 45]. However, although less pronounced, the same effect was also observed by Bilston et al. [4] with a centered rotational configuration.

3.2 Constitutive Modelling of Brain Tissue

The ability of a constitutive model to accurately describe the mechanical response of brain tissue is important for reliable simulations of head injury. Therefore, a model is required that describes the response to large deformations in different deformation modes. The ability of a model to describe the anisotropic behaviour of brain tissue may be important for some regions of the brain, although the degree of anisotropy observed for brain tissue is relatively low [20, 36, 40, 46].

A number of constitutive models have been developed for brain tissue. Many researchers used an integral viscoelastic model, often in combination with Ogden hyperelasticity to describe the viscoelastic behaviour of brain tissue [11, 27, 28, 36, 40, 45], and some propose a differential constitutive equation [4, 8, 13, 18].

Arbogast et al. [1] fitted an integral viscoelastic model with a third order polynomial for the instantaneous elastic response to data from stress relaxation tests. Darvish et al. [10–12] investigated the use of two integral constitutive models, that is the QLV model and a third order Green–Rivlin viscoelastic non-linear model (GR). The latter model was shown to be superior at frequencies above 44 Hz. The elastic response of the QLV model showed strain hardening for shear strains above 10% which is in contradiction with QLV models proposed by Prange et al. [39] and Mendis et al. [27]. Also Takhounts et al. [44, 45] used linear, quasi linear and GR non-linear viscoelastic constitutive models that were fitted to results from stress relaxation experiments. Mendis et al. [27] proposed a QLV constitutive equation with a Mooney–Rivlin elastic function, where the viscoelastic properties were characterised by time-dependent coefficients. They fitted this constitutive equation to the measured data from constant strain rate compression tests. A QLV model with a Mooney–Rivlin function was presented by Miller and Chinzei [31] and was fitted to the results from unconfined compression experiments on porcine brain tissue. The model agreed well with the experimental data for compression levels reaching 30%. Miller [28] fitted a viscoelastic model with a polynomial strain energy function to the same results and obtained similar results as with the previous model. Miller [29] modified the non-linear model presented in Miller and Chinzei [31] and then compared the prediction with the coefficients from in-vitro tests with results from in-vivo indentation experiments. Predicted forces were approximately 31% lower than those recorded during the experiment. Based on the results from tensile experiments Miller [30] proposed a non-linear viscoelastic model based on the Ogden strain energy hyperelastic function. The model was found to apply well for tissue properties in compression as well as in tension for strains up to 30%. In Prange et al. [39, 41] and Prange and Margulies [40], the non-linear material properties were modelled with a QLV model in combination with a first order Ogden hyperelastic function. This model predicted data from shear measurements well and the result was validated with compression data. Nicolle et al. [35] fitted a five mode viscoelastic model in a QLV framework to dynamic frequency data, where an Ogden hyperelastic function was used to describe the equilibrium moduli from relaxation tests at different strain levels.

Bilston et al. [4] concluded that a quasi linear viscoelastic (QLV) integral model is not sufficient and used a multi-mode upper convected Maxwell model in combination with a Mooney–Rivlin hyperelastic response, multiplied by a strain dependent function and a stress-dependent viscosity. This model performed well, comparing results from stress relaxation, constant strain rate and dynamic measurements in shear. However it was not validated in other deformation modes. Two drawbacks of this non-linear constitutive equation were mentioned, that is computational expensiveness and complexity. Brands et al. [8] proposed a non-linear viscoelastic constitutive model in a differential framework in combination with a second order Mooney–Rivlin model. However, because of the higher order terms present in this model, application for strains outside the range for which the model was characterised should be done with care. In the model by Hrapko et al. [18], a non-linear elastic mode is used in combination with a number of viscoelastic modes showing viscous shear thinning behaviour. The most characteristic aspect of this model is however the non-linear equilibrium response. The model is formulated in a general differential viscoelastic framework that is appropriate for large deformations [38]. This model will be discussed in more detail in Sect. 4.

In summary, several authors [11, 27, 28, 36, 40, 45] have used an Ogden model in a viscoelastic framework to describe the non-linear response of brain tissue. In this model, the elastic strain energy is given by:

$$W = \frac{\mu}{\alpha}\left(\lambda_1^\alpha + \lambda_2^\alpha + \lambda_3^\alpha\right), \tag{15}$$

where λ_i are the principal stretch ratios and μ and α are material parameters. For incompressible materials, this leads to the following elastic stress–strain relation:

$$\boldsymbol{\sigma} = \mu\left(\lambda_1^\alpha \boldsymbol{n}_1\boldsymbol{n}_1 + \lambda_2^\alpha \boldsymbol{n}_2\boldsymbol{n}_2 + \lambda_3^\alpha \boldsymbol{n}_3\boldsymbol{n}_3\right), \tag{16}$$

with $\boldsymbol{n}_i$ the principal stretch directions. Note that an additional hydrostatic term should be added to this stress tensor. Other authors [e.g., 4, 8, 18, 27, 31] have used a Mooney–Rivlin model instead. In some cases, additional higher order terms or an additional non-linear function have been used [4, 8, 18]. In most studies [1, 11, 27, 30, 35, 40, 45], an integral constitutive equation was used to model viscoelastic effects in the mechanical behaviour of brain tissue. However, Bilston et al. [4], Darvish and Crandall [10, 11], and Takhounts et al. [45] have pointed out the incapability of the QLV integral models to describe the large strain response of this material. Differential formulations provide a general framework for describing viscoelastic material behaviour with the appropriate choice of the (linear or non-linear) constitutive relations for the elastic and viscous components and were used by several authors [4, 8, 13, 18].

Constitutive models for brain tissue should be validated in different deformation modes such as shear and compression. Some studies have presented combined shear and compression measurement results and subsequent model predictions [19, 40, 43]. Prange and Margulies [40] have shown a good model prediction of shear stress relaxation; however the compressive response was validated only with the equilibrium stress obtained from stress relaxation. Shen et al. [43] have validated a simplified version of a constitutive model developed by Bilston et al. [4] with constant strain rate measurements in compression only up to a strain of 5%. However, in both of these studies, the model tends to underpredict the tissue response in compression deformation, which is similar to results obtained by Hrapko et al. [19], where compression and shear results obtained for the same sample and on the same setup were used.

Sofar, most finite element models of the head have considered isotropic material behaviour only. However, in particular white matter in some regions consists of highly oriented axons. In these axons, neurofilaments (i.e. fibers) contribute to the strength of the axons in their axial directions. As a consequence, the mechanical properties of this material will be anisotropic, with the properties being dependent on the axonal orientation. For this reason, an anisotropic model of brain tissue may be based on a strain energy function with a term $W_\mathrm{f}(\lambda_\mathrm{f})$ that depends on the stretch ratio λ_f in the direction of the axons, see Eq. 10.

Transversely isotropic models for brain tissue of this type have been considered by for example Meaney [26] and Velardi et al. [46].

The models discussed above have been developed predominantly for the prediction of traumatic brain injury as a result of impact conditions. Under these high rate loading conditions, the brain tissue can be assumed to be nearly incompressible with a bulk modulus that is several orders of magnitude higher than the shear modulus. Under slow (quasi-static) loading conditions, however, the volumetric deformation of the tissue may be considerably larger and some authors have suggested a porosity model for these conditions [e.g., 14].

4 Non-linear Viscoelastic Model for Brain Tissue

A non-linear viscoelastic model for brain tissue was developed based on rheological shear measurements [18]. This model is formulated in a general differential viscoelastic framework that is applicable to large deformations and consists of multiple Maxwell elements. In this section, a detailed overview of a compressible form of this model is given, as well as a numerical implementation scheme that is based on an explicit integration procedure.

4.1 Constitutive Model

A parallel arrangement of N (deviatoric) viscoelastic modes, denoted by the subscript "ve", and a non-linear elastic mode, denoted by the subscript "e", is used to model the mechanical behaviour of brain tissue. The total Cauchy stress in this model is then written as:

$$\boldsymbol{\sigma} = \boldsymbol{\sigma}_{\mathrm{e}} + \sum_{i=1}^{N} \boldsymbol{\sigma}^{\mathrm{d}}_{\mathrm{ve}_i}, \tag{17}$$

in which the superscript "d" indicates a deviatoric tensor. The viscoelastic model is schematically represented in Fig. 6. It shows the elastic mode ($\boldsymbol{\sigma}_{\mathrm{e}}$) and the viscoelastic modes ($\boldsymbol{\sigma}^{\mathrm{d}}_{\mathrm{ve}_i}$ with $i = 1, \ldots, N$).

The elastic stress is derived from a generalised compressible strain energy function of the form:

$$W_{\mathrm{e}} = \tilde{W}_{\mathrm{e}}(\tilde{I}_1, \tilde{I}_2) + W^{\mathrm{h}}(I_3), \tag{18}$$

where $\tilde{I}_1 = I_3^{-\frac{1}{3}} I_1$, $\tilde{I}_2 = I_3^{-\frac{2}{3}} I_2$ are the invariants of the isochoric Finger tensor $\tilde{\boldsymbol{B}} = I_3^{-\frac{1}{3}} \boldsymbol{B}$ and where for the term $\tilde{W}_{\mathrm{e}}(\tilde{I}_1, \tilde{I}_2)$ the following function is used:

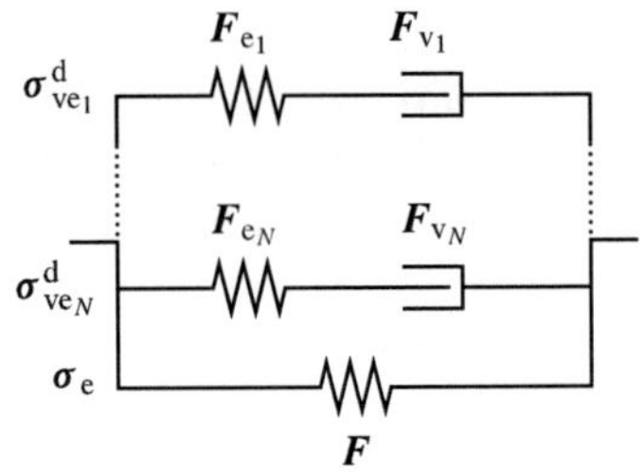

Fig. 6 Schematic illustration of the viscoelastic model. The viscoelastic modes ($\boldsymbol{\sigma}^d_{ve_i}$) are represented by elastic springs (elastic deformation gradient tensor $\boldsymbol{F}_{e_i}$) and dashpots (viscous deformation gradient tensor $\boldsymbol{F}_{v_i}$) and a non-linear elastic mode ($\boldsymbol{\sigma}_e$) is represented by an elastic spring

$$\tilde{W}_e = G_\infty \left\{ -\frac{(1-A)}{C^2} [(C\tilde{x}+1)\exp(-C\tilde{x}) - 1] + \frac{1}{2} A\tilde{x}^2 \right\}, \tag{19}$$

with

$$\tilde{x} = \sqrt{b\tilde{I}_1 + (1-b)\tilde{I}_2 - 3}. \tag{20}$$

The stress in the elastic mode can be written as:

$$\boldsymbol{\sigma}_e = \boldsymbol{\sigma}^d_e + \boldsymbol{\sigma}^h_e, \tag{21}$$

where "h" indicates the hydrostatic part. The deviatoric part of the elastic stress is then obtained as:

$$\begin{aligned} \boldsymbol{\sigma}^d_e &= \frac{2}{\sqrt{I_3}} \left[\frac{\partial \tilde{W}}{\partial \tilde{I}_1} \tilde{\boldsymbol{B}}^d - \frac{\partial \tilde{W}}{\partial \tilde{I}_2} (\tilde{\boldsymbol{B}}^{-1})^d \right] \\ &= \frac{G_\infty}{\sqrt{I_3}} \left[(1-A)\exp\left(-C\sqrt{b\tilde{I}_1 + (1-b)\tilde{I}_2 - 3} \right) + A \right] \end{aligned} \tag{22}$$

$$\left[b\tilde{\boldsymbol{B}}^d - (1-b)(\tilde{\boldsymbol{B}}^{-1})^d \right], \tag{23}$$

with G_∞ the elastic shear modulus. The parameter b distinguishes between the first and the second invariant of $\tilde{B}$ and is often taken as 1. The parameters A and C determine the non-linearity of the response. The shapes of the partial derivatives of the strain energy function were chosen based on the non-linear strain dependence of the equilibrated response to shear relaxation tests and resulted in the strain energy function of Eq. 19. For the hydrostatic part, the following strain energy function is assumed:

$$W^h = \frac{1}{2} K (\sqrt{I_3} - 1)^2, \tag{24}$$

with K the bulk modulus and $\sqrt{I_3} = \det(\boldsymbol{F})$ the volume ratio. Then, the hydrostatic part of the elastic stress is given by:

$$\boldsymbol{\sigma}^{\mathrm{h}} = 2\sqrt{I_3}\frac{\partial W}{\partial I_3}\boldsymbol{I} \tag{25}$$

$$= K(J-1)\boldsymbol{I} \qquad \text{with} \qquad J = \sqrt{I_3}. \tag{26}$$

The second term on the right hand side of Eq. 17 consists of a summation of the viscoelastic modes. A general differential framework is used to describe each viscoelastic mode. In this framework, the deformation gradient tensor $\boldsymbol{F}$ is partitioned into an elastic deformation gradient tensor $\boldsymbol{F}_{\mathrm{e}_i}$ and a viscous deformation gradient tensor $\boldsymbol{F}_{\mathrm{v}_i}$ [37]. The partitioning of the viscoelastic deformations is depicted in Fig. 7. For simplicity, the subscript i indicating the viscoelastic mode i is omitted throughout the remaining part of this section. A multiplicative decomposition of the deformation gradient tensor $\boldsymbol{F}$ is assumed:

$$\boldsymbol{F} = \boldsymbol{F}_{\mathrm{e}} \cdot \boldsymbol{F}_{\mathrm{v}}. \tag{27}$$

The decomposition involves a fictitious intermediate state, which could exist after application of only the viscous deformation gradient tensor $\boldsymbol{F}_{\mathrm{v}}$. After application of the elastic deformation tensor $\boldsymbol{F}_{\mathrm{e}}$, this stress-free state transforms into the final state. By using Eq. 27, with the definition of the velocity gradient tensor $\boldsymbol{L} = \dot{\boldsymbol{F}} \cdot \boldsymbol{F}^{-1}$, it follows that

$$\boldsymbol{L} = \boldsymbol{L}_{\mathrm{e}} + \boldsymbol{L}_{\mathrm{v}} \quad \text{with} \quad \boldsymbol{L}_{\mathrm{e}} = \dot{\boldsymbol{F}}_{\mathrm{e}} \cdot \boldsymbol{F}_{\mathrm{e}}^{-1} \quad \text{and} \quad \boldsymbol{L}_{\mathrm{v}} = \boldsymbol{F}_{\mathrm{e}} \cdot \dot{\boldsymbol{F}}_{\mathrm{v}} \cdot \boldsymbol{F}_{\mathrm{v}}^{-1} \cdot \boldsymbol{F}_{\mathrm{e}}^{-1}. \tag{28}$$

The velocity gradient tensor $\boldsymbol{L}$ can also be decomposed as

$$\boldsymbol{L} = \boldsymbol{D} + \boldsymbol{\Omega} \quad \text{and} \quad \boldsymbol{L}_{\mathrm{e}} = \boldsymbol{D}_{\mathrm{e}} + \boldsymbol{\Omega}_{\mathrm{e}} \quad \text{and} \quad \boldsymbol{L}_{\mathrm{v}} = \boldsymbol{D}_{\mathrm{v}} + \boldsymbol{\Omega}_{\mathrm{v}}. \tag{29}$$

in which $\boldsymbol{D} = \frac{1}{2}(\boldsymbol{L} + \boldsymbol{L}^{\mathrm{T}})$ is the symmetric rate of deformation tensor and $\boldsymbol{\Omega} = \frac{1}{2}(\boldsymbol{L} - \boldsymbol{L}^{\mathrm{T}})$ is the skew-symmetric spin tensor. To obtain a unique intermediate state, the viscous deformations are chosen to be spin-free, so that $\boldsymbol{\Omega}_{\mathrm{v}} = \boldsymbol{0}$.

For the response of the viscoelastic modes, in analogy to the elastic mode, the elastic strain energy is assumed to be given by

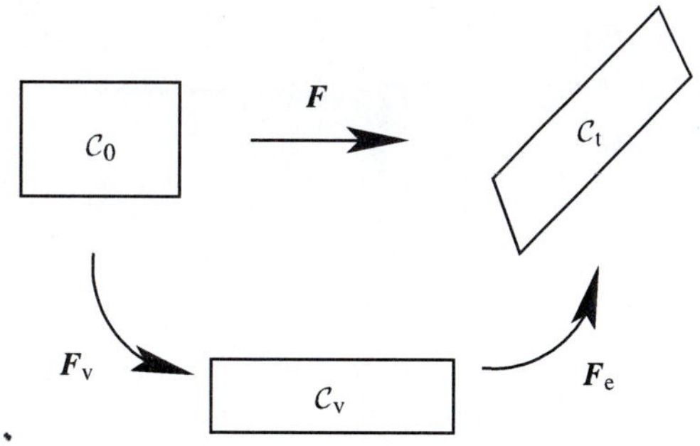

Fig. 7 Partitioning of the deformation gradient tensor $\boldsymbol{F}$ into an elastic deformation gradient tensor $\boldsymbol{F}_{\mathrm{e}}$ and a viscous deformation gradient tensor $\boldsymbol{F}_{\mathrm{v}}$ in order to obtain a fictitious stress-free state C_{v} between the initial configuration C_0 and the current configuration C_{t}

$$W_{\mathrm{ve}} = \frac{1}{2}G\left[a\tilde{I}_{\mathrm{e}_1} + (1-a)\tilde{I}_{\mathrm{e}_2} - 3\right], \tag{30}$$

with G the shear modulus, $\tilde{I}_{\mathrm{e}_1}$ and $\tilde{I}_{\mathrm{e}_2}$ the invariants of the isochoric elastic Finger tensor $\tilde{\boldsymbol{B}}_{\mathrm{e}}$, and a a fitting parameter. As a result, the viscoelastic stress is given by:

$$\boldsymbol{\sigma}_{\mathrm{ve}}^{\mathrm{d}} = \frac{G}{\sqrt{I_3}}\left[a\tilde{\boldsymbol{B}}_{\mathrm{e}}^{\mathrm{d}} - (1-a)(\tilde{\boldsymbol{B}}_{\mathrm{e}}^{-1})^{\mathrm{d}}\right]. \tag{31}$$

The viscous deformation $\boldsymbol{F}_{\mathrm{v}}$ is assumed to be volume-invariant, so that $\det(\boldsymbol{F}_{\mathrm{v}}) = 1$ and $\det(\boldsymbol{F}_{\mathrm{e}}) = \det(\boldsymbol{F})$. This justifies the use of the third invariant of $\boldsymbol{B}$ instead of $\boldsymbol{B}_{\mathrm{e}}$ in Eq. 31.

The viscous rate of deformation tensor is given by the following flow rule, assuming viscous incompressibility:

$$\boldsymbol{D}_{\mathrm{v}} = \frac{\boldsymbol{\sigma}_{\mathrm{ve}}^{\mathrm{d}}}{2\eta(\tau)}, \tag{32}$$

where the dynamic viscosity η is a function of the scalar equivalent stress measure $\tau = \sqrt{\frac{1}{2}\boldsymbol{\sigma}^{\mathrm{d}} : \boldsymbol{\sigma}^{\mathrm{d}}}$. It is described by the Ellis model, which states

$$\eta(\tau) = \eta_\infty + \frac{\eta_0 - \eta_\infty}{1 + \left(\frac{\tau}{\tau_0}\right)^{n-1}}. \tag{33}$$

This model considers a powerlaw dependence, with a plateau value η_0 for small stress levels and a value of η_∞ for infinite values of τ. The initial value for viscosity is defined as $\eta_0 = G\lambda$, whereas the infinite viscosity is written as $\eta_\infty = k\eta_0$. In Fig. 8, predictions with an incompressible version of this model are compared with experimental results obtained for shear [18]. This model was also validated for compressive loading in Hrapko et al. [19].

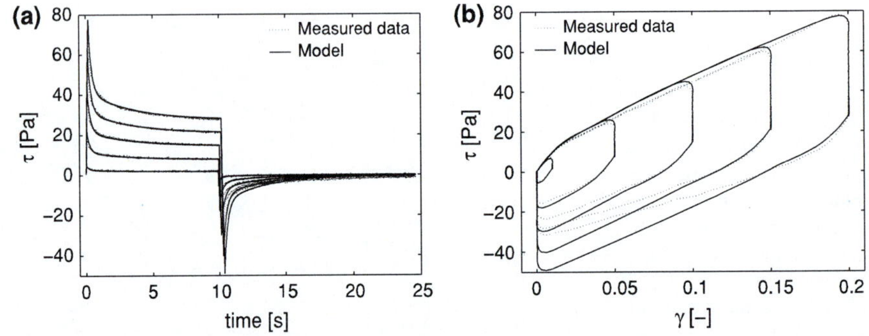

Fig. 8 Model fit and prediction for stress relaxation experiments. Reprinted from Hrapko et al. [18], with permission from IOS Press

4.2 Numerical Implementation

In order to use the previously presented constitutive model in a finite element head model, an incremental integration scheme is required. In this section, a framework is given that is based on an explicit time integration scheme for the viscous right Cauchy–Green deformation tensor. This tensor is defined as

$$\boldsymbol{C}_{\mathrm{v}} = \boldsymbol{F}_{\mathrm{v}}^{\mathrm{T}} \cdot \boldsymbol{F}_{\mathrm{v}}. \tag{34}$$

In order to update the tensor $\boldsymbol{C}_{\mathrm{v}}$ in the time integration scheme use is made of its time derivative, which can be written as:

$$\dot{\boldsymbol{C}}_{\mathrm{v}} = \boldsymbol{F}^{\mathrm{T}} \cdot \boldsymbol{B}_{\mathrm{e}}^{-1} \cdot [(\boldsymbol{L} - \boldsymbol{L}_{\mathrm{e}}) \cdot \boldsymbol{B}_{\mathrm{e}} + \boldsymbol{B}_{\mathrm{e}} \cdot (\boldsymbol{L}^{\mathrm{T}} - \boldsymbol{L}_{\mathrm{e}}^{\mathrm{T}})] \cdot \boldsymbol{B}_{\mathrm{e}}^{-1} \cdot \boldsymbol{F}. \tag{35}$$

Using the assumption of a spin-free viscous deformation, the time derivative of the viscous right Cauchy–Green deformation tensor for a unique intermediate state yields

$$\dot{\boldsymbol{C}}_{\mathrm{v}} = 2 \cdot \boldsymbol{F}^{\mathrm{T}} \cdot \boldsymbol{B}_{\mathrm{e}}^{-1} \cdot \boldsymbol{D}_{\mathrm{v}} \cdot \boldsymbol{F}, \tag{36}$$

where the elastic Finger tensor is calculated by

$$\boldsymbol{B}_{\mathrm{e}} = \boldsymbol{F} \cdot \boldsymbol{C}_{\mathrm{v}}^{-1} \cdot \boldsymbol{F}^{\mathrm{T}}. \tag{37}$$

This evolution law can then be integrated explicitly.

For the computation of the Cauchy stress (Eq. 17), for a given deformation $\boldsymbol{F}(t)$, at the end of an increment from $t - \Delta t$ to t, several steps are taken (see Fig. 9).

The elastic stresses $\boldsymbol{\sigma}_{\mathrm{e}}^{\mathrm{h}}(t)$ and $\boldsymbol{\sigma}_{\mathrm{e}}^{\mathrm{d}}(t)$ can be computed directly. For the time integration of the viscoelastic stress modes, Heun's method, also known as the improved Euler method, is used [7]. First, a prediction (indicated by the superscript "*") of the time derivatives of the viscous right Cauchy–Green deformation tensors $\boldsymbol{C}_{\mathrm{v}_i}^{*}(t)$ are computed using the Euler forward method. These are used to calculate $\boldsymbol{\sigma}_{\mathrm{ve}_i}^{*\mathrm{d}}(t)$ and $\dot{\boldsymbol{C}}_{\mathrm{v}_i}^{*}(t)$. After that, the tensors $\boldsymbol{C}_{\mathrm{v}_i}(t)$ are more accurately determined by applying a trapezoidal rule, after which $\boldsymbol{\sigma}_{\mathrm{ve}_i}^{\mathrm{d}}(t)$ and subsequently $\boldsymbol{\sigma}$ are calculated. For each time step, the values of tensors $\boldsymbol{C}_{\mathrm{v}_i}(t)$ and $\dot{\boldsymbol{C}}_{\mathrm{v}_i}(t)$, are stored for the next time increment. For the initial values, no deformation is assumed, which yields $\boldsymbol{F}(0) = \boldsymbol{I}$, $\boldsymbol{C}_{\mathrm{v}_i}(0) = \boldsymbol{I}$, and $\dot{\boldsymbol{C}}_{\mathrm{v}_i}(0) = \boldsymbol{0}$. The complete integration scheme is given in the appendix.

5 Influence of Constitutive Modelling in a 3D Head Model

A number of constitutive models have been developed to describe the mechanical behaviour of brain tissue. Important for the use of these models to predict injury

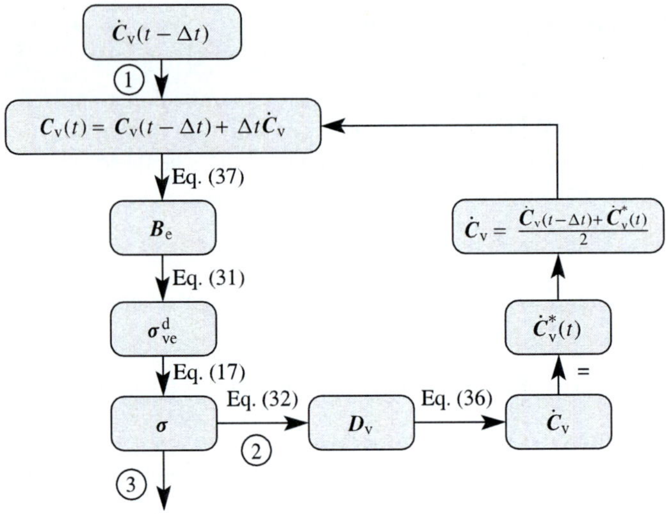

Fig. 9 Flow diagram of integration scheme. Each increment starts at *1*. Subsequently, routes *2* and *3* are followed

may be the ability to correctly describe the non-linear behaviour for complex loading histories and large deformations in different deformation modes. To be able to use uniform tissue level injury criteria, i.e. injury criteria that can be transferred between different head models, accurate stress and strain levels should be predicted, for which the correct description of the constitutive response of brain tissue is required.

In this section, the consequences of using different constitutive descriptions of the mechanical response of brain tissue in FE head models for injury prediction are demonstrated. For this purpose, a non-linear viscoelastic constitutive model for brain tissue is used, see Sect. 4. This model has shown to provide a good description of the response to both shear and compression [18, 19]. Numerical simulations using the constitutive model in a 3D head model are compared with predictions using a simplified version of this model and the constitutive model developed by Brands et al. [8]. These constitutive models are used within a 3D head model with a sliding interface between the brain and the skull. This sliding interface model was compared to a previously validated 3D head model with a tied interface [7, 9] in Hrapko et al. [21].

5.1 Model Description

A 3D numerical head model with a sliding interface was used to investigate the consequences of constitutive non-linearities by comparing the performance of different constitutive models. For these simulations the explicit FE code MADYMO,

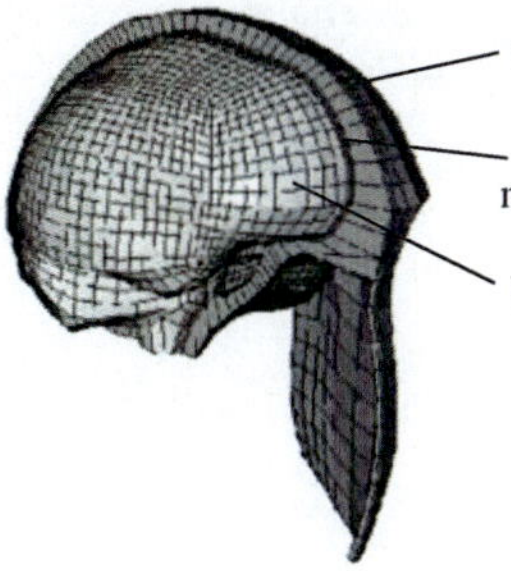

Fig. 10 3D FE head model [9, 7]

version 6.3.2 was used. The finite element head model (see Fig. 10) is based on a model developed by Claessens et al. [9] and Brands et al. [7]. In their model, the brain, dura mater and skull were tied, not allowing sliding and separation of each other. This model was validated [9] using the experimental data of Nahum et al. [34]. All anatomical components are described by solid, reduced integration, eight node brick elements. The skull consists of 3,212 elements, the dura mater consists of 3,178 elements, and the brain is modelled by 7,478 elements. The skull is assumed to be rigid and the dura mater is modelled as a linear elastic material. A skull–brain interface is used for which the original model is extended with a sliding interface [5, 32]. The interface between the skull and the brain was modelled by a 0.5 mm gap between the dura mater and the pia mater. The brain is not restrained in the foramen magnum in the current model. This was previously compared with the brain being restrained in the foramen magnum, showing only small differences in strains and stresses [5]. The pia mater is composed of 3,210 shell elements with 0.5 mm thickness which completely envelope the brain. Sliding and separation is enabled between the dura mater and pia mater by a frictionless contact condition. The pia mater is modelled as a linear elastic material. Material parameters of various components are summarised in Table 1.

The brain tissue was assumed to be homogeneous, isotropic and was described by three different constitutive models:

- NHVE: a simplified version of the constitutive model presented in Sect. 4, obtained with $k = 1, A = 1$. The resulting model is a visco-hyperelastic model in which the elastic behaviour is neo-Hookean. The model is also known as the Upper Convected Maxwell (UCM) model.
- NLVE-A: the non-linear viscoelastic model presented in Sect. 4 which is an extension of the UCM and was described in the previous section. The model is

Table 1 Material parameters of the various anatomical components of the FE head model

	Young's modulus E (MPa)	Poisson's ratio ν (−)	Mass density ρ (kg/m^3)
Skull	–	–	2,070
Dura mater	31.5	0.45	1,130
Pia mater	31.5	0.45	1,130

Table 2 Material parameters of the NLVE-A model

Elastic and viscous parameters	Viscoelastic parameters
$G_\infty = 182.9$ Pa	$G_1 = 9{,}884$ Pa $\lambda_1 = 0.00013$ s
$A = 0.73$	$G_2 = 835.5$ Pa $\lambda_2 = 0.012$ s
$C = 15.6$	$G_3 = 231.2$ Pa $\lambda_3 = 0.35$ s
$n = 1.65$	$G_4 = 67.1$ Pa $\lambda_4 = 4.62$ s
$\tau_0 = 9.7$ Pa	$G_5 = 3.61$ Pa $\lambda_5 = 12.1$ s
$k = 0.39$	$G_6 = 2.79$ Pa $\lambda_6 = 54.3$ s
$K = 2.5$ GPa	

of a multi-mode Maxwell type and consists of a non-linear elastic mode in combination with a number of viscoelastic modes. The material parameters of the model (see Table 2) were determined in [18] and extended with an extra viscoelastic mode for higher frequencies based on experimental data by Shen et al. [19] and data obtained in [46].

- NLVE-B: the non-linear constitutive model developed in Brands et al. [7], using the material parameters obtained from the same study. The model is a non-linear extension of a multi-mode UCM model. The elastic behaviour is modelled by a hyperelastic, higher order Mooney–Rivlin formulation. The inelastic, time-dependent behaviour is modelled using a Newtonian law, acting on the deviatoric part of the stress only.

For each constitutive model, the skull was subjected to translational and rotational accelerations in three directions each, to study the model response in a total of six different loading directions, see Fig. 11. In addition to anterior–posterior (AP) translation and posterior–anterior (PA) rotation, translation in lateral and superior–inferior (SI) direction and rotation in lateral and axial direction were applied. In all rotational cases, the axis of rotation was chosen to coincide with the anatomical origin of the model, corresponding to the ear hole projected to the sagittal plane.

The skull was subjected to sinusoidal acceleration pulses for translation and rotation with six different acceleration pulse amplitudes with a duration of 5 ms. The translational acceleration amplitudes were chosen according to HIC values of 10, 100, 500, 1,000, 1,500 and 2,000. These values were chosen in order to study

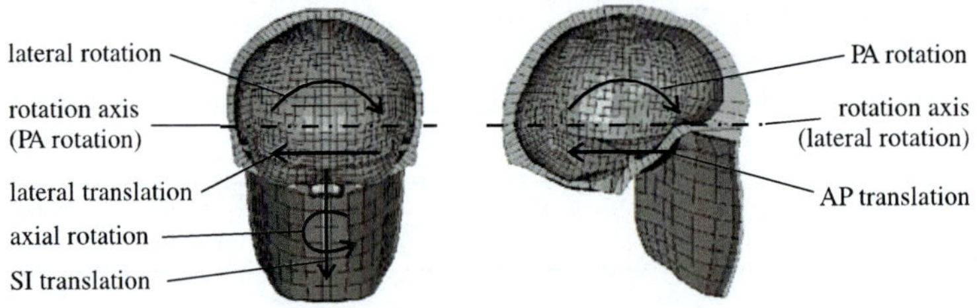

Fig. 11 Loading directions for translational and rotational accelerations applied to the skull. *PA* posterior–anterior, *AP* anterior–posterior, *SI* superior–inferior)

the model response in a range of loading conditions covering the non-injurious to the injurious regime. The acceleration amplitudes for rotation were chosen to produce translational acceleration in the upper brain–skull interface approximately corresponding to the previously chosen translational acceleration levels.

Results will be presented in terms of equivalent Von Mises stress and strain levels which are defined as: $\bar{\sigma} = \sqrt{\frac{3}{2}\boldsymbol{\sigma}^{\mathrm{d}} : \boldsymbol{\sigma}^{\mathrm{d}}}$ and $\bar{\varepsilon} = \sqrt{\frac{3}{2}\boldsymbol{\varepsilon}^{\mathrm{d}} : \boldsymbol{\varepsilon}^{\mathrm{d}}}$, respectively, with $\boldsymbol{\sigma}$ the Cauchy stress tensor and $\boldsymbol{\varepsilon} = \frac{1}{2}(\boldsymbol{I} - \boldsymbol{B})$ the left Green–Lagrange strain tensor.

5.2 Results

The patterns of Von Mises stress and Von Mises strain predictions with the FE head model using different constitutive models for brain tissue are compared in Fig. 12. Stress and strain predictions obtained with the different constitutive models show similar patterns. However, the magnitudes of stress and strain concentrations differ. During simulations with translational acceleration, the highest deformations were in all cases caused by the brain being obstructed by relatively stiff parts of the dura mater, the falx cerebri or the tentorium cerebelli. For all models, during AP translation, the maximum deformations were observed in the

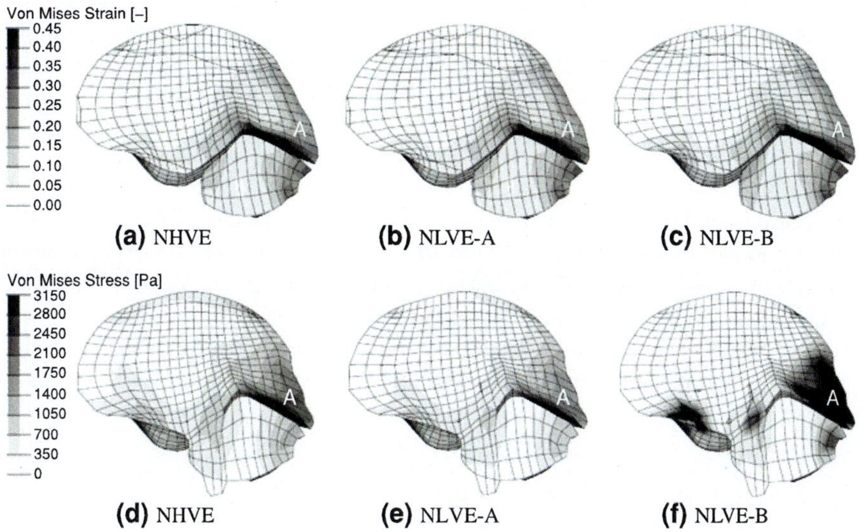

Fig. 12 FE model predictions of (**a–c**) Von Mises strain in a parasagittal cross-section (17.5 mm off-centre) during AP translation with an acceleration of 730 m/s^2 and (**d–f**) Von Mises stress in a parasagittal cross-section (10 mm off-centre) during AP translation with an acceleration of 2,400 m/s^2 at 5 ms. **a**, **d** NHVE model, **b**, **e** NLVE-A model, **c**, **f** NLVE-B model. Reprinted from Hrapko et al. [21], with permission from Taylor & Francis Ltd

posterior and inferior side of the region corresponding to the occipital lobe of the brain at approximately 5 ms, indicated by A in Fig. 12.

For all models during PA rotation, the maximum deformations were observed in the anterior, superior and inferior side of the region corresponding to the frontal lobe and anterior side of the region corresponding to the temporal lobe of the brain at approximately 9 ms. For lateral rotation, the maximum deformations were observed in the posterior side of the corpus callosum of the brain which is caused by the falx cerebri at 10–13 ms [21]. The second major deformation area was found to be the inferior side of the region corresponding to the frontal lobe at the same time. For a discussion of results for other loading conditions, see [21].

Von Mises stress and Von Mises strain values that are exceeded by various amounts of elements using different constitutive models for brain tissue are shown in Fig. 13 for translation. For simplicity, only maximum values obtained during the first 15 ms of the simulation are shown. The entire time-dependent response and similar results for rotational loading can be found in [21].

The differences found between the NLVE-A and NHVE models are slightly dependent on the applied acceleration amplitude in both translation and rotation. Also the differences found between predictions with the NLVE-A and NLVE-B model for translation are slightly dependent on the applied acceleration amplitude. However, the differences found between the NLVE-A and NLVE-B model predictions in rotation are strongly depending on the applied acceleration amplitude. In the case of translational acceleration, the relative differences between predictions with different constitutive models are up to 55%. When assuming the NLVE-A model to be the reference model, the NHVE constitutive model predicts up to 10% smaller strains and up to 40% larger stresses, depending on the acceleration amplitude (Fig. 14). The strain predictions from the NLVE-B constitutive model are up to 15% larger than the strains obtained with the NLVE-A constitutive model, and the stresses predicted are up to 55% smaller than the stresses predicted when using the NLVE-A constitutive model. In the case of rotational acceleration, the relative differences between predictions of different constitutive models are up to 630% in the case of 24 krad/s^2. The NHVE constitutive model predicts up to 17% smaller strains and up to 27% larger stresses than the NLVE-A model. The strain predictions from the NLVE-B constitutive model are up to 450% smaller than the stress obtained with the NLVE-A constitutive model, and the strain predictions are up to 630% larger than the strains predicted when using the NLVE-A constitutive model.

5.3 Discussion and Conclusions

To obtain a reliable prediction of a mechanical response from any FE model, the constitutive models used for the materials involved have to be chosen carefully. In this section, the consequences of using different material models for brain tissue in 3D FE head models for injury prediction were analysed. For this purpose,

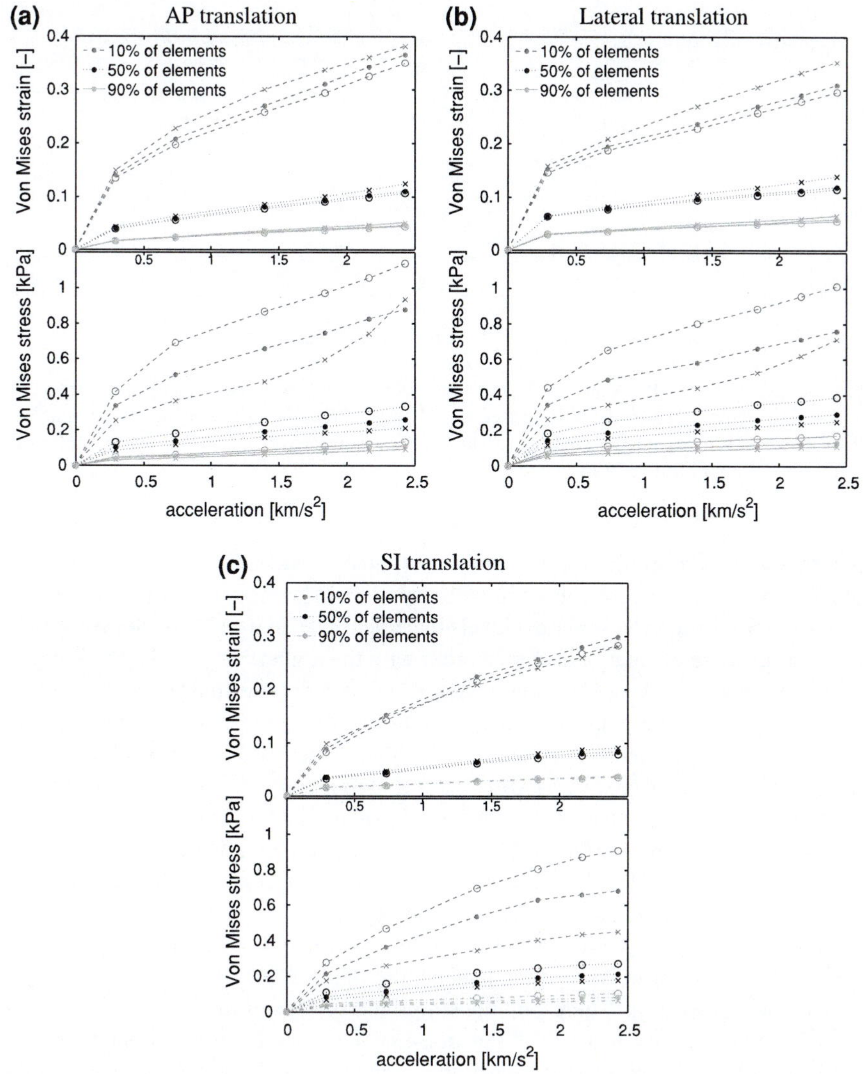

Fig. 13 FE model predictions of Von Mises strains and Von Mises stress in (**a**) anterior–posterior translation, (**b**) lateral translation, and in (**c**) superior–inferior translation. *Open circle* NHVE, *filled circle* NLVE-A, *multiple symbol* NLVE-B. Reprinted from Hrapko et al. [21], with permission from Taylor & Francis Ltd

a non-linear viscoelastic constitutive model for brain tissue, that was presented in Sect. 4, a simplified version of this model, and another non-linear model [8] were compared.

Different constitutive models show similar patterns for the stress and strain predictions, however, the magnitudes differ. Regions exhibiting maximum

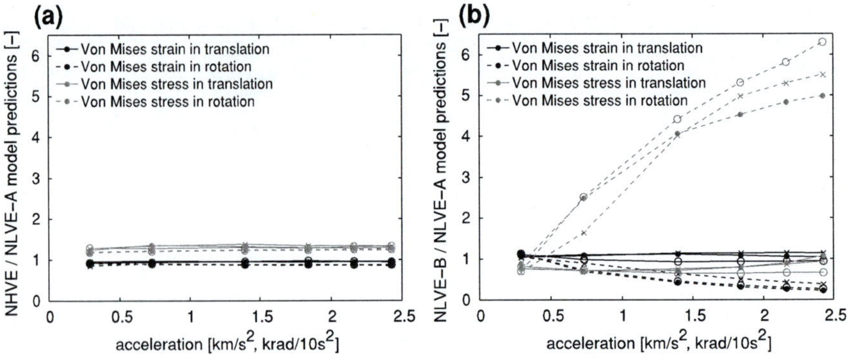

Fig. 14 Difference in the Von Mises strain and Von Mises stress levels that are exceeded by 10% of elements. *Filled circle* PA rotation, AP translation, *multiple symbol* lateral rotation, translation, *open circle* axial rotation, SI translation. Reprinted from Hrapko et al. [21], with permission from Taylor & Francis Ltd

deformations were similar for different constitutive models. Comparing predictions of the two non-linear constitutive models, relative differences of 4.5 for Von Mises strain and 6.3 for Von Mises stress levels that are exceeded by 10% of elements were found in the case of rotational acceleration with the highest amplitude. Differences found in translational acceleration were more than a decade smaller. An important observation is that these differences were non-linearly increasing with an increasing amplitude of the acceleration applied. This is due to the (hardening) non-linearity of the stress response of the NLVE-B constitutive model above shear strains of 0.25. This non-linearity is considered to be non-physical.

Contrary, the variation in the Von Mises stress and Von Mises strain response between the simplified and the non-linear version of the constitutive model of [18] which was found to accurately describe the non-linear response of brain tissue in both shear and compression, was found to be only up to 17 and 40%, respectively. These differences were not dependent on the applied acceleration level in both translation and rotation. Therefore, the simplified version of the recently developed model could be used instead of the non-linear model to obtain reliable injury predictions with FE simulations, since the response can easily be scaled according to the constitutive model used.

It was shown that choosing a different constitutive model for brain tissue to be used in a FE model can have large consequences, depending on the presence of non-linearities in the model. However, in the case of a simplified and non-linear version of a model that has been shown to match the non-linear response of brain tissue, the response predicted with a numerical head model for different conditions (i.e. severity and type of loading) varies consistently with the constitutive behaviour used. Therefore, still a reliable assessment of injury can be made with the less accurate simplified constitutive model by using a model-specific injury criterion that then is not a true threshold for injury of brain tissue.

6 Conclusions

In order to further improve traffic safety, it is required to be able to assess the likelihood of traumatic brain injury due to a mechanical load to the head. By using a finite element model of the head, this assessment can be based on local, tissue level, injury criteria based on stress and strain values. However, to obtain a reliable prediction of these mechanical quantities, the constitutive response of brain tissue should be well understood and modelled.

The mechanical response of brain tissue has been found to be viscoelastic and non-linear by many rheological studies. Moreover, some regions of the brain may show anisotropic behaviour as a result of the oriented microstructure of the tissue. Models aimed at describing this viscoelastic behaviour are based on either an integral approach or a differential approach. The non-linearity is often captured in a non-linear hyperelastic model that is part of the viscoelastic framework. Hyperelastic models that have been used for brain tissue are often based on the Ogden form or on a Mooney–Rivlin model, where for the latter sometimes a higher order formulation or a non-linear pre-factor is used. A specific model based on this approach in a differential framework was discussed in more detail.

The influences of constitutive non-linearities of brain tissue in 3D numerical head model simulations were investigated by comparing the performance of the model of Hrapko et al. with a simplified version, based on neo-Hookean elastic behaviour, and another non-linear constitutive model. In the case of the simplified and fully non-linear version of the model, the response predicted with a head model for varying conditions (i.e. severity and type of loading) varied consistently with the constitutive behaviour. Consequently, when used in a finite element head model, the response can be scaled according to the constitutive model used. However, the differences found when using the non-linear model of Brands et al. [7] were dependent on the loading conditions.

Acknowledgments The authors gratefully acknowledge the support from the European Integrated Project APROSYS and the Dutch Technology Foundation STW, applied science division of NWO and the Technology Program of the Ministry of Economic Affairs.

1 Appendix

In Sect. 4, a constitutive model for brain tissue was presented. In this appendix, a numerical integration scheme for this model based on explicit integration of Eq. 36 is given. For each increment, the state of the material, resulting from a given deformation and the solution at the end of the previous increment, is obtained from the following procedure:

1. Compute the deformation $\boldsymbol{F}(t)$ from the nodal displacements.
2. Compute

$$\boldsymbol{\sigma}_{\mathrm{e}}^{\mathrm{h}}(t) = K(J(t) - 1)\boldsymbol{I}$$

3. Compute

$$\boldsymbol{\sigma}_{\mathrm{e}}^{\mathrm{d}}(t) = \frac{G_\infty}{\sqrt{I_3(t)}}\left[(1 - A)\exp\left(-C\sqrt{b\tilde{I}_1(t) + (1 - b)\tilde{I}_2(t) - 3}\right) + A\right]$$
$$\left[b\tilde{\boldsymbol{B}}^{\mathrm{d}}(t) - (1 - b)(\tilde{\boldsymbol{B}}^{-1}(t))^{\mathrm{d}}\right]$$

4. a. Retrieve $\boldsymbol{C}_{\mathrm{v}_i}(t - \Delta t)$ and $\dot{\boldsymbol{C}}_{\mathrm{v}_i}(t - \Delta t)$ from the previous time increment for each mode $i = 1$ to N.

   b. For mode $i = 1$ to N, predict $\boldsymbol{C}_{\mathrm{v}_i}(t)$, $\boldsymbol{B}_{\mathrm{e}_i}(t)$, and $\boldsymbol{\sigma}_{\mathrm{ve}_i}^{\mathrm{d}}(t)$

$$\boldsymbol{C}_{\mathrm{v}_i}^*(t) = \boldsymbol{C}_{\mathrm{v}_i}(t - \Delta t) + \dot{\boldsymbol{C}}_{\mathrm{v}_i}(t - \Delta t)\Delta t$$

$$\boldsymbol{B}_{\mathrm{e}_i}^*(t) = \boldsymbol{F}(t) \cdot \boldsymbol{C}_i^{*-1}(t) \cdot \boldsymbol{F}^{\mathrm{T}}(t)$$

$$\boldsymbol{\sigma}_{\mathrm{ve}_i}^{*\mathrm{d}}(t) = \frac{G_i}{\sqrt{I_3(t)}}\left[a\tilde{\boldsymbol{B}}_{\mathrm{e}_i}^{*\mathrm{d}}(t) - (1 - a)(\tilde{\boldsymbol{B}}_{\mathrm{e}_i}^{*-1}(t))^{\mathrm{d}}\right]$$

   c. Predict $\boldsymbol{\sigma}^{\mathrm{d}}(t)$ and $\tau(t)$

$$\boldsymbol{\sigma}^{*\mathrm{d}}(t) = \boldsymbol{\sigma}_{\mathrm{e}}^{\mathrm{d}}(t) + \sum_{i=1}^{N} \boldsymbol{\sigma}_{\mathrm{ve}_i}^{*\mathrm{d}}(t)$$

$$\tau^*(t) = \sqrt{\frac{1}{2}\boldsymbol{\sigma}^{*\mathrm{d}}(t) : \boldsymbol{\sigma}^{*\mathrm{d}}(t)}$$

   d. For mode $i = 1$ to N, predict $\eta_i(t)$, $\boldsymbol{D}_{\mathrm{v}_i}(t)$, and $\dot{\boldsymbol{C}}_{\mathrm{v}_i}(t)$

$$\eta_i^*(t) = \eta_{\infty_i} + \frac{\eta_{0_i} - \eta_{\infty_i}}{1 + \left(\frac{\tau^*(t)}{\tau_0}\right)^{(n_i - 1)}}$$

$$\boldsymbol{D}_{\mathrm{v}_i}^*(t) = \frac{\boldsymbol{\sigma}_{\mathrm{ve}_i}^{*\mathrm{d}}(t)}{2\eta_i^*(t)}$$

$$\dot{\boldsymbol{C}}_{\mathrm{v}_i}^*(t) = 2\boldsymbol{F}^{\mathrm{T}}(t) \cdot \boldsymbol{B}_{\mathrm{e}_i}^{*-1}(t) \cdot \boldsymbol{D}_{\mathrm{v}_i}^*(t) \cdot \boldsymbol{F}(t)$$

   and determine

$$\boldsymbol{C}_{\mathrm{v}_i}(t) = \boldsymbol{C}_{\mathrm{v}_i}(t - \Delta t) + \frac{1}{2}\left(\dot{\boldsymbol{C}}_{\mathrm{v}_i}(t - \Delta t) + \dot{\boldsymbol{C}}_{\mathrm{v}_i}^*(t)\right)\Delta t$$

$$\boldsymbol{B}_{\mathrm{e}_i}(t) = \boldsymbol{F}(t) \cdot \boldsymbol{C}_i^{-1}(t) \cdot \boldsymbol{F}^{\mathrm{T}}(t)$$

$$\boldsymbol{\sigma}^{\mathrm{d}}_{\mathrm{ve}_i}(t) = \frac{G_i}{\sqrt{I_3(t)}}\left[a\tilde{\boldsymbol{B}}^{\mathrm{d}}_{\mathrm{e}_i}(t) - (1-a)(\tilde{\boldsymbol{B}}^{-1}_{\mathrm{e}_i}(t))^{\mathrm{d}}\right]$$

and store $\boldsymbol{C}_{\mathrm{v}_i}(t)$ for the next time increment.

e. Determine

$$\boldsymbol{\sigma}^{\mathrm{d}}(t) = \boldsymbol{\sigma}^{\mathrm{d}}_{\mathrm{e}}(t) + \sum_{i=1}^{N} \boldsymbol{\sigma}^{\mathrm{d}}_{\mathrm{ve}_i}(t)$$

$$\tau(t) = \sqrt{\frac{1}{2}\boldsymbol{\sigma}^{\mathrm{d}}(t) : \boldsymbol{\sigma}^{\mathrm{d}}(t)}$$

f. For mode $i = 1$ to N, determine

$$\eta_i(t) = \eta_{\infty_i} + \frac{\eta_{0_i} - \eta_{\infty_i}}{1 + \left(\frac{\tau(t)}{\tau_0}\right)^{(n_i - 1)}}$$

$$\boldsymbol{D}_{\mathrm{v}_i}(t) = \frac{\boldsymbol{\sigma}^{\mathrm{d}}_{\mathrm{ve}_i}(t)}{2\eta_i(t)}$$

$$\dot{\boldsymbol{C}}_{\mathrm{v}_i}(t) = 2\boldsymbol{F}^{\mathrm{T}}(t) \cdot \boldsymbol{B}^{-1}_{\mathrm{e}_i}(t) \cdot \boldsymbol{D}_{\mathrm{v}_i}(t) \cdot \boldsymbol{F}(t)$$

and store $\dot{\boldsymbol{C}}_{\mathrm{v}_i}(t)$ for the next time increment.

5. Compute

$$\boldsymbol{\sigma}(t) = \boldsymbol{\sigma}^{\mathrm{h}}_{\mathrm{e}}(t) + \boldsymbol{\sigma}^{\mathrm{d}}(t)$$

References

1. Arbogast, K.B., Meaney, D.F., Thibault, L.E.: Biomechanical characterization of the constitutive relationship for the brainstem. In: Proceedings of the 39th Stapp Car Crash Conference, SAE 952716, pp. 153–159 (1995)
2. Bain, A.C., Meaney, D.F.: Tissue-level thresholds for axonal damage in an experimental model of cerebral nervous system white matter injury. J. Biomech. Eng. Trans. ASME **122**(6), 615–622 (2000)
3. Bandak, F.A., Eppinger, R.H.: A three-dimensional finite elements analysis of the human brain under combined rotational and translational accelerations. In: Proceedings of the 38th Stapp Car Crash Conference, SAE 942215, pp. 148–163 (1994)
4. Bilston, L.E., Liu, Z., Phan-Thien, N.: Large strain behavior of brain tissue in shear: some experimental data and differential constitutive model. Biorheology **38**(3), 335–345 (2001)
5. Brands, D.W.A.: Implementation of Sliding Interface in the tu/e fe Head Model. Technical Report, TNO Automotive, The Netherlands (2002)
6. Brands, D.W.A., Bovendeerd, P.H.M., Peters, G.W.M., Wismans, J.S.H.M.: The large shear strain dynamic behavior of in-vitro porcine brain tissue and the silicone gel model material. In: Proceedings of the 44th Stapp Car Crash Conference, SAE 2000-01-SC17, pp. 249–260 (2000)

7. Brands, D.W.A., Bovendeerd, P.H.M., Wismans, J.S.H.M.: On the potential importance of non-linear viscoelastic material modelling for numerical prediction of the tissue response: test and application. Stapp Car Crash J. **46**(SAE 2002-22-0006), 103–121 (2002)
8. Brands, D.W.A., Peters, G.W.M., Bovendeerd, P.H.M.: Design and numerical implementation of a 3-d non-linear viscoelastic constitutive model for brain tissue during impact. J. Biomech. **37**(1), 127–134 (2004)
9. Claessens, M.H.A., Sauren, F., Wismans, J.S.H.M.: Modelling of the human head under impact conditions: a parametric study. In: Proceedings of the 41th Stapp Car Crash Conference, SAE 973338, pp. 315–328 (1997)
10. Darvish, K.K., Crandall, J.R.: Investigating Nonlinear Viscoelastic Properties of Brain Tissue Using the Forced Vibration Method. American Society of Biomechanics, 24th Annual Meeting (1999)
11. Darvish, K.K., Crandall, J.R.: Nonlinear viscoelastic effects in oscillatory shear deformation of brain tissue. Med. Eng. Phys. **23**(9), 633–645 (2001)
12. Darvish, K.K., Takhounts, E.G., Crandall, J.R.: A dynamic method to develop nonlinear viscoelastic model of brain tissue. Advances in Bioengineering. In: Proceedings of the ASME International Mechanical Engineering Congress, vol. 39. Anaheim, California (1998)
13. Donnelly, B.R., Medige, J.: Shear properties of human brain tissue. J. Biomech. Eng. Trans. ASME **119**(4), 423–432 (1997)
14. Franceschini, G., Bigoni, D., Regitnig, P., Holzapfel, G.A.: Brain tissue deforms similarly to filled elastomers and follows consolidation theory. J. Mech. Phys. Solids **54**(12), 2592–2620 (2006)
15. Fung, Y.: Biomechanics: Mechanical Properties of Living Tissues. Springer, New York (1981)
16. Gurdjian, E.S., Lissner, H.R., Patrick, L.M.: Protection of the head and neck in sports. J. Am. Med. Assoc. **182**, 502–512 (1962)
17. Horgan, T.J., Gilchrist, M.D.: The creation of three-dimensional finite element models for simulating head impact biomechanics. Int. J. Crashworthiness **8**(3), 1–14 (2003)
18. Hrapko, M., van Dommelen, J.A.W., Peters, G.W.M., Wismans, J.S.H.M.: The mechanical behaviour of brain tissue: large strain response and constitutive modelling. Biorheology **43**(5), 623–636 (2006)
19. Hrapko, M., van Dommelen, J.A.W., Peters, G.W.M., Wismans, J.S.H.M.: Characterisation of the mechanical behaviour of brain tissue in compression and shear. Biorheology **45**, 663–676 (2008)
20. Hrapko, M., van Dommelen, J.A.W., Peters, G.W.M., Wismans, J.S.H.M.: The influence of test conditions on characterisation of the mechanical properties of brain tissue. J. Biomech. Eng. Trans. ASME **130**(3), 031003 (2008)
21. Hrapko, M., van Dommelen. J.A.W., Peters, G.W.M., Wismans, J.S.H.M.: On the consequences of non linear constitutive modelling of brain tissue for injury prediction with numerical head models. Int. J. Crashworthiness **14**, 245–257 (2009)
22. Iwata, A., Stys, P.K., Wolf, J.A., Chen, X.H., Taylor, A.G., Meaney, D.F., Smith, D.H.: Traumatic axonal injury induces proteolytic cleavage of the voltage-gated sodium channels modulated by tetrodotoxin and protease inhibitors. J. Neurosci. **24**(19), 4605–4613 (2004)
23. Kleiven, S.: Evaluation of head injury criteria using a finite elementmodel validation against experiments on localized brain motion, intracerebralacceleration, and intracranial pressure. Int. J. Crashworthiness **11**(1), 65–79 (2006)
24. Langlois, J.A., Rutland-Brown, W., Thomas, K.E.: Traumatic Brain Injury in the United States: Emergency Department Visits, Hospitalizations, and Deaths. Technical Report, Centers for Disease Control and Prevention, National Center for Injury Prevention and Control (2004)
25. Macosko, C.W.: Rheology: Principles, Measurements, and Applications. VCH Publishers, Berlin (1994)
26. Meaney, D.F.: Relationship between structural modeling and hyperelastic material behavior: application to cns white matter. Biomech. Model. Mechanobiol. **1**, 279–293 (2003)

27. Mendis, K.K., Stalnaker, R.L., Advani, S.H.: A constitutive relationship for large deformation finite element modeling of brain tissue. J. Biomech. Eng. Trans. ASME **117**(3), 279–285 (1995)
28. Miller, K.: Constitutive model of brain tissue suitable for finite element analysis of surgical procedures. J. Biomech. **32**(5), 531–537 (1999)
29. Miller, K.: Biomechanics of soft tissues. Med. Sci. Monit. **6**(1), 158–167 (2000)
30. Miller, K.: How to test very soft biological tissue in extension. J. Biomech. **34**(5), 651–657 (2001)
31. Miller, K., Chinzei, K.: Constitutive modeling of brain tissue: experiment and theory. J. Biomech. **30**(11, 12), 1115–1121 (1997)
32. Moerman, K., Herlaar, K.: Finite Element Modelling of the Human Head to Predict and Analyse Brain Injury due to Blast Induced Acceleration. Technical Report. TNO-DV2 2005 IN017, TNO Defense, Security and Safety, The Netherlands (2006)
33. Morrison, B. III., Cater, H.L., Wang, C.C.B., Thomas, F.C., Hung, C.T., Ateshian, G.A., Sundstrom, L.E.: A tissue level tolerance criterion for living brain developed with an in vitro model of traumatic mechanical loading. Stapp Car Crash J. **47**(SAE 2003-22-0006), 93–106 (2003)
34. Nahum, A.M., Smith, R.W., Ward, C.C.: Intracranial pressure dynamics during head impact. In: Proceedings of the 21st Stapp Car Crash Conference, SAE 770922, pp. 339–366 (1977)
35. Nicolle, S., Lounis, M., Willinger, R.: Shear properties of brain tissue over a frequency range relevant for automotive impact situations: new experimental results. Stapp Car Crash J. **48**(SAE 2004-22-0011), 239–258 (2004)
36. Nicolle, S., Lounis, M., Willinger, R., Palierne, J.F.: Shear linear behaviour of brain tissue over a large frequency range. Biorheology **42**(3), 209–223 (2005)
37. Peters, G.W.M., Baaijens, F.: Modelling of non-isothermal viscoelastic flows. J. Non-Newt. Fluid Mech. **68**(2, 3), 205–224 (1997)
38. Peters, G.W.M., Meulman, J.H., Sauren, A.H.J.: The applicability of the time/temperature superposition principle to brain tissue. Biorheology **34**(2), 127–138 (1997)
39. Prange, M.T., Margulies, S.S.: Directional properties of gray and white brain tissue. In: Symp. Proc. Center for Disease Control, Wayne State University (1998)
40. Prange, M.T., Margulies, S.S.: Regional, directional, and age-dependent properties of the brain undergoing large deformation. J. Biomech. Eng. Trans. ASME **124**(2), 244–252 (2002)
41. Prange, M.T., Meaney, D.F., Margulies, S.S.: Defining brain mechanical properties: effects of region, direction, and species. In: Proceedings of the 44th Stapp Car Crash Conference, SAE 2000-01-SC15, pp. 205–213 (2000)
42. Ruan, J.S., Prasad, P. Head injury potential assessment in frontal impacts by mathematical modeling. In: Proceedings of the 38th Stapp Car Crash Conference SAE 942212, pp. 111–121 (1994)
43. Shen, F., Tay, T.E., Li, J.Z., Nigen, S., Lee, P.V.S., Chan, H.K.: Modified bilston nonlinear viscoelastic model for finite element head injury studies. J. Biomech. Eng. Trans. ASME **128**(5), 797–801 (2006)
44. Takhounts, E.G., Crandall, J.R., Matthews, B.T.: Shear properties of brain tissue using non-linear green-rivlin viscoelastic constitutive equation. In: Injury Biomechanics Research, Proceedings of the 27th International Workshop, pp. 141–156 (1999)
45. Takhounts, E.G., Crandall, J.R., Darvish, K.K.: On the importance of nonlinearity of brain tissue under large deformations. Stapp Car Crash J. **47**(SAE 2003-22-0005), 107–134 (2003)
46. Velardi, F., Fraternali, F., Angelillo, M.: Anisotropic constitutive equations and experimental tensile behavior of brain tissue. Biomech. Model. Mechanobiol. **5**(1), 53–61 (2006)
47. Versace, J.: A review of the severity index. In: Proceedings of the 15th Stapp Car Crash Conference, SAE 710881, pp. 771–796 (1971)
48. Willinger R, Baumgartner D (2003) Human head tolerance limits to specific injury mechanisms.Int. J. Crashworthiness **8**, 605–617

Modeling of the Brain for Injury Prevention

King H. Yang, Haojie Mao, Christina Wagner, Feng Zhu, Clifford C. Chou and Albert I. King

Abstract From an ethical point of view, it is extremely difficult to propose a well-controlled human subject study aimed at understanding brain injury mechanisms and establishing the associated tolerance values. For this reason, many numerical models of the human and animal head or brain have been developed over the past several decades in an attempt to obtain in-depth insights into brain injury biomechanics, minimizing the need for human subject research. This chapter highlights and contrasts the essence of human and animal head numerical models developed for studying blunt impact and blast-induced brain injuries. Even with the vast amount of literature produced by these investigations and studies, the precise mechanisms of brain injury have not yet been fully established to date.

Through this review, it is clear that a lot of information can be garnered by numerical brain modeling but few efforts have been devoted so far to using these numerical models to provide guidelines in the discovery of brain injury mechanisms. Based on the brain models reported in the current literature, there are some inherent deficiencies. However, with further revisions and improvements to the currently available models, as opposed to developing new models from scratch, these issues can be overcome, and the state of the art can be advanced. More research effort into brain injury mechanisms, especially under in vivo conditions, is needed for computational model improvements so that the injury mechanisms can be thoroughly understood and effective countermeasures for protecting human from traumatic brain injury can be developed.

K. H. Yang (✉) · H. Mao · C. Wagner · F. Zhu · C. C. Chou · A. I. King
Wayne State University, Detroit, USA
e-mail: aa0007@wayne.edu; di8797@wayne.edu

Stud Mechanobiol Tissue Eng Biomater (2011) 3: 69–120
DOI: 10.1007/8415_2010_62

Published Online: 5 March 2011

1 Introduction

Traumatic brain injury (TBI) resulting from vehicular collisions, contact sports, falls, or blasts can have devastating consequences. Brain injuries not only pose a serious disability for those involved, but also place an enormous burden on society, often exacting a heavy economical, social, and emotional price. To reduce the likelihood and effects of these injuries, impact biomechanics, also referred to as the science of injury control, has been established as the field which aims to protect humans through the application of established engineering and medical research methods. The four general areas in the field of impact biomechanics involve the study of injury mechanisms, mechanical response to impact, injury tolerance, and assessment of the effectiveness of countermeasures using human surrogates. Although numerous research projects have been conducted, established scientific knowledge in all four areas of head impact biomechanics remains limited and an important area of modern research.

Modern day impact biomechanics research can be traced back to 1939 when Professor H. R. Lissner (a professor in Engineering) and Dr. E. S. Gurdjian (a neurosurgeon) began studying the mechanism of skull fracture at Wayne State University (WSU). They placed human skulls at the bottom of an elevator shaft and dropped steel balls onto these specimens from as high as the 12th floor. Since then, numerous investigations have been conducted to further our understanding of injury mechanisms, impact response, and injury tolerance. Experimental studies include the use of animals, physical models, volunteers, and cadavers and have subsequently led to the companion development of mathematical and computational models to enhance the understanding of brain injury. These virtual models can allow for more in-depth biomechanics studies, if properly developed and validated.

Based on globally measured parameters in these studies, numerous textbooks and articles affirmed that brain injury is due to one or more of the following mechanisms: (1) positive pressure, (2) negative pressure, (3) pressure gradient, and/or (4) rotational effects. Positive pressure, typically associated with the so-called coup injury mechanism, is assumed to be the result of the moving skull towards the stationary brain, producing a compressive wave in the brain at the time of impact or direct compression of the brain due to in-bending of the skull. Negative pressure, which has been associated with the so-called contrecoup injury mechanism, is hypothesized to be the result of tension generated by skull moving away from the brain that is lagging behind skull. Alternately, the negative pressure could be due to a tensile wave that was formed by the reflection of the original compression wave off the skull. Cavitation (collapse of a vapor bubble) occurs if negative pressure is lower than the vapor pressure of water and may also damage brain tissues. Additionally, shear stresses resulting from high pressure gradients in areas such as junctions between gray matter and white matter have been associated with diffuse axonal injury (DAI), but despite this, no techniques have been developed to physically measure intracranial shear stress experimentally. This shear mechanism has been postulated to explain why massive loss of neuronal function is seen in the central areas of the brain in some injuries. Lastly, rotational acceleration or velocity has been associated with several types of brain injury, such as surface contusions due to frictional contact of the smooth brain on the rough bony cranial vault, pulling of the brainstem through the foramen magnum, acute subdural hematoma (ASDH) as a result of ruptured bridging veins due to large relative motion between the brain and skull, and concussion or DAI due to high shear deformation of the brain owing to the high bulk modulus with low shear modulus. The reader is cautioned that not all hypothesized statements made above have been scientifically validated or proven. In particular, the shearing of axons is difficult to imagine considering that it is not a very stiff material. Also, it can be shown from first principles of fluid mechanics that rupture of the bridging vein is not the cause of ASDH. This highlights why further research in TBI is necessary.

In addition to blunt injury mechanisms, recent conflicts in the Middle East have ignited a debate regarding blast wave induced TBI. As early as the 1950s, Gurdjian and colleagues observed that direct impacts to the head could generate pressure waves in the brain. They performed the first fluid percussion experiment on dogs to recreate this phenomenon. Without any global head acceleration, brain injuries were observed, although it remains uncertain as to whether these injuries included DAI or not. In any case, this mechanism was "forgotten" for some time and is being rekindled due to the many mild TBI injuries attributed to blast overpressure in which global head motion may not be occurring.

To research these and other brain injury mechanisms, experimental animal models are used as human surrogates to provide an opportunity to monitor the brain's physiologic response over time, which can rarely be done in human studies. However, methods for extrapolating or scaling animal biomechanical data or tolerances to the human are often unreliable, and some researchers have raised ethical concerns on the use of animal subjects. As an alternative, physical models may provide good control over the experimental setup but the mechanism of injury cannot be delineated from such models due, in part, to poor biofidelity. Data from volunteers, which removes biofidelity issues, are very sparse, and experiments can only be performed under uncontrolled conditions (such as real world athletic events) or under controlled conditions in a laboratory at impact severities well below the injury level. Although human cadavers allow for well-controlled tests and have the same anatomical features as a living person, the lack of muscular and physiological responses limits the scientist's ability to assess acute as well as secondary sequelae. That is to say, pathophysiological responses of the brain that are vital to the understanding of functional brain injury cannot be fully assessed. The ideal surrogate for brain injury research has yet to be discovered.

Possible surrogate candidates can be found in numerical modeling techniques, some of which have been in development for decades. Mathematical modeling of the head is a powerful tool for the study of head injury and associated head impact protection. In early models, simplified intracranial pressures and stresses in the skull were calculated based on lumped-mass-spring-damper models consisting of only a handful of degrees-of-freedom (e.g. [3, 31, 50, 51, 90, 131]). There were also models of fluid filled shells, mathematically represented by a set of partial differential equations that were solved using either the infinite series method (e.g. [17, 29]) or the finite difference method (e.g. [91]). Readers are referred to the review article by King and Chou [66] for more detail on these early numerical models. In recent years, technological advances have allowed for the development of complex finite element (FE) head models, which mimic the irregular geometry and anatomical features of the head, with more than 1 million degrees-of-freedom (DOFs). Such computational models have been used to study brain impact response, mimicking the complex boundary and loading conditions. Simulated tissue level stress, strain and/or deformation distributions for a given biomechanical input, such as a direct impact or a non-impact inertial loading, are available for correlation with data obtained experimentally or from real-world incidents to establish injury mechanisms and thresholds. Based on the promise of these techniques, numerous FE head/brain models have been developed all over the world, but it is important to note that not every computational model will yield the same biomechanical conclusions, an issue that will be discussed further in concluding this review.

In this chapter, human and animal head FE models developed over the past several decades are reviewed and several directions for future research are highlighted. Due to length limitations, computational models developed for purposes other than injury research, such as those for studying brain-electromagnetic (EM) field interaction phenomena (e.g. [18]), neurosurgical procedures (e.g. Wittek et al. [154] on needle insertion into the brain; Saberi et al. [124] on effect of hematoma; Gao et al. [40] on decompressive effect of craniotomy; Hagemann et al. [44], Soza et al. [132] Gao et al. [39], Wittek et al. [153], Hu et al. [55], Chakrabarty and Hanson [15], on brain shift calculation), hematoma size and shape (e.g. [136]) are not included in this review. This review is divided into four major: Sects. 2.1, 10.1, 10.2, and 10.3.

2 Literature Review

2.1 Blunt Impact Brain Models

Blunt impact induced TBI has caused significant numbers of death and disability among children and young adults in the United States [139]. It is estimated that 1.5 million Americans sustain a TBI annually, and 50,000 die each year from these injures. Among the affected, 80,000 to 90,000 patients suffer permanent disability from their injuries [79]. In addition to being a major public health problem, TBI is also a major socioeconomic problem. The direct medical costs and indirect costs (such as lost productivity) of TBI totaled an estimated $60 billion in the United States in 2000 [33]. As discussed previously, there are limited methods that can be used to study the mechanism of injury and evaluation of appropriate countermeasures. Computational models of the head have been and continue to be used to assist in this research.

The basic principle behind the FE method is that any structural system can be subdivided into a finite number of discrete elements and its response to loading approximated using nodal interconnections. Mechanical properties, such as elastic modulus, density, Poisson's ratio, and tangent modulus are assigned to each element to govern the material behavior. The geometric information and material properties for each element are used to form an element stiffness matrix. Static or dynamic structural deformations can then be calculated through the assembly of all element stiffness matrices, definition of appropriate boundary conditions, and application of loading conditions using physical energy conservation laws and differential equations. Therefore, this is a numerical method that is limited by computational resources and underlying approximations.

3 History of FE Head Modeling

Early FE models of the head generally assumed a simplified geometry due to the lack of computational power needed to solve the equations mentioned above. To the best of the authors' knowledge, the first three-dimensional (3D) FE head model with actual skull geometry was reported by Hardy and Marcal [45]. This model was used to simulate static frontal and lateral loading, although the model did not include representation of the brain. Chan [16] reported an axi-symmetric model with the skull and brain represented by viscoelastic materials and used the model to investigate the hypothesis that large shear stress could rupture the cerebral blood vessels and injure brain matter. Other earlier models used simplified spherical, spheroidal, or ellipsoidal shells to represent the skull and inviscid fluid, viscoelastic, or elastic materials to form the intracranial contents (e.g. [62, 63]).

A more realistic 3D FE model developed by Shugar [130] followed. In this model, the skull and brain were assumed to be linearly elastic, and the skull was represented by a layered structure to incorporate the inner and outer tables. This model was later refined by Ruan et al. [117] and Zhou et al. [163] with changes in material properties and the differentiation of gray and white matter in the brain model, discussed in the following subsections. Concurrently with Shugar's paper, Ward and colleagues at the Naval Construction Battalion Center and University of California at San Diego developed 3D human half-brain models which included the cerebrum, cerebellum, brainstem, ventricles, dural membrane, and a rigid skull [145, 147, 148] for studying frontal impacts. Compared to the Shugar model, this model was much more thoroughly validated. In the publication by Ward and Thompson [148], roller-supported linear springs were used to simulate the tethering of the brain to the skull, and the model was validated against data obtained from static measurements of the brainstem during flexion and extension of the head, as well as experimentally determined mode shapes and natural frequencies. The model was also validated against intracranial pressure time history data from a single test reported by Nahum et al. [97]. Based on results from these simulations, the authors noted that a model without the falx cerebri and tentorium could not predict the superior brainstem deflections correctly. This conclusion has been echoed by a newer 2D model developed by Li et al. [82] at the Medical College of Wisconsin with the aim of evaluating the effect of falx with the head subjected to lateral impact. It is worth noting that the FE model reported by Hosey and Liu [54] used the same brain geometry as Ward and Thompson [148] and was reasonably complete, but was too large to be run on the computers available at that time, underscoring the difficulties in FE model development in the early history of the field. Utilizing animal and human studies, Ward et al. [147] further developed monkey and baboon brain models for calculating the differences in intracranial responses among animals and humans subjected to the same loading conditions. In 1979, Nahum et al. continued to investigate closed head injury and studied the effectiveness of helmets on protecting the brain [98]. In 1982, Ward reported that the predicted cortical displacement by their model under dynamic loading was too high [145]. Restraint conditions were added to the brain–skull interface and the Young's modulus selected to represent the brain was increased to ten times that used in Ward's study in order to lower the cortical displacement.

Eventually, Nahum et al. [99] expanded the half-brain model developed by Ward and Thompson [148] to a whole brain model, which was then used to predict subdural pressure under lateral impacts. Tests conducted by Nahum et al. [100] yielded multiple impact events along the lateral direction on one pressurized cadaver, with or without a helmet, and provided head linear and angular accelerations along with corresponding subdural pressures. The FE model simulations revealed a less distinct pressure gradient pattern than that seen in frontal impacts, and the authors suggested that this difference was due to the combination of all three components of linear acceleration not seen in frontal impacts. The authors also reported that the falx and tentorium played a significant role in compartmentalizing brain motion within each compartment, yet again accenting the importance of anatomical complexity in biomechanical models.

Besides the consequence of anatomy, these early head models also generated some insight into tissue material representation. In 1982, based on past experiences in brain modeling, Ward concluded that the brain should not be modeled as an incompressible material. A selective reduced integration scheme was suggested [145]. Specifically, a reduced integration scheme was used for the dilatational (volumetric) component, whereas a full integration scheme for the distortional (shear) component was used in conjunction with a nearly incompressible material representing the brain. The findings were summarized as follows: (a) brain models could accurately predict brain stresses or displacements if the model included the dural folds, falx, tentorium, foramen magnum, and effective compressibility, (b) stresses and strains developed in the brain lagged those in the skull, (c) coup and contrecoup contusions were caused by pressure associated with translational acceleration, (d) subdural hematomas were caused by high shear strains, (e) the occurrence and duration of concussion were related to the magnitude of the brain response, and (f) the largest shear strains were predicted along the brain–skull interface, brainstem, and cerebellum. It is suggested that readers should refer to a report by Khalil and Viano [64] for a discussion of the differences, similarities, and the deficiencies of the FE models developed by the two groups led by Shugar and Ward.

Very few FE models were developed to study head and brain injury in the late-1970s or 1980s. One of few examples showing the use of FE models to investigate injury countermeasures, Saczalski et al. [125] developed a spherical brain model covered with a linear elastic dura, a linear elastic spherical skull, and a nonlinear scalp. The model-predicted pressure response was first qualitatively correlated with experimentally measured results before a helmet model was added to determine the effect of different liners. Despite the dearth of literature publications from this time, significant advancement on computational power, commercially available nonlinear finite element codes, meshing software, and graphical and animation tools became available during this period, indirectly facilitating the development of more complex numerical models of head and brain injuries. Since the 1990s, numerous FE head models have been developed around the globe for blunt impact analyses, and some iteration of many are still in use today. In the following subsections, these models are reviewed in accordance with the institutions where the model was first developed. As each institution developed their model independently, it is difficult to draw comparisons based on model predictions. Instead, one can assume it is more relevant to critique how the models were developed and how their predictions compare to experimentally measured data. It should be noted that tolerance values should not be compared between models, as these data are model-dependent, although trends can be seen throughout the brain modeling field.

4 Wayne State University

To investigate the dynamic response of the brain during side impact, Ruan et al. [117] developed an axi-symmetric model and two plane strain head models, which were subjected to a triangular pulse loading with a peak pressure of 40 kPa to determine the effects of mechanical properties of the skull, brain, and membrane. As these were 2D models, future refinements were made by the same authors to incorporate the geometry of the 3D FE head model reported by Shugar [130] to include most essential components of the head including the scalp, a three-layered skull, cerebrospinal fluid (CSF), dura mater, falx cerebri, and brain [118, 119]. The model-predicted time histories of the impact force, head acceleration, and intracranial pressure compared favorably with data from a single cadaveric subject reported by Nahum et al. [97]. Although one set of experimental data is obviously insufficient to validate a FE model, this was the only dataset reported by the authors in which time histories of the intracranial pressure were provided as opposed to peak pressures. For direct impact, a correlation was found between the head injury criterion (HIC) and intracranial pressure, and head acceleration and intracranial pressure. Biomechanical responses to varying impact locations (side, occipital and vertex), changes in impact velocity, and mass of the impactor were studied parametrically using the model. A higher contrecoup pressure was predicted from an occipital impact than from a frontal impact. This finding supports clinical findings of contrecoup injury being more likely to result from an occipital impact than from a frontal impact. It was also reported that the maximum shear stress occurred in the brainstem.

The mesh of this model was refined and inhomogeneous material properties of the gray and white matter were added by Zhou et al. [163]. Additionally, ventricles and ten pairs of parasagittal bridging veins were incorporated into the model in an attempt to predict subdural hematoma when stretching of these veins exceeded their experimentally-determined tolerance limit. By prescribing different material properties of the gray and white matters, the authors found larger variations in the shear stress distribution patterns, without affecting the intracranial pressure, compared to the model with homogeneous properties. This finding is qualitatively similar to porcine experimental results in which diffuse axonal injuries were found in the white matter at its boundary with gray matter and near the ventricles [112] and underscores tissue differentiation as having a possible role in brain injury.

This model was coupled to a multilink rigid body model of the Hybrid III dummy, which is an anthropomorphic representation of the 50th percentile human male, to simulate human head responses during automotive barrier crashes [121]. Skull–isostress and brain–isostrain response curves were established based on these kinematics. The head model was further refined and exercised by Al-Bsharat et al. [2] through validation against intracranial pressure time histories reported by Nahum et al. [97]. While Nahum et al. [97] did not report time history curves for more than one test, they did report the peak pressure values for other test conditions although one of the peak values appeared to be an outlier. After excluding that data point, Al-Bsharat et al. were able to match their model predictions against all the other data points. Unfortunately, the relative displacements between the brain and the skull were far below the measured data reported by King et al. [67] using a high-speed X-ray system. To overcome this discrepancy, Al-Bsharat et al. [2] studied various types of numerical definition of the sliding interface between the brain and skull to determine the best scheme to represent the movement of the cerebral spinal fluid (CSF) layer. It was found that by adding a sliding interface between the brain and skull—more specifically, the interface between the pia and the arachnoid—model predictions of brain motion were found to match the measured displacements better.

With this improved model, Zhang et al. [160] investigated differences in brain response due to frontal and lateral impacts under identical impact and boundary conditions. Simulation results suggested that skull deformation and internal partitions of the brain may be responsible for the directional sensitivity of the head in terms of intracranial pressure and shear stress response. This conclusion is qualitatively consistent with experimental findings using subhuman primates in which a lateral impact was more injurious than a frontal impact of the same severity. Zhang et al. [160] also found that this version of the head model was numerically unstable when simulating large rotational impact.

To resolve this issue, a much more refined FE head model consisting almost eight times the number of elements was developed by Zhang et al. [161]. The new model had an anatomically detailed facial structure and used tri-linear solid elements and bi-linear membrane/shell elements to ensure numerical stability. Results indicated that this new model could be used to simulate direct and indirect impacts with combined translational and rotational accelerations as high as 200 g and 12,000 rad/s^2 or higher [161]. The model, referred to as the Wayne State University Head Injury Model (WSUHIM), has been subjected to extensive validation using published cadaveric test data including the intracranial and ventricular pressure data reported by Nahum et al. [97] and Trosseille et al. [140], the relative displacement data between the brain and the skull by King et al. [67] and Hardy et al. [46], and the facial impact data by Nyquist et al. [102] and Allsop et al. [4].

Although it is known that human brain tissue consists of a network of neurons, axons, arterioles, capillaries, and venules interspersed within a matrix of supporting cells; none of the aforementioned models had sufficient mesh resolution to replicate this complex architecture. In order to consider the effect of tethering due to the blood vessels and other structures, material properties of the brain used in head models have typically assumed a higher value than those reported from direct measurements of in vitro specimens dissected from cadaveric or animal subjects. However, to include all of the major vasculature in a 3D brain model is a significant modeling challenge. Zhang et al. [159] developed a 2D FE model of the human head, consisting of the skull, dura matter, CSF, tentorium, brain tissue, and the parasagittal bridging veins, and a second 2D FE model which also included major branches of the cerebral arteries. The authors found that the maximum principal strain/stress in the brain was lower in the model that included simulated blood vessels. The inclusion of the cerebral vessels added regional strength to the brain substance, and thereby contributed to the load bearing capacity of this composite brain model during head impact, analogous to reinforcing bars in a reinforced concrete structure. Unfortunately, incorporation of blood vessels in a 3D FE head model is not practical at this stage due to the lack of computing power.

Up to this point, data generated by all head models described above had never been correlated with injuries seen in living humans. Because the testing of volunteers cannot be taken to an injurious level, contact sports and real world car crash events present a unique opportunity to overcome this shortcoming. Through collaboration with the US National Football League (NFL), 24 head-to-head field collisions were simulated using the WSUHIM. Several injury predictors and injury levels were analyzed by correlating brain tissue responses with the site and occurrence of mild traumatic brain injury (concussion) seen in the field. Results indicated that the shear stress around the brainstem region could be an injury predictor for concussion. Viano et al. [144] further analyzed strain responses occurring at specific time points during and after impact and compared with the signs and symptoms of concussion seen in these players. They found that locations with the largest strains significantly correlated with the removal from play, cognitive and memory problems, and loss of consciousness observed in injured players. Additionally, concussive injuries occur during the rapid displacement and rotation of the cranium, after peak head acceleration and momentum transfer in helmeted impacts.

In another study, Franklyn et al. [35] recreated four real-world full-vehicle automotive side impact cases at a proving ground. The cases were selected from an Australian crash sampling database. Head kinematics (three translational and three angular velocities) measured during the crashes were used as inputs to both the WSUHIM and the National Highway Traffic Safety Administration (NHTSA) Simulated Injury Monitor (SIMon)—detailed in the next section—to determine model-predicted injury outcomes, which were then compared with known location of severity of brain injuries received by the real-world occupants. The results demonstrated that both models were capable of predicting varying injury severities (i.e., varying AIS injury levels) in the real-world cases. The WSUHIM predicted a slightly higher injury threshold than the SIMon, probably due to the finer mesh and different software used for the simulations. Additionally, the WSUHIM could be used to determine regions of the brain which had been injured, although the computer resources needed to run this model were much more than those required by the SIMon model.

Furthermore, the WSUHIM was also used to improve automotive hood design. In order to reduce the risk of brain injury during head-to-car-hood impact in pedestrian crash, a collaborative study between WSU and Autoliv was conducted to investigate the influence of impact speed on head and brain injury risk (Fredriksson et al. [37]. Head kinematics generated from headform-hood contact was used as input to the WSUHIM. Results of this study showed that: (1) a lower head injury criterion (HIC) value did not always reduce the risk for brain injury and (2) a under-hood clearances of 60 mm in 20 km/h and 80 mm in 30 km/h impacts reduced the risk of skull fracture and brain injury significantly.

5 National Highway Traffic Safety Administration

Brain models are also being considered by NHTSA for potential rule making in the US. To upgrade the HIC-based head injury standard, NHTSA worked on a brain model which could calculate brain strain based on head kinematics measured from crash test dummies, thereby giving a more complex picture of brain injuries from non-accelerative mechanisms. A simple model "SIMon I" with a non-deformable skull was developed to minimize run time so that it can be used by original equipment manufacturers who routinely conduct a large number of crash tests. The NHTSA model was originally developed by DiMasi et al. [27] and later modified by Bandak and Eppinger [30] and DiMasi et al. [28]. This model contained only the cerebrum and had a very thick falx (average 7 mm) to avoid poor quality meshes typically associated with a low aspect ratio. A slip interface with a low coefficient of friction was introduced between the interior dura and external surface of the cortex to facilitate brain motion. The authors further hypothesized that the risk of sustaining a diffuse axonal injury was proportional to the fraction of the brain volume which exceeded a preset injury threshold. Simulation results indicated that HIC was sensitive to translational kinematics only, whereas cumulated strain damage measure (CSDM) responded to rotation or combined translational and rotational conditions. Bandak [7] and Bandak et al. [8] further developed techniques to allow users to develop FE head models from biomedical images with ease. Their model also emphasized the simulation of skull fracture.

To improve the mesh quality of this model, Takhounts et al. [134] extended and rounded the lower part of the cerebrum of the previous SIMon model to avoid stress concentrations. The authors tested several nonlinear material models, including the Ogden rubber and Mooney–Rivlin rubber, and concluded that the linear viscoelastic material law was the best available approximation at the time. Based on a logistic regression of the model-predicted responses and animal injury outcomes, the authors derived three injury measures to estimate the risk associated with three different injury types: (a) a cumulative strain damage measure (CSDM) exceeding 55% of the brain volume represented serious risk of sustaining diffuse axonal injury (DAI), (b) a dilatation damage measure (DDM) of lower than −100 kPa (tension) represented an occurrence of brain contusion and focal lesions, and (c) a relative motion damage measure (RMDM) of one corresponded to the occurrence of an acute subdural hematoma (ASDH). The authors also found that side impact was potentially more injurious than frontal impact due to the more severe rotational kinematics.

A new geometrically detailed FE head model comprised of the cerebrum, cerebellum, falx, tentorium, pia-arachnoid complex with CSF, ventricles, brainstem, and parasagittal blood vessels was developed based on the SIMon concept [135]. The new model represented the brain of a 50th percentile male and can be used to simulate combined translational and rotational accelerations of up to 400g and 24,000 rad/s^2. The model was used to simulate mild TBI cases in American football players at the collegiate level to derive injury thresholds before it was used to investigate brain injury potential in NHTSA conducted side impact tests.

6 Université Louis Pasteur of Strasbourg

In France, Trosseille et al. [140] conducted five cadaveric head impact tests to provide much needed cadaveric data that can be used for model validation. They also used a 2D sagittal plane model developed by Lighthall et al. [83] to determine the effect of material properties on model-predicted responses. Other French human head models came mostly from Université Louis Pasteur of Strasbourg (ULP) where the initial emphasis was on finding the natural frequencies of the head in order to design protective devices that can prevent the head from being exposed to these resonant frequencies [149, 150]. Constant energy shocks of varying duration were used as input to the model and the resulting compressive, tensile, and shear stresses, along with the relative brain–skull displacement and skull deformation were evaluated. Turquier et al. [142] used the head model developed by Willinger et al. [151] based on MRI scans to simulate the experiments conducted by Trosseille et al. [140]. The model-predicted responses (in terms of three intracranial accelerations measured at the lenticular nucleus as well as frontal and occipital lobes, three epidural pressures measured at the frontal, occipital, and temporal regions, and two intracranial pressures measured at the third and lateral ventricles) matched experimental results in terms of trend but showed significant oscillations. After introducing damping into the model, numerical oscillations were only slightly reduced. The authors called for more investigation into the assumptions used to model the subarachnoid space.

A short time later, Kang et al. [61] reported another model, which corrected the simplified geometry used in modeling the temporal lobe developed previously by Willinger et al. [149], and validated it against the intracranial pressure data reported by Nahum et al. [97]. A helmeted Hybrid II headform was used in a series of drop tests to reconstruct a real-world head impact of a motorcyclist into a Range Rover. The motorcyclist suffered a severe contusion to the right temporal lobe, tentorial contusion and subarachnoid hematoma over the occipital lobe, and laceration of the brainstem. Kinematics from the best correlated test—defined as the best match of the impact location and helmet damage—were used as input to the model to predict intracranial responses. The authors reported that shear stress correlated the best with the reported right temporal contusion observed in autopsy. Willinger et al. [152] further used the model to simulate cadaver test results. Although the model-predicted responses matched well against data from the short duration test, they did not match those obtained from a long duration impact. The authors emphasized the need to validate the FE head models against a variety of impact conditions to minimize errors.

Raul et al. [110] extended the application of the ULP head model from simulating automotive-induced head injuries to fall-induced head injuries of a 63-year-old male who fell backward to a wooden floor after being pushed by another person and then fell again from an emergency cart during lifting of the victim to the fire truck. The estimated impact velocity was 6 and 1.5 m/s for the first and second fall, respectively. The authors believed that the model-predicted responses could be used to exclude those injury mechanisms that were not likely in causing the injury seen in this case. Raul et al. [111] further suggested that FE human head models should be used routinely in forensic medicine. Marjoux et al. [89] reconstructed 61 real-world head injury cases experimentally and numerically to compare the predictive performance of using HIC, head impact power (HIP) proposed by Newman et al. [101], the CSDM, DDM, RMDM computed by the SIMon model [134], and intracranial responses predicted by the ULP model [61] as the injury mechanism. The authors concluded that moderate and severe brain injuries can only be distinguished with numerical model-based criteria and not with external (global) head acceleration.

7 Royal Institute of Technology (KTH) of Sweden

The geometries of most FE head models are taken from an idealized or a representative single subject's anatomy. Kleiven and von Holst [68] developed a parameterized human head and a simplified neck model to allow rapid changes in the model geometry. The model was validated against intracranial pressure data measured by Nahum et al. [97]. Six models of different sizes, representing a small female, a 5th percentile female, a 50th percentile female, a 50th percentile male, a 95th percentile male, and a large male were created to study the effect of head size in terms of von Mises stresses and HIC for padded frontal impacts and inertial loading. The authors found more than a fourfold increase in the peak stress while the HIC decreased 43% when the head size was increased from the smallest to the largest. They suggested that the size dependency was not reflected by HIC and recommended that any new head injury criterion should include the variations in head size. Ruan and Prasad [123] in their comments on this KTH model indicated that the negative pressure was too low (nearly −280 kPa) which may invalidate the findings of this study.

Again considering global response parameters, Kleiven [70, 71] compared FE model predicted intracranial responses to HIC and HIP from nine different applied pure translational and rotational acceleration pulses. He found that HIC responded in accordance with the magnitude of translational acceleration, but was not influenced by pure rotational impulse, while HIP required individual scaling coefficients for the different terms to account for different loading directions. Additionally, the largest bridging vein motion occurred in the shortest bridging veins, specifically those that were oriented in the plane of the motion and were angled in the direction of motion during rotation of the head in the sagittal plane. Following this train of thought, bridging vein geometry would significantly affect the strain in the bridging veins and may present a substantial obstacle for researchers to create subject-specific model suitable for predicting bridging vein rupture and associated injuries.

A variation on these models, Kleiven and Hardy [69] presented a FE head model which appeared to resemble the model reported by Kleiven and von Holst [68], but was "substantially different" from it according to Kleiven and von Holst (2006). The model used a hyperelastic Mooney–Rivlin linear viscoelastic constitutive law to simulate the brain and a selective reduced integration (SRI) scheme instead of the commonly used reduced integration scheme. Their model predicted a smaller relative motion between the brain and skull due to a lateral impact as compared with frontal or occipital impacts, consistent with the experimental data obtained by Hardy et al. [47]. A tied interface definition to represent the CSF resulted in model predictions which best correlated with measured intracranial pressures. However, the combination of a SRI scheme, Mooney–Rivlin material model, and a tied interface was later challenged by Takhounts et al. [134]. These authors speculated that the SRI scheme was susceptible to shear locking and was less stable than the reduced integration methods. Additionally, the

Mooney–Rivlin model did not match material properties obtained experimentally. Perhaps these numerical procedures were model specific, thus making it more difficult for researchers to decide from the literature what to follow when attempting to develop a reliable brain injury model.

To investigate localized anatomical effects, Ho and Kleiven [48] developed a 3D head model in which major arteries and veins were represented by "beam" elements to study the effect of cerebral vasculature on brain response. Differences in maximum principal strain predicted by models with and without vasculature were no more than 4% of the peak maximum principal strain. The predicted peak differences in strain between the model with and without vasculature were lower than those reported by Zhang et al. [159] using their 2D model which included Explicit modeling of the cerebral arteries. Ho and Kleiven [48] believed that the difference was due to the high vessel-to-brain volume ratio used in Zhang's model as well as the high vascular density in the para-sagittal plane selected by Zhang and the low vascular density in their 3D study. Nevertheless, the authors found that vessel-induced strain reduction in regions representing the thalamus and corpus callosum were higher than other regions, indicating that the effect of cerebral vasculature could be highly localized, rendering it more difficult to model brain vasculature accurately. Additionally, Ho and Kleiven [49] developed two highly detailed FE models, one with and one without sulci (a network of folds that cover the brain surface), with an average element size of about 1 mm^3. The models were loaded in three configurations: translational acceleration in the sagittal plane, rotational acceleration in the coronal plane, and rotational acceleration in the sagittal plane. Except for the parietal lobe in sagittal and coronal rotation accelerations, the model with sulci predicted a lower peak maximum principal strain when compared to the model without. The authors suggested that future FE brain model should include the sulci because this anatomical feature could significantly alter the strain distribution pattern.

For comparison to real-world injury, Kleiven [72] simulated NFL collisions, in which 25 players had concussion and 23 without, and a motorcycle accident, in which the rider sustained severe hematomas. The model used to reconstruct these cases was the same as that published previously [69, 72]. A series of parametric studies was conducted to show the effect of different material properties selected to represent brain tissues. Aside from intracranial responses predicted by the model, global injury measures (e.g. peak angular and translational acceleration, angular velocity change, HIP, and HIC) were also calculated. He found that the maximal pressure in the gray matter correlated best with injury while the maximal principal strain could be used to identify the location of injury observed in the motorcyclist. After testing several combined injury measures, the author found that a linear combination of HIC and peak change in rotational velocity correlated best with model-predicted maximum principal strain.

8 Eindhoven University of Technology (TUE)

Claessens et al. [22] created a coarse mesh model with geometry based on the Visible Human Data available at the U.S. National Library of Medicine. Because the brain and skull were directly coupled, the model failed to predict the intracranial pressures reported by Nahum et al. [97]. Better correlation with experimental data was achieved after adding a layer of friction-free contact interface between the brain and the skull. A more detailed model was also developed to include the falx, tentorium, and brainstem in the same report. Although experimental data on relative motion between the brain and the skull were not available at that time, the authors concluded that the junction between the brain and skull was somewhere between the rigid coupling and the free interface they simulated. However, they suggested that the coupling would be closer to a free interface. Brands et al. [11] continued this modeling effort by transforming Claessens's model into a MADYMO (version 5.4.1) model which used a reduced integration scheme. Before using this nonlinear model, a physical gel model was tested and modeled. After the authors matched the gel experimental results against model predictions, nonlinear "strain softening" viscoelastic properties for the human brain were derived and incorporated into the model. The motion of brain tissue through the foramen magnum was suppressed. Still, the model predicted a maximum shear strain which was ten times that reported by Bandak and Eppinger [6] when the same loading condition was applied. The authors attributed this increase in strain to the low shear modulus and strain softening effect they assumed in their model simulations.

A 2D plane strain model was developed from MRI data by Kuijpers et al. [75] to study the influence of linear elastic or viscoelastic brain material properties, different contact conditions at the skull–brain interface, and incorporation of a neck constraint. They found that intracranial pressures were more sensitive to the type of skull–brain interface condition than to the presence or absence of a force-free foramen magnum. Additionally, constraints in the neck affected the deformation of the brain but different time-dependent deviatoric material parameters did not significantly change the response. Cloots et al. [23] speculated that the convolutions of the cerebral cortex could affect the cortical stresses and strains. Several highly detailed substructure 2D models containing gyri and sulci were developed and loaded by the boundary conditions generated from the model reported by Brands et al. [11]. Results indicated that inclusion of the gyri and sulci had no significant effect on the mean von Mises stress, but had a significant effect on the maximum value.

9 Other Institutions

Numerous institutions throughout the world also developed human head models for impact simulations. Idealized models and simplified physical surrogates with accompanying models have also been reported. At the Medical College of

Wisconsin (MCW), Pintar et al. [104] developed a simplified spherical FE model to represent the human skull and brain and then used it to simulate penetrating head injuries by two different types of projectiles. Five years later, Zhang et al. [157] applied head kinematics measured from Hybrid III and EuroSID-2 dummies to the NHTSA SIMon model to estimate the risk of brain injuries. A total of 13 frontal and 20 side impacts without head contact and with low HIC (less than 1,000) were simulated. The authors concluded that more than 80% of the brain strains were contributed by rotational acceleration and recommended that rotational accelerations should be quantified in future crash tests to improve occupant safety. Additional research was performed by Yoganandan et al. [156], who used the same simplified 2D human brain model reported by Li et al. [82] and mentioned previously to subject the brain to four different angular acceleration pulses. The brain was subdivided into 17 regions and the average model-predicted maximum principal strains for each region was presented. The highest strains were found in corpus callosum while the lowest strains were found in the lower parietal lobes and the shape of acceleration pulses had a profound effect on the regional intracranial strain.

In Taiwan, Chu et al. [21] reported a 2D plane-strain model and validated it against pressure data by Nahum et al. [97] and resonant frequencies reported in the literature. The authors identified that shear strain better predicted cerebral contusion than intracranial pressure. Kurosawa et al. [77] in Japan reported a simple cylindrical model consisting of an acrylic container (skull), water (CSF), and agar (brain). However, the pressure they predicted was too high. Johnson and Young [60] in UK segmented the skull from high-resolution MR images of a volunteer and then used the data to rapidly prototype a plastic skull. A pendulum was used to impact the plastic skull filled with water to represent the brain. A FE model representing the same skull was also created and simulation results matched well with those obtained experimentally. Although their study proved that a FE model could indeed duplicate experimental results, the plastic-skull and water-brain was far too simple to represent a human head. Sarron et al. [126] in France tested 30 human skulls filled with silicone gel to study the "rear effects" head injury, which is defined as injury induced by non-penetrating indentation of the military helmet during ballistic impact. A 3D FE model was used to conduct a parametric study, in which results showed that the diploë layer played a role in protecting the skull from fracture. In India, Kumaresan and Radhakrishnan [76] reported a 3D head model and calculated the first five modal frequencies and the maximum shear stress due to an occipital impact. Xu and Yang [155] in Changsha, China also reported a FE human head model in an article written in Chinese. This model also used experimental data reported by Nahum to validate the model.

Some 3D models have considered the role of the neck in brain kinematics. Zong et al. [166] in Dalian, China developed a simplified 3D head-and-neck model and validated it against intracranial pressure data obtained by Nahum et al. [97] and Trosseille et al. [140]. The model geometry appeared to have been adapted from Shugar [130] but the authors did not provide a clear indication regarding how the model geometry was obtained. A "structural intensity" method was used to indicate the power flow within the loaded structure. Although the title of this study was related to head impact, the authors concluded that the spinal cord was vulnerable during the three impact scenarios simulated. Huang et al. [56] in Taiwan developed a 3D head model to simulate indirect impact due to flexion and extension about the upper cervical spine and found that intracranial pressures were lower in indirect impact compared to direct impact. They assumed that brain injuries also occurred in indirect impact (even though researchers have never produced any direct evidence of brain injury in real world cases), the authors concluded that shear strain is a better injury predictor compared to pressure. In Japan, Kimpara et al. [65] developed a 3D head–neck model consisting of major components of the head and cervical spine to investigate the biomechanical responses of the brain–spinal cord complex. The head–neck model was validated against three sets of brain test data obtained by Nahum et al. [97], Trosseille et al. [140], and Hardy et al. [46] and two sets of neck test data obtained from Thunnissen et al. [141] and Pintar et al. [104]. A series of parametric studies was conducted and results showed that a soft head contact or no head contact could reduce the CSDM predicted diffused axonal injury. The authors used the ultimate strains of the spinal cord and pia mater to estimate the risk of neck injury.

Automotive impacts have also been investigated in models not yet described. Horgan and Gilchrist [52] in Ireland developed a 3D head model for simulating pedestrian accidents. The intracranial pressure was validated against Nahum et al.'s experiment [97]. Parametric studies regarding the effect of different mesh densities and influence of material properties were performed. The authors found that the short-term shear modulus of the brain tissue had the biggest effect on intracranial frontal pressure, and on the Von-Mises response. The bulk modulus of the CSF had a significant effect on the contre-coup pressure when the CSF was modeled. The coarse mesh model was fine for pressure prediction compared to a finer mesh. The authors also concluded that careful modeling of the CSF and skull thickness is necessary for correctly predicting intracranial pressure. Also, the FE head model is better to be scaled to the particular head being simulated for accurate prediction. A version of this model was also used by Rousseau et al. [116] to investigate how changes in neck kinematics affect brain tissue strain. It was found that increased neckform compliance increased maximal principal strains in the cerebellum, but no effect was seen in other brain regions when increasing or decreasing compliance. The authors stated difficulties in applying their findings to brain injury due to a lack of established threshold. Krabbel and Appel [74] in Germany developed a 3D human cranium model from CT scans. Kinematics of a Hybrid III head due to frontal impact and a EuroSID I due to lateral impact were calculated from MADYMO and used as input to the skull model. No injury assessment was reported.

10 Summary of Blunt Impact Brain Models

Based on the literature review discussed above, a summary of key FE adult human head/brain models developed for blunt impact simulations is presented in Table 1.

Table 1 Summary of FE adult human head/brain models

First author, year	Ruan et al. [117]	Ruan et al. [118, 119]	Zhou et al. [163]	Al-Bsharat et al. [2]	Zhang et al. [160, 161]	Zhang et al. [159]
Institute	WSU					
Mesh	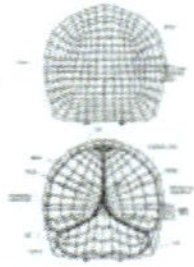		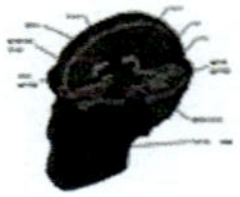	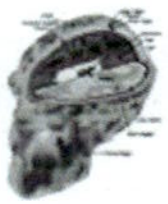	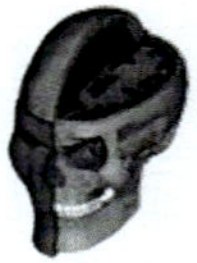	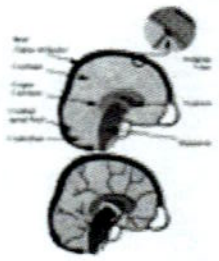
Part	Head	Head	Head	Head	Head	Head
Solving software	MARC, ANSYS	PAM-CRASH	PAM-CRASH	PAM-CRASH	PAM-CRASH	PAM-CRASH
2D/3D	2D	3D	3D	3D	3D	2D
Geometry	50th percentile human head geometry	Average adult male head geometry	50th percentile adult male head geometry	50th percentile adult male head geometry	50th percentile male head, anatomical drawings	Anatomical drawings
Elements	Shell	Hexahedral elements: brain 1,760, skull 2,800, CSF 896 Shell elements: scalp 864, dura 896, falx 135.	22,995 elements. Total mass 4.37 kg with brain being 1.41 kg	41,354 elements. Total mass of 4.3 kg.	314,500 elements with a mass of 4.5 kg	Model I and II, 4,501 elements with a mass of 41.07 g

(continued)

Table 1 (continued)

First author, year	Ruan et al. [117]	Ruan et al. [118, 119]	Zhou et al. [163]	Al-Bsharat et al. [2]	Zhang et al. [160, 161]	Zhang et al. [159]
Material law	Elastic material for brain and skull	Linear viscoelastic material for brain, Elastic–plastic material for bone, elastic material for the remaining	Similar to Ruan [119]	Similar to Zhou [163]	Linear viscoelastic material for brain, elastic–plastic material for bones, elastic material for membrane, skin, cartilage	Linear viscoelastic material for brain, linear elastic elemens for vessels
Validation	No	Intracranial pressure	Intracranial pressure	Intracranial ressure, Perform comparison of skull–brain relative displacements from FE model and experiment	Intracranial pressure, brain/skull relative motion	Intracranial pressure, brain/skull relative motion

First author, year	Ward [148, 147, 145]	Dimasi et al. [27] Bandak et al. [6] Dimasi et al. [28]	Takhounts et al. [135]	Willinger et al. [149]	Kang et al. [61]	Claessens et al. [22]
Institute	Biodynamics/ Engineering Inc.	NHTSA		ULP		TUE
Mesh			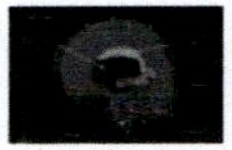	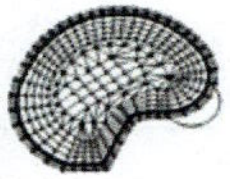	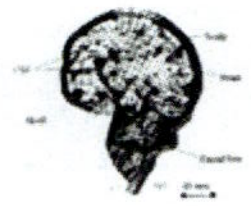	
Part	Head	Head	Head	Head	Head	Head

(continued)

Table 1 (continued)

First author, year	Ward [148, 147, 145]	Dimasi et al. [27] Bandak et al. [6] Dimasi et al. [28]	Takhounts et al. [135]	Willinger et al. [149]	Kang et al. [61]	Claessens et al. [22]
Solving software	SAPIV, SAPV	LS-DYNA	LS-DYNA	NA	NA	MARC
2D/3D	3D	3D	3D	2D	3D	3D
Geometry	Atlas	Anatomical slice data	CT scan of 50th percentile male	NA	Lab obtained skull geometry and atlas	CT and MRI
Elements	Coarse hexahedral elements	Approximately 5,900 elements	5153 shell, 40,708 hexahedral elements, 14 bean elements	573 plane strain elements	10,395 hexahedral elements, 2,813 shells, with total mass of 4.7 kg	1,756 hexahedral elements, with total mass 3.1 kg
Material law	Elastic material	Linear viscoelastic material for brain, elastic material for membrane, rigid skull	Linear viscoelastic material for brain and PAC-CSF, elastic fluid for ventricle, elastic material for membrane, cable discrete beam for vessels, rigid skull	Elastic material for brain, bone. Membrane neglected	Linear viscoelastic material for brain, elastic material for CSF, scalp, membrane, and facial bone, elastic brittle property for composite skull	Linear elastic material for brain and skull
Validation	Intracranial pressure	Intracranial pressure	Intracranial pressure, Brain/skull relative motion	No	Intracranial pressure	Intracranial pressure

(continued)

Table 1 (continued)

First author, Year	Kleiven [68]	Ho et al. [49]	Chu et al. [21]	Huang et al. [56]	Kimpara et al. [65]	Horgan et al. [52]
Institute	KTH		National Yang-Ming Medical College, Taiwan	Taipei Medical College, Taiwan	Toyota Central R&D Labs., Inc	University College Dublin, Ireland
Mesh		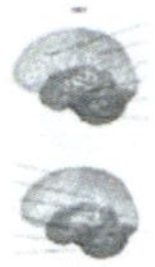		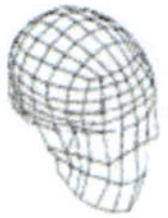		
Part	Head	Head	Head	Head	Head	Head
Solving software	LS-DYNA	LS-DYNA	ANSYS	ANSYS	LS-DYNA	ABAQUS
2D/3D	3D	3D	2D	3D	3D	3D
Geometry	NA	MRI	Atlas	Modified from Shugar's model	THUMS-AM50	CT
Elements	11,454 hexahedral elements, 6,940 four-node elements, 22 two-node truss elements	Average element size 1 mm^3, with brain mass around 1.2 kg	1,261 quadral and triangle elements	1,328 solid elements (inlcuding 60 four-node pyramids and 24 five-node tetrahedral elements), with brain mass 1.307 kg and total mass 2.932 kg	49,579 solid elements, 25,119 shell, and 364 seatbelt element, with a total mass as 4.39 kg	Element densities varying from 9,000 elements to 50,000 elements

(continued)

Table 1 (continued)

First author, Year	Kleiven [68]	Ho et al. [49]	Chu et al. [21]	Huang et al. [56]	Kimpara et al. [65]	Horgan et al. [52]
Material law	Hyperelastic and viscoelastic material for brain tissue, linear elastic for bone, scalp, and membrane	Ogden hyperelastic material for brain, elastic fluid for CSF, quasilinear viscoelastic shell for pia, simplified rubber/foam material for falx and tentorium, rigid skull,	Linear elastic mateiral for brain and skull	Linear elastic mateiral for brain and skull	Elastic–plastic material for bone, linear viscoelastic material for brain, elastic material for membrane	Linear viscoelastic for brain tissue, hybrid elements for CSF, elastic material for bone and scalp
Validation	Intracranial pressure, Brain/skull relative motion	Brain/skull relative motion	Intracranial pressure	Intracranial pressure	Intracranial pressure, Brain/skull relative motion	Intracranial pressure

First author, year	Zong et al. [166]	Bandak [7]	Johnson et al. [60]	Krabbel et al. [74]	Cloots et al. [23]	Li et al. [82]
Institute	Dalian University of Technology, China	NHTSA	University of Exeter, UK	Technical University Berlin, Germany	TUE	MCW
Mesh		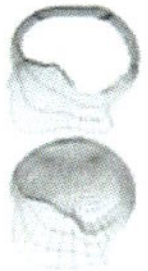			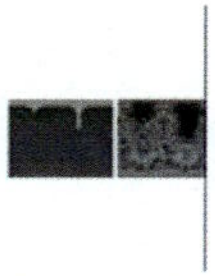	
Part	Head	Head bone	Head bone	Head bone	Cortex slice	Simple phycical model

(continued)

Table 1 (continued)

First author, year	Zong et al. [166]	Bandak [7]	Johnson et al. [60]	Krabbel et al. [74]	Cloots et al. [23]	Li et al. [82]
Solving software	MSC/DYTRAN	LS-DYNA	LS-DYNA	PAM-CRASH	ABAQUS	ABAQUS
2D/3D	3D	3D	3D	3D	2D	2D
Geometry	NA	CT	MRI	CT	Topological study	Sphere
Elements	Minimum side length of the solid elements is 0.64 mm. Average side length is about several centimeters	Projection method used to generate high resolution and medium resolution models	Slightly over 200,000 heahedral and tetrahedral elements (head bone)	1,342 solid elements (head bone)	4,243–4,533 elements for heterogeneous models, 3,072 elements for homogeneous model	4,650 elements
Material law	All materials assured as linear elastic	Lineat elastic material for bone, brain, scalp	NA	Elastic–plastic material for bone	Non-linear viscoelastic material for brain, low shear modulus elastic solid for CSF	Linear elastic for skull and falx, hydrodynamic material model for CSF, linear viscoelastic material for gel
Validation	Intracranial pressure	No	No	No	No	Motion and strain

WSU Wayne State University, *NHTSA* National Highway Traffic Safety Administration, *ULP* Université Louis Pasteur of Strasbourg, *KTH* Royal Institute of Technology of Sweden (Swedish: Kungliga Tekniska högskolan), *MCW* Medical College of Wisconsin, *TUE* Eindhoven University of Technology

The table lists the first author of the model developers and the year when the model reported. It also compares mesh, software, element type, and material laws used, model geometry and what data were used for model validation. Several key conclusions from this review are:

First, the complexity of human head models has been augmented significantly over the past two decades and each new version of the model was able to address additional clinical phenomena or to make site-specific correlation between the model results and real world injury. Even though all available data have been used for validation of the latest models, they can only be considered partially validated because the quantity and quality of these experimental data are still relatively low. Additionally, the shortage of quantitative material property data continues to prevent total validation. Nevertheless, useful information has been derived from model predictions. Two common conclusions can be drawn from these FE modeling studies: shear strain is mainly due to rotational impact, while intracranial pressures are affected mostly by translational acceleration and stiffness of the skull.

However, there are many unanswered issues which may hinder continued development of the human head model. For example, modeling of the pia-arachnoid complex, within which the CSF flows, continues to be an unresolved issue for many researchers. Among the techniques used to model this interface are a direct connection with no slip, direct coupling at the junction, sliding interface with different coefficients of friction, or tie-break with a preset threshold. Unfortunately, the method that best represents the real-world situation cannot be determined with confidence until the relative motion at this junction is actually measured. Nevertheless, most researchers reported that representing the CSF layer by a gap cannot be used to generate tension in the contre-coup site, thus making it unsuitable to model the contre-coup phenomenon reported by clinicians. Experimental data reported by Jin et al. [57–59] on bovine pia-arachnoid complex showed that the trabeculae in the CSF layer offers finite in-plane, traction, and shear resistance, thus it would be a mistake to model this layer as an incompressible fluid.

Second, variations in skull thickness are generally neglected in the simulation of closed head injuries. However, skull thickness is an important issue for direct impacts to the head because one cannot simply input head kinematics to the center of gravity of the head and expect to model skull fracture. Both its thickness and material properties have an effect on how or where a fracture would initiate. According to Ruan and Prasad [122], human skull thickness varied from 4 to 9 mm in five published studies, but all models so far have assumed a uniform thickness. Such information should be incorporated into the model for direct-impact simulations to reflect how local bone thickness affects bone fracture and transfer of energy to the brain. Also, it is necessary to include the scalp in direct-impact simulations because the scalp dampens an impact by increasing the duration and contact area over the skull.

Third, tissue level injury thresholds are unknown. It is a common trend to use the computed maximum principal strain or stress and pressure as an estimate for the risk of head injury. Ruan and Prasad [120] suggested the use of a computed stress threshold (pressure or shear) to derive the head acceleration and impact duration for risk estimation. A maximum von Mises stress of 100 MPa in the skull was suggested as the threshold for skull fracture while a maximum shear stress of 22 kPa was the proposed threshold of a reversible concussion. With this method, the automotive designers can use measured dummy head kinematics to predict the risk of brain injury. However, it should be mentioned parenthetically that the use of von Mises stresses is a convenience but not necessarily a biomechanically valid parameter for predicting injury. Additionally, the CSDM, DDM, and RMDM proposed by NHTSA make physical sense because occurrence of high stress in a single element may not be a meaningful measure of injury to the brain due to numerical issues as well as physiological reasons. However, more research is needed to establish different injury thresholds for different brain tissues. For example, a 15% strain may be injurious to the brainstem, but not to the gray matter.

Fourth, high strain rate material properties of the brain remained the largest road block en route to accurate modeling of brain response. Selection of constitutive laws and their associated material constants varies greatly among different research groups. Until a set of realistic and commonly accepted properties becomes available, readers should be aware that a combination of incorrect input data can still yield a "correct" model prediction. This quest for proper material properties was identified by Goldsmith [41] more than 40 years ago and was reflected in the simplified FE model of skull and brain by Chan [16] that remotely resembled a human head. This issue has not been resolved as of today. For example, shear properties of the brain reported by Galford and McElhaney [38] and Prange et al. [108] differed by more than two orders of magnitude. Additionally, several researchers have reported a "strain softening" effect of the brain (i.e., shear stiffness decreases as the strain increases). This matter has been discussed by Brands et al. [10], Prange et al. [108] and Bilston et al. [9]. This phenomenon probably bears further study because the concept is counterintuitive and the phenomenon is not seen in other biological tissues. Nevertheless, the trend predicted by all models seems to be similar. In particular, intracranial pressure appears to be the easiest parameter to match by all groups, even though only a limited number of experimental datasets are available.

Finally, the predominant deficiency is the lack of experimental data to properly validate the model-predicted results against the impact responses of the head and brain. In terms of intracranial pressure, only a total of three datasets—two reported by Trosseille et al. [140] and one from the often-cited Test 37 reported by Nahum et al. [97]—are available to validate forehead impact. Similarly, only a few tests reported by Hardy et al. [46, 47] are available for the validation of the motion of the brain relative to the skull. Additionally, for a comparison of model response to human injury, there are some 30 cases of mTBI reconstructed from NFL games [144] and four sets of graded AIS scale head injury derived from real world crashes reported by Franklyn et al. [35]. Other real world situations are not well-documented enough for this purpose. New and higher quality experimental or real world data are needed for the continued improvement of simulation models to enhance their capability in accurately predicting the risk of brain injury under a given blunt impact condition.

10.1 Blast-Induced Brain Injury Models

Blast neurotrauma has become the "signature wound" of the current Mideast conflict due to exposure to improvised explosive devices [42, 103, 167]. Victims of blast-related TBI suffer from complex neuropsychiatric symptoms such as dyspraxia, dysphasia, executive dysfunctions, paralysis, deficits, and dysfunctions of special senses, and mood disorders [32, 84]. This type of blast-related brain injury is categorized as primary because it is related directly to the shock wave itself, as opposed to subsequent insults. Brain injuries resulting from fragments caused by blast are classified as secondary, and those caused by impacts with objects when the individual is propelled through space are termed tertiary [32, 84]. Secondary and tertiary TBI are the result of blunt trauma routinely seen in falls, vehicle crashes or contact sports, and have been studied with some degree of success over the past seven decades, and related computational brain models are summarized in the Blunt Impact section. In this section, the focus is on primary blast injury, as its mechanism is poorly understood, and the injury threshold remains unknown. In recent years, efforts have been made to study the effects of primary blast brain injury using FE models.

Compared with blunt head impacts, blast loading has a very short duration and very rapid pressure change. Because the magnitude of strains generated by these extremely short duration impacts is minuscule, many researchers hypothesize that pressure or pressure gradient is the mechanism of primary blast-induced brain injury. Currently, there exist several unvalidated blast related brain injury models, and the current status of their development is briefly presented below. Because this is a current above-the-horizon topic, this review also includes reports published in conference proceedings that may not have been rigorously peer-reviewed.

Unlike other hallmark blast injuries (such as blast lung), primary brain injury mechanism and tolerance have not been clearly established in terms of exposure. To investigate blast-induced brain injury mechanisms, Chafi et al. [12] developed a LS-DYNA head model consisting of all essential anatomical features (i.e. brain, falx and tentorium, CSF, dura matter, pia mater and skull) and used it to simulate primary blast brain injury. Material properties were taken from the published literature and the input was a blast created by 13 kg of C4 explosive at a distance of 3 m. High compressive stresses were observed in the frontal and parietal regions, while tensile stresses were large in the posterior fossa and occipital regions. They hypothesized that the pressure gradient might be responsible for brain damage. The same authors [13] also investigated the effect of shear stress using the same head model which was exposed to smaller charges (0.5 and 1.0 lb TNT). They compared the predicted stress level with published impact injury thresholds and concluded that the deviatoric or shearing stresses developed at localized regions in the brain surrounding the ventricles might be the injury mechanism. The same authors [14] also reported the intracranial pressure gradient, shear stress, and principal strain when the model was exposed to overpressures in the range of 2.4 to 8.7 atmospheres. They suggested that it would be more appropriate to use an overall evaluation for injury prediction using various measurements instead of one single injury criterion. Additionally, they reported that dynamic responses of the brain were better injury predictors than head input accelerations.

Several other research groups have explored multiple mechanisms to determine which criteria may be most appropriate to establish blast loading tolerance. For example, Moss et al. [95] hypothesized that blast-induced head injury might be caused by secondary factors, such as forced relative motions between the skull and brain. An in-house software package was used to calculate the interaction between the shock wave and head, which was represented by a simplified elliptical solid with description of the skull, CSF, brain, and face. The model was exposed to a blast created by 2.3 kg of C4 at a distance of 4.6 m. They concluded that the likely cause of blast induced injuries was localized skull flexure with displacements in the order of 50 μm which generated a large pressure gradient in the underlying brain. Another model was developed by Zhang et al. [158] to study biomechanical responses, such as strain and displacement, within the brain. This idealized FE head model was exposed to a blast created by 10 kg of TNT at a distance of 1 m. Brain strains predicted in the coup and contre-coup regions were 4 to 7 times higher than that in the central region. Additionally, high brain strain (15%) and large deformations (4 mm) occurred in the brainstem region. From these results, the authors stated that there may be a higher probability of injury in the peripheral brain and brainstem regions due to blast overpressure loading. No consensus has been reached on the role of skull and brain displacement in blast-induced brain injury at this time.

Moore et al. [96] used an in-house fluid/solid coupling tool to simulate the detonation of 0.0648 and 0.324 kg TNT. A head model consisting of only the upper part of the head was used to calculate intracranial responses. The behavior of the brain tissue was described by a neo-Hookean elastic model with the Tait and Mie-Gruneiesen equation of state Meyer et al. [92]. Results indicated that intracranial responses of an unprotected head with an intensity of 50% lethal dose blast lung injury were comparable to impact-induced mild TBI.

Other researchers have taken different approaches to simulate the blast event. Instead of using explosive simulations to produce shock waves, Taylor and Ford [137] applied an energized air flow with a peak pressure of 1.3 MPa as input to a head model developed from the Visible Human Female data set using a self-developed FE solver. Simulation results revealed that the blast-induced high pressure, shear stress and volumetric tension occurred within the first 2 ms of the blast exposure and was much too soon for any significant global motion of the head to take place. Consequently, the authors concluded that injury criteria based on linear and rotational accelerations were not suitable for evaluating blast-induced TBI, indicating that this conclusion holds true whether the blast is simulated by explosives or by air flow alone.

Considering that there is currently no blast-related intracranial pressure data obtained from human cadavers for model validation, an FE model of the rat head subjected to air shock loading was developed and validated against the experimental data [165]. The rat head model was taken from a previously developed rat brain model for simulating blunt controlled cortical impacts [85] and described in more detail in the Animal Brain Model section. An FE model representing gas flow in a 0.305-m diameter shock tube was formulated using an Eulerian approach to provide input (incident) blast overpressures to the rat model. An arbitrary Lagrangian–Eulerian (ALE) fluid–structure coupling algorithm was then utilized to simulate the interaction between the shock wave and the rat head. The model-predicted pressure–time histories at the cortex and in the lateral ventricle of the rat were in reasonable agreement with those obtained experimentally. Further examination of the FE model predictions revealed that pressure amplification, caused by shock wave reflection at the interface of the materials with distinct wave impedances, was found in the skull. The overpressures in the anterior and posterior regions were 50% higher than those at the vertex and central regions, indicating a higher possibility of injuries in the coup and contre-coup sites. At an incident pressure of 85 kPa, the shear stresses and principal strains in the brain were at low levels, implying that they are not the main mechanism of injury in this particular scenario.

Table 2 briefly summarizes blast-related brain models developed and published to date. The table compares the mesh and geometry of the models, loading conditions applied, solvers, and parameters assessed. There are several key techniques involved include the mathematical description of the shock wave, fluid/solid coupling, and very high strain rate constitutive modeling, the details of which are beyond the scope of this chapter. Generally, it can be seen from Table 2 that at least three techniques were used in the simulation of fluid–structure interaction between the blast wave and the head, namely the multi-material ALE/Lagrangian coupling formulation, empirical pressure functions, and a pure Eulerian computational fluid dynamics algorithm. Shock wave/head coupling was implemented either by using commercially available software (i.e. LS-DYNA) or in-house computational tools. Intracranial pressure gradients, stress (compressive, tensile or shear), as well as shear strain were used to quantify the response of the head under shock loading. Despite the fact that these modeling efforts are aimed at providing relevant information in using computer models to study blast-induced primary brain injury, these reported models are far from perfect. All studies applied quasi-static material properties from the literature without consideration of the high strain rates involved in blast loading. Except the rat brain model reported by Zhu et al. [165], no other models were validated against experimental data pertaining to simulated blast loadings due to a paucity of such data. Based on aforementioned modeling results, several key parameters need to be measured during blast experiments in order to fully validate model predictions: (a) strain or deformation

Table 2 Brief summary of FE models developed to simulate blast-induced brain injuries

First author, year	Chafi et al. [12]	Chafi et al. [13]	Chafi et al. [14]	Moore et al. [96]	Moss et al. [95]	Zhang et al. [158]	Taylor et al. [137]
Mesh	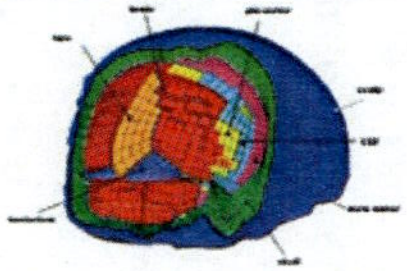			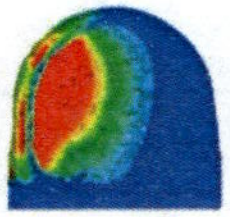	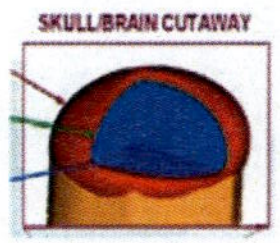	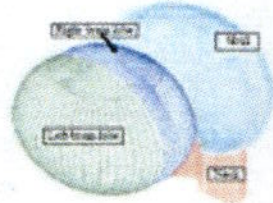	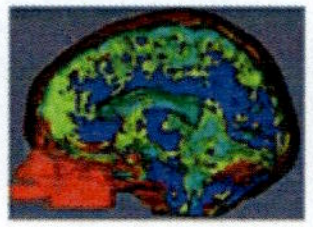
Loading condition	13 kg C4 at 3 m	0.227 and 0.454 kg TNT at 0.8 m	0.038, 0.093 and 0.227 kg TNT at 0.8 m	0.0648 kg and 0.324 kg TNT at 0.6 m	2.3 kg C4 at 4.6 m	10 kg TNT at 1 m	Energized air flow with the peak pressure of 1.3 MPa
Loading modeling method	Multi-material ALE/Lagrangian coupling			Multi-material ALE/Lagrangian coupling	Multi-material ALE/ Lagrangian coupling	Empirical pressure equation (ConWep)	Pure Eulerian (CFD)
Geometry	Human head with detailed anatomy: Brain, falx and tentorium, CSF, dura mater, pia mater, skull bone, scalp			Human head with detailed anatomy (only upper part): Skull, ventricle, glia, white and gray matters, eyes, venous sinus, CSF, air sinus, muscle, skin/fat	Simplified model: skull, CSF, brain and face	Simplified model: Skull, left and right lobes, and neck	Human head with detailed anatomy: Skull, white and gray matters, CSF
Solver	LS-DYNA			In-house software	In-house software	LS-DYNA	In-house software
Parameters assessed	Compressive and tensile stresses	Shear stress	Intracranial pressure gradient, shear stress and shear strain	Pressure, and tensile and compressive stresses	Intracranial pressure gradient	Pressure and maximum principal strain	Volumetric tension, shear stress and pressure

of the skull and (b) pressures at multiple locations such as the coup, central, and contre-coup sites.

Additionally, there is no standard set of experimental data using a phantom (physical surrogate) to verify the biofidelity of software developed to calculate blast-induced intracranial response. While some researchers have commented the over-simplification and associated inaccuracies of certain commercially available software for blast simulations, none have provided a common dataset to check the accuracy of such software. Besides, there is a need to use a very fine mesh model to correctly simulate the thickness and density of the shock front and the solid–fluid coupling algorithms are very computationally expensive. Unfortunately, a model incorporated these features cannot be run with reasonable turnaround time at present. Improvements in modeling techniques, determinations of material properties at appropriate loading rates, accurate measurements of biomechanical responses in skull and brain, and faster computer are needed to comprehensively verify and validate FE models as a reliable tool in blast-induced brain injury research in future.

10.2 Pediatric Head and Brain Models

Although many lack the sophistication of adult FE head/brain models, several pediatric head and brain models have been developed to investigate biomechanical response with particular relevance to children. Major obstacles in this field are the lack of accurate material properties obtained from pediatric specimens and appropriate datasets for model validation. Each of these models was developed for distinctly different purposes, even though some age groups overlap, and as such, it is not relevant to directly compare predicted response values, especially in computational models which cannot be properly validated. Nevertheless, parametric studies using these models and comparative analyses of simulated results from these models can still provide useful information on pediatric head/brain injuries.

Some pediatric head models have been developed to study the effects of the more compliant infant skull on brain response. The first such model by Thibault et al. [138] utilized a simplified geometrical representation and different techniques for modeling the cranial sutures. They showed that, in a geometrically simple model meant to represent the infant brain as solid elements and the infant skull as shell elements, the maximum principal strain distributions of the intracranial contents were affected by the material properties of a thin, linearly elastic skull. A similar semi-ellipsoidal model by Kurtz et al. [78] used 1D springs to simulate sutures without fontanels (which may not be anatomically correct for the 3-month-old material property data used) and expanded the skull behavior to include plasticity. The linear springs allowed the sutures to support tension, but not bending, which seems to be biomechanically counterintuitive. This model was not validated, but posterior impacts were reported to cause the highest brain strains. Lateral impacts showed skull deformation remote to the impact area and instigated diffuse patterns of strain in the brain.

Another model with idealized geometry was published by Margulies and Thibault [88] in which the sutures were represented by elastic shell elements. As with the previously mentioned simplified models, no CSF layer was considered. Identical brain properties were assigned to both models but the skull and sutures were assumed to have either adult or 1-month-old infant material properties. An oblique impact load was applied to investigate the effect of different cranial bone properties on the intracranial tissue deformation pattern. Results indicated that increased skull deformation in the infant head might lead to changes in strain distribution within the brain. More specifically, a similar impact scenario could yield diffuse bilateral strain in the infant but focal unilateral strain in the adult. Again, no validation was performed in either model. The use of boundary conditions along the flat edge to prevent any global translation or rotation in simulations might have affected the conclusions reached by the authors.

Prange et al. [107] created an infant (2-week-old) and an adult FE skull–brain models from 2D MRI slices, each representing a 2 mm thick coronal section, to study brain injury induced by rotational loading applied at a fixed pivot point approximately at level of the mid-cervical spine (C4–C5). For the infant model, the brain was assigned either adult or infant material properties while only the adult properties were used in the adult model. Maximum principal strains at five anatomical regions were analyzed, and it was concluded that the brain size had a greater effect on response than material properties, though both played a role. No statistical significance tests were performed. It should be pointed out that projecting a 2D slice to create a uniformly thick 3D model can produce results of questionable validity as far as 3D response is concerned.

More advanced, anatomically accurate models have also been developed for various purposes. Desantis-Klinich et al. [26] reconstructed real-world automotive frontal impact cases, in which concomitant skull and brain injuries were attributed to deploying passenger-side airbags, using a child restraint air bag interaction (CRABI) dummy. The authors also developed an anatomically realistic 6-month-old infant head FE model from digitized skull CT contours for the purpose of predicting skull fractures through a stress-based criterion without consideration of intracranial response. When simulating the real-world scenario based on bilateral loading from the CRABI reconstruction, the highest stresses were observed at the point of impact, while real-world injuries suggested fractures tend to occur at locations remote from the impact site. This inconsistency was not resolved even though a von Mises stress-based fracture tolerance was presented. Although the model was sensitive to changes in modulus of the elastic skull, different moduli were not assigned to different bones. The brain was modeled as linearly viscoelastic with a CSF layer, and a parametric study found that decreasing the bulk modulus of the brain affected total skull responses, while increasing long-term shear modulus increased head acceleration. Based on comparative responses of a CRABI and the infant FE models, the authors proposed alternative values that were different from the accepted injury assessment reference values (IARV) for the CRABI dummy (scaled from adult data). However, given the lack of validation in the infant FE model, the applicability of the proposed values is questionable.

The predicted corridors for injury measures such as HIC and acceleration were wider for the infant FE model and higher than the CRABI values.

Another FE model, developed for the purpose of investigating skull fracture from drops and falls, was published by Coats et al. [24]. The geometry of the model was obtained from MRI and CT image sets of a 5-week-old infant. The interaction between the 8-noded skull shell elements and the tetrahedral solid brain elements was defined as a frictional interface with no CSF. The sutures were modeled as membrane elements. The parameters assigned for the brain were based on Ogden formulation, the sutures were linearly elastic, and different orthotropic parameters were given to the occipital and parietal bones based on bending data. Occipital loading was applied to simulate a short fall, and several brain and suture properties were parameterized and compared to responses from baseline. Brain stiffness (μ) was investigated due to the fact that Prange and Margulies [109] found infant porcine brain tissue to be twice as stiff as adult porcine tissue. Therefore, the baseline μ-value was defined as two times that of the human adult value (257 Pa) measured by this research group. It was observed that using 257 or 559 Pa did not significantly affect skull stresses, but increasing the stiffness to four times baseline (another adult value from the literature) increased peak principal stress, peak force, and contact area by greater than 15%. The authors showed that variations in material incompressibility also affected the skull response. However, no analysis was presented on how to choose appropriate values for the pediatric brain. Parametric studies of the suture indicated that decreasing suture thickness had little to no effect on the skull, but widening the suture to an unrealistic 10 mm affected peak stress and contact area. Removing the suture entirely did not change the predictions of this model, which may be a consequence of suture morphology simplification. Qualitative validation was performed based on using ultimate stress as an indicator of fracture, but the exact location and orientation of the fracture was not considered, nor were measurable quantitative values such as impact force.

The previously mentioned models incorporated a completely homogeneous brain mass, with or without CSF. However, a more detailed 6-month-old head model developed at ULP [113, 114] included several brain structures such as the falx and tentorium. As the geometry was derived from CT imaging, the anatomical accuracy of these structures is assumed to be estimated from an anatomical atlas. Nodal connectivity was employed to define interaction between skull, CSF, and brain. The brain was linearly viscoelastic, while all other parameters were linearly elastic. Several studies have been performed using versions of this head model, but no validation has been presented. The first publication [113] attempted to reconstruct the two phases of shaken baby syndrome (SBS) separately by the oscillatory shaking and the blunt occipital impact when the child was released. Model-predicted pressure and von Mises stresses were found to be much higher in the impact case, while strain predicted in the bridging veins through relative motion was similar for both loading conditions. The 2008 paper [114] illustrated the differences in biomechanical responses between the infant head model and a structurally scaled version of the adult ULP head model with similar material definitions. Results from the two models indicated visibly different predictions of skull fracture location using a stress-based criterion for a single case history. However, due to rudimentary nature of this model, it does highlight the fact that children are not small adults, and that the biomechanical responses are both material and shape dependent.

The same researchers have also developed a 3-year-old head model using similar techniques to those of their 6-month-old model [115]. This 3-year-old head model was used to develop tolerance criteria for neurological lesions in real-world fall cases and two sets of brain material parameter values from the literature. In the absence of proper data for validation, 25 cases were used as a basis for statistical analysis of peak intracranial von Mises stress, peak internal pressure, peak angular and linear accelerations, and HIC using logistic regression risk curves. Based on an incomplete analysis in this study, HIC was shown to be the best predictor among different injury predictors. Variance in impact location was shown to affect the peak stress in the brain, as did a 30% error in fall height. However, it was concluded by the authors that the quantitative effect was within acceptable limits.

In addition to the study performed by Roth et al. [113], Couper and Albermani [25] also investigated SBS through FE modeling. In their study, a 2D model of a hemi-spherical brain MRI slice from a 3-month-old was developed, and more complex material properties compared to other published pediatric brain models were assigned. The constitutive model employed a nonlinear μ coefficient for the Odgen portion and various dissipation modes in the Maxwell element, with gray matter and three different myelinated white matters being assigned different properties. These values were estimated from published human and animal data for both adults and infants. Various subarachnoid space representations were also considered. The input to the model was an oscillatory acceleration, and the different simulation conditions were shown to have dissimilar biomechanical response. Simulated results from models without validation made it difficult for the authors to draw conclusions. Nevertheless, the authors postulated that using solid Lagrangian elements for the CSF layer is not appropriate for cyclic loading schemes based on phase differences. The volume of CSF is shown to have a large effect on brain response and is therefore an important factor in predicting brain injury. Although material properties are important, it was shown that the stress path may be more sensitive to the brain interface than to material stiffness based on this unvalidated model.

Table 3 lists a summary of the pediatric head and brain models discussed above, in terms of mesh, age, solving software, geometry, elements and material used, and validation status. Although the lack of proper validation in these models and the scarcity of appropriately documented injury criteria for children negates the possibility of drawing concise conclusions on injury risk, these models may point scientists and engineers in the direction of future pediatric brain injury research. It is interesting to note that brain material behavior often has been considered only secondary to the braincase itself, and that many of the FE model results show less effect from suture stiffness than might be expected. A recent review of directly measured pediatric material properties [36] indicates that such

Table 3 Summary of FE pediatric head/brain models

First author, year	Kurtz et al. [78]	Margulies and Thibault [88]	(DeSantis) Klinich, [26]	Roth et al. [113, 114, 115]	Coats et al. [24]	Couper et al. [25]	Prange et al. [107]
Mesh			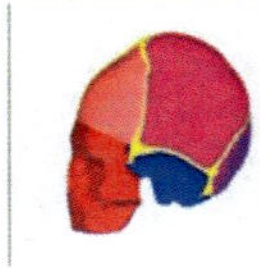	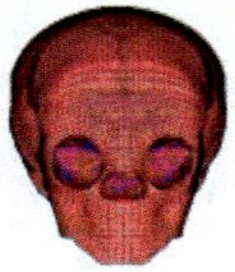			
Age	3 months	1 month	6 months	6 months and 3 years (2009)	1.5 months	3 months	2 weeks
Solving software	LS-DYNA3D	ANSYS/LS-DYNA3D	LS-DYNA	Radioss	ABAQUS/Explicit	MSC MARC	PATRAN
2D/3D	3D	3D	3D	3D	3D	2D	Pseudo 3D
Geometry	Idealized	Idealized	CT of 27 week subject; facial geometry from Zygote model	CT of 6 month and 3 year subjects	CT and MRI of 5 week old subject; suture geometry idealized	MRI of 3 month subject	4 mm thick MRI of a 2 week subject
Elements	25,279 solid elements, 5,514 shell elements, 137 springs	12,772 elements	NA	6 MO: 69,324 solid elements, 9,187 shell elements 3 YO: 23,000 solid elements, 3,500 shell elements	11,066 tetrahedral solid elements, 624 hexagonal solid elements, 18,706 shell elements, 2,485 membrane elements	4,000 plane strain shell elements	5,834 elements

(continued)

Table 3 (continued)

First author, year	Kurtz et al. [78]	Margulies and Thibault [88]	(DeSantis) Klinich, [26]	Roth et al. [113, 114, 115]	Coats et al. [24]	Couper et al. [25]	Prange et al. [107]
Material law	Linear viscoelastic brain, elastic–plastic bone, springs for suture	Linear viscoelastic brain, linear elastic bone/suture	Linear viscoelastic brain, linear elastic bone/suture/CSF/scalp	Linear viscoelastic brain, linear elastic bone (interpolated)/suture/CSF/scalp	Ogden brain, orthotropic linear elastic bone, linear elastic suture and scalp	Ogden with Maxwell for infintesimal strains (various dissipation modes)	Ogden brain, rigid skull and falx
Validation	None	None	None	None	Skull fracture location as predicted by ultimate stress	None	None

shortcomings exist not only in the computational modeling field, but are also indicative of a general lack of experimental data.

10.3 Animal Head and Brain Models

In vivo TBI experiments animal surrogates are extensively used to study neurological responses, which cannot be investigated with cadavers or volunteers, as described in the introductory section to this chapter. Because there are many types of TBI in the human, no single published in vivo TBI experimental model can reproduce the entire spectrum of human TBI. To study individual brain injury types, different animal TBI experimental models have been developed. These experiments offer controlled and measurable external impact parameters which are uniquely suited to rigorous validation of FE simulations in multiple loading conditions, often an impediment to human FE model validation. Furthermore, the detailed brain internal injuries revealed through histological and/or imaging techniques can be compared against FE-predicted intracranial tissue response maps to study brain injury mechanisms as well as tolerance. In conjunction with human FE models, parallels can be drawn between human and animal injury, which may allow researchers to develop scaling laws to facilitate application of tolerance values from animal studies to human injury.

11 Rodent Brain FE Models

Among all the animals, rodents are the most frequently used animal in the laboratory for TBI experiments. Because of the minute nature of the rodent brain certain difficulties are encountered in the development of rodent FE brain models. Despite the extensive use of the rodent as a TBI experimental model, only five rodent FE brain models have been published to date, generally to investigate a specific experimental setup, limiting the ability to compare simulation predictions between different rat brain models based on what has been reported in the literature.

Shreiber et al. [129] developed 3-D FE rat brain model to study the mechanical threshold for blood–brain-barrier (BBB) injury. The model consisted of a homogeneous brain and a rigid skull which served as the boundary condition. In experimental studies, negative pressure pulses with magnitudes of 2, 3, and 4 psi (each at three different durations of 25, 50, and 100 ms) were applied to the exposed brain tissue to induce local brain deformation. The FE brain model calculated cortical displacement was then validated against experimentally measured brain surface deformation using a laser displacement transducer. The model-predicted maximum principal logarithmic strain, maximum principal stress, strain energy density, and von Mises stress were compared against the BBB injury observed at specific locations within the brain. The authors reported that the maximum principal logarithmic strain was the best predictor with a strain value of 0.1888 for a 50% probability of BBB injury in rats.

Pena et al. [106] first attempted to characterize displacements, mean stress, and shear stress using a 2D FE brain model representing a single coronal section due to controlled cortical impact (CCI). Levchakov et al. [81] developed a 3D coarsely meshed FE model of a rat brain using tetrahedral elements to predict intracranial strain/stress for both neonate and mature rat brains in closed head CCI. Each of these two models assumed homogeneous material properties with no consideration given to the detailed anatomical organization of the brain. As such, these models are limited in their capability in predicting region-specific responses to TBI. The biofidelity of the above models are also questioned because neither has been validated against experimental data.

Mao et al. [85] developed a 3D FE rat brain model representing all essential anatomical features of a rat brain, including the olfactory bulb, cortex, hippocampus, thalamus, hypothalamus, corpus callosum, brainstem (midbrain, pons and medulla oblongata), cerebellum, lateral ventricle, third ventricle, fourth ventricle, internal capsule, external capsule and part of the spinal cord, based on histological studies of a rat brain [105]. The brain model consisted of 255,700 hexahedral elements with a typical spatial resolution 200 μm. The FE model was first validated against cortical tissue deformation measured during dynamic cortical deformation experiments conducted by Shreiber et al. [129]. The biomechanically validated rat brain model was then used to simulate four different series of CCI using unilateral craniotomy [19, 73, 127, 133]. Simulation results indicated that the peak maximum principal strain (MPS) with a threshold of 0.30 correlated with contusion volumes experimentally measured between 7 and 14 days post-injury. For predicting contusion measured at 24 h post-injury, 0.265 MPS is suggested [86]. To demonstrate the convergence of this rat brain model, Mao et al. [87] used five simplified 3D rat brain models with a spatial resolution of 1.6, 0.8, 0.4, 0.2, and 0.1 mm. Results demonstrated that continued decrease in element size resulted in less and less variation in the average MPS value. For example, the difference between the 1.6 and 0.8 mm spatial resolution models was 33.9% but the difference between the 0.2 and 0.1 mm resolution model was only 4.4% for the region close to impact area. Similarly, the differences for ventral region remote from impact were 42.7% between 1.6 and 0.8 mm spatial resolution model, and 1.1% between 0.2 and 0.1 mm resolution model. These results indicated that a model with 0.2 mm spatial resolution reasonably balanced between computational accuracy and efficiency at current stage.

Using the same rat brain model [85], Mao et al. [86, 87] proposed a new injury metrics: cumulative strain damage percentage measure (CSDPM). The CSDPM concept is based on the hypothesis that the element-level peak strain magnitude is related to the injury intensity within that element. In particular, the percentage of cell loss is found to be related to the magnitude of element strain. The difference between CSDPM and CSDM proposed by NHTSA researchers (see Sect. 2.1) is that CSDM counted the total volume of brain elements exceeding a certain threshold strain value. In CSDPM, neuronal loss percentage was calculated for each brain element ($\sim$0.2 mm resolution for the FE rat brain model) first before the average percentage of cell loss for a region was determined. Equation 1 shows the proposed relationship between MPS and CSDPM [87]

$$\text{CSDPM} = \sum_{i=1}^{N} (1992 * \text{MPS} - 0.028) * [\text{volume ratio}\,(i)]$$

$$\text{Volume ratio}\,(i) = \frac{\text{volume of element}\,(i)}{\text{total brain volume}} \tag{1}$$

where i represents the brain element number and N is the total number of elements in the FE brain model.

The external impact parameters for the in vivo TBI models subjected to CCI tended to vary considerably among different laboratories, which make the comparison of research findings among different institutions very difficult. Mao et al. [87] adopted the design of experiments (DOE) method to investigate the effect of external impact parameters and the potential of using FE rat brain model previously developed by Mao et al. [85] to aid in the design of animal TBI model of desired injury intensity. A five-factor two-level factional factorial DOE was performed. Results demonstrated that the impact depth was the leading factor affecting the predicted brain internal responses. Interestingly, impactor shape ranked as the second most important factor, surpassing impactor diameter and velocity, which were commonly reported in the literature as indicators of injury severity along with the impact depth. The differences in overall brain responses due to a unilateral craniotomy or bilateral craniotomies were small, but there were significant regional differences. The interaction effects of any two external parameters were found to be not significant. Such analysis demonstrates that FE rat brain model can be used to assist the engineering of better experimental TBI models in the future.

Shafieian et al. [128] developed a FE rat brain stem model to validate the assumption of force–displacement relationship during an in vitro indentation. The FE-predicted force–deflection curve agreed with the linear portion of the experimental results if the brainstem was modeled as a linear elastic material. At an impact depth of 0.5 mm, the FE model predicted a 20% effective strain at the ponto-medullary junction (PmJ) and a 38% effective strain at the pyramidal decussation (PDx). Quantification of axonal injuries yielded means of 1.7 ± 0.3 injured axons at PmJ, and 16.8 ± 1.1 at PDx per 100,000 μm^2. The FE-predicted strains seemed to qualitatively match the trend of axonal injuries.

12 Primate Brain FE Models

Ward et al. [146] used both human and monkey finite element brain models developed by the same group [147] to predict intracranial pressure (ICP) in the human and monkey during blunt impact. The FE brain model was validated against experimentally observed ICP data. By comparing ICP and brain injury in the form of contusion and hematoma/hemorrhage, Ward concluded that pressure above 234 kPa (34 psi) could induce severe brain injuries. However, due to the limited computational capabilities at that time, the brain meshes were relatively coarse. Tissue strains at both the coup and contre-coup sites were not investigated and their contributions to brain contusion remained unclear in the study.

In order to study the subdural hematoma injury mechanism, Lee et al. [80] developed a 2D rhesus monkey model to simulate the animal experiments designed and conducted by Abel et al. [1]. The brain was assumed to be an isotropic elastic material with a shear modulus of 80 kPa. The authors stated that this high shear modulus was selected to account for the absence of the dura, tentorium, and blood vessels. Structural damping was added in some simulations to remove the high frequency response components in the simulations. A no-slip boundary condition was assumed between the brain and skull. Because the authors were interested in establishing an injury threshold, which included both the peak linear and angular accelerations, they applied purely linear and purely angular loading to their model and calculated the brain deformation. The amount of linear loading required to generate the same deformation as a purely rotational loading was then determined to generate "equal deformation" lines on an angular versus tangential acceleration plot. From this point of view, a combined injury threshold was proposed.

13 Pig and Sheep Brain FE Models

In 1994, Zhou et al. developed three 2D finite element models representing three coronal sections of the porcine brain [162]. The models consisted of a three-layered skull, the dura, CSF, white matter, gray matter and ventricles. Model I was a section at the septal nuclei and anterior commissure level and contained 490 solid elements and 108 membrane elements. Model II was a section at the rostral-thalamic level and contained 644 solid elements and 130 membrane elements. Model III was a section at the caudal hippocampal level and contained 548 solid elements and 104 membrane elements. Plane strain conditions were assumed for all models. The input was the angular velocity time history used in the animal experiments performed at the University of Pennsylvania to determine the distribution of diffuse axonal injury (DAI) in porcine brain [112]. Regions of high shear stresses predicted by the model agreed qualitatively with experimental findings when the shear modulus of white matter was assumed to be 60% higher than that of gray matter. If the white and gray matter were assumed to possess the same properties, regions of DAI predicted by the model did not match those observed experimentally.

Miller's study [93] focused on FE modeling approaches for the simulation of the relative motion between the skull and the cerebral cortex in miniature pigs.

They performed rotational acceleration tests along an axis normal to the axial plane (the plane perpendicular to the brainstem in a miniature pig), which, according to the authors, was equivalent to the coronal plane in the human. Diffuse axonal injury data from five experiments were used to validate their models which assumed that the skull and brain were either connected directly by a virtually incompressible low shear modulus material or by a frictional interface. The numerical responses of the two approaches were compared using the model across two-axial planes. Model I was partitioned through the dorsal part of the frontal, parietal and occipital lobes, while Model II was partitioned through the brain stem. The brain was assumed to be a Mooney–Rivlin hyperelastic material. When predicting the distribution of diffuse axonal injuries, results of their simulations demonstrated that the frictional interface was a better representation of the subarachnoid space than the virtually incompressible solid that had a low shear modulus. It was also found that the maximum principal nominal strain and von Mises stress were good predictors of axonal and macroscopic hemorrhagic cortical contusions, while negative pressure was a poor predictor for both forms of injury. A year later, Miller et al. [94] noted that most animal models provided only qualitative injury maps instead of quantitative ones for comparisons with FE model predictions, which could provide both temporal and spatial metrics for correlations with injury. They graded the severity of axonal injury from the animal experiments based on histological analyses. The two sectional models described in their 1998 paper [93] were separated into 20 regions for comparisons with experimentally observed graded injuries. They concluded that omitting the dura mater in the FE model of miniature pig could best approximate the graded experimental results they observed.

Anderson [5] developed a coarse-mesh 3D sheep brain model to simulate the blunt impact induced on the sheep head. He found that high von Mises stresses best correlated with their own axonal injury scores ($R^2 = 0.296$) among all parameters examined including strain and pressure. However, even in the best correlation case, the majority of the variation in axonal injury score was still unaccounted for. This seems to indicate numerical improvements of the coarse-mesh 3D brain model are required. Lastly, Zhu et al. [164] applied a 3D 3- to 5-day-old piglet brain model developed at the University of Pennsylvania to predict duration of unconsciousness. Using a published strain threshold developed previously with the same brain model, the length of unconscious time could be accurately predicted.

14 Other Animal Models

Ueno et al. [143] developed 3D FE ferret brain models that simulated the controlled cortical impacts performed at General Motors by Lighthall et al. [83]. The pressure predicted by the FE model compared favorably with experimental pressure data. The authors found that the pressure predicted by the FE model propagated to the skull–brain boundary. High shear deformation was generated at the impact site and was similar to the contusion hemorrhage observed experimentally.

15 Summary of Animal FE Models

It can be seen that each of the animal brain FE models was developed for a specific purpose, and rarely were the computational models exercised fully. This makes it challenging to contrast the findings in any meaningful way. Instead, a brief summary of the aforementioned animal models is provided in Table 4 to illustrate how geometry, material laws, and validation compared. Except for the 3D rat brain model developed by Mao et al. [85] the rest of the animal models lacked detailed anatomical structures critical to the prediction of regional tissue responses for correlation with regional injury patterns and severities. Some of the models were not biomechanically validated and most of these animal FE models were developed to simulate a specific injury scenario without investigating other types of injury scenarios. Theoretically, a brain FE model with good biofidelity should be able to reasonably predict different types of TBI. In other words, it is better to evaluate a FE brain model for various types of TBI rather than for one specific condition so that its universal applicability and reliability can be assessed. FE model predictions need to be fully validated against multiple scenarios before they can be considered trustworthy and care must be taken when determining injury mechanisms and thresholds through comparisons of FE model predictions with experimentally obtained injury data.

16 Concluding Remarks

Any numerical model requires validation against multiple scenarios of experimental data before it can be used to predict model responses under conditions where experiments are difficult to conduct. In TBI impact biomechanics research, numerous researchers have emphasized the need for better characterizations of material laws and associated properties of brain and skull as they play important influential roles in simulation results. Additionally, numerical models developed thus far are in general under-validated due to lack of high fidelity data for model validation. For these reasons, it is recommended that head model developers should always conduct convergence study on the effect of mesh size before the simulated results are overused in predicting the injury outcomes. Limitations of models also need to be reported clearly. Nevertheless, a number of FE models have been developed and validated to a limited extent for advancing our understanding of head and brain biomechanics in hope to develop useful countermeasures to reduce the incidence of TBI.

Table 4 Summary of FE animal brain models

First author, Year	Ward et al. [146]	Lee et al. [80]	Zhou et al. [162]	Ueno et al. [143]	Miller et al. [93, 94]	Anderson [5]
Mesh						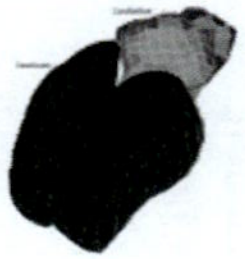
Species	Monkey	Monkey	Pig	Ferret	Pig	Sheep
Solving software	NA	Nastran	PAM-CRASH	ABAQUS	ABAQUS	LS-DYNA
2D/3D	3D	2D	2D	3D	2D	3D
Geometry	Exact geometry	Mid-sagittal section	Atlas	Brain surface contour	T2 MRI	MRI, CT
Elements	NA	291 elements, 317 grid points	Model 1, 2, 3: 490, 644, 548 solid elements, and 108, 130, 104 membrane elements, respectively	1920 solid elements for brain, 883 shell element s for skull	NA	3936 solid elements, 1699 shell ele ments
Material law	Various elastic modulus and Poisson's ratio for brain	Brain as isotropic linear elastic material	Distinct gray and white matter, elastic material for brain, CSF, and skull	Elastic property for brain, rigid skull	Rigid skull, Mooney Rivlin for brain, distinct gray and white matter, elastic dura	Mooney-Rivlin for brain
Validation	Pressure	No	No	Pressure	No	No

(continued)

Table 4 (continued)

First author, Year	Zhu et al. [164]	Shreiber et al. [129]	Pena et al.[106]	Levchakov et al. [81]	Mao et al. [85, 86, 87]	Shafieian et al. [128]
Mesh	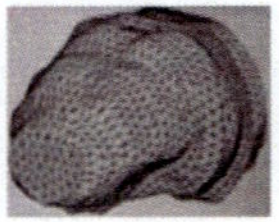				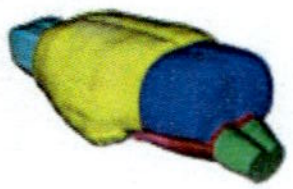	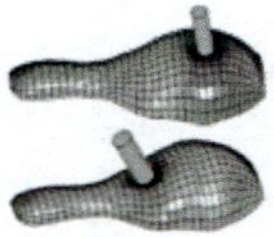
Species	Pig	Rat	Rat	Rat	Rat	Rat brainstem
Solving software	ABAQUS	ABAQUS	FEMLab 2.0	Nastran	LS-DYNA	LS-DYNA
2D/3D	3D	3D	2D	3D	3D	3D
Geometry	MRI	Atlas	T2 MRI	Atlas	Atlas	Optical digitizer
Elements	66,445 Tetrahedral elements	36,664 solid elements	NA	About 30,000 tetrahedral elements	255,700 hexahedral elements	Hexahedral elements
Material law	Ogden material for brain	Modified hyperelastic material for brain, rigid skull	Elastic property for brain	Viscoelastic brain, elastic skull	Linear viscoelastic	Linear elastic
Validation	No	Cortical deformation	No	No	Cortical deformation	No

To overcome these difficulties is a monumental task which is probably why usable experimental data for validation are still lacking even after 70 years of TBI research. However, difficulties in conducting animal TBI research may not be as profound as using the human as research subjects. It is our belief that carefully designed animal experimental models with the aim of providing needed data for accurate material properties characterization and precise documentation of tissue-level impact responses will be the key to determining which techniques are most suitable for modeling the skull and brain tissues. Using these data, high resolution FE model can be developed to establish injury mechanisms and region-specific injury tolerances. If a number of high fidelity FE models representing a couple of small and large animals are successfully developed and rigorously validated, scaling laws and methodologies may be established to extrapolate animal responses to the human. The validated human head model will, in turn, be used to develop helmets or other countermeasures to mitigate the injury severity, reduce the number of TBI incidents, or even completely eliminate TBI.

Before this chapter is concluded, it is worth noting that none of the published numerical models addresses secondary brain injury, defined as a progressive cascade and evolvement of primary injury or injury that is independent of the primary injury [20, 34, 43, 53]. This is because the passageway and progression from primary to secondary injury is mostly pathophysiological and therefore beyond the predictive capability of FE models at this time.

References

1. Abel, J.M., Gennarelli, T.A., Segawa, H.: Incidence and severity of cerebral concussion in the rhesus monkey following sagittal plane angular acceleration. In: 22nd Stapp Car Crash Conference. Ann Arbor, Michigan, USA (1978)
2. Al-Bsharat, A., Hardy, W.N., Yang, K.H., Khalil, T.B., King, A.I., Tashman, S.: Brain/skull relative displacement magnitude due to blunt head impact: new experimental data and model. In: 43rd Stapp Car Crash Conference, SAE Paper No. 99SC22. Society of Automotive Engineers, Warrendale (1999)
3. Alem, N.M., Simulation of head injury due to combined rotation and translation of the brain. In: Proceedings of the 18th Stapp Car Crash Conference, Ann Arbor, Michigan, USA, SAE 741192 (1974)
4. Allsop, D.L., Warner, C.Y., Wille, M.G., Scheider, D.C., Nahum, A.M.: Facial impact response—a comparison of the hybrid iii dummy and human cadaver. In: Proceedings of the 32nd Stapp Car Crash Conference, SAE Paper No. 881719. Society of Automotive Engineers, Warrendale (1988)
5. Anderson, R.W.: A study on the biomechanics of axonal injury. In: Road Accident Research Unit and the Department of Mechanical Engineering. The University of Adelaide (2000)
6. Bandak, F.A., Eppinger, R.H.: A three-dimensional finite element analysis of the human brain under combined rotational and translational accelerations. In: 38th Stapp Car Crash Conference, SAE 942215. Ft. Lauderdale, Florida, USA (1994)
7. Bandak, F.A.: On the mechanics of impact neurotrauma: a review and critical synthesis. J. Neurotrauma **12**(4), 635–649 (1995)

8. Bandak, F.A., Vander Vorst, M.J., Stuhmiller, L.M., Mlakar, P.F., Chilton, W.E., Stuhmiller, J.H.: An imaging-based computational and experimental study of skull fracture: finite element model development. J. Neurotrauma **12**(4), 679–688 (1995)
9. Bilston, L.E., Liu, Z., Phan-Thien, N.: Large strain behaviour of brain tissue in shear: some experimental data and differential constitutive model. Biorheology **38**(4), 335–345 (2001)
10. Brands, DW., Bovendeerd, PH., Peters, GW., Wismans, JS.: The large shear strain dynamic behaviour of in-vitro porcine brain tissue and a silicone gel model material. Stapp Car Crash J. **44**:249–260 (2000)
11. Brands, D.W.A., Bovendeerd, P.H.M., Wismans, J.S.H.M.: On the potential importance of non-linear viscoelastic material modelling for numerical prediction of brain tissue response: test and application. In: Proceedings of the 46th Stapp Car Crash Conference, pp. 103–121 (2002)
12. Chafi, M.S., Karami, G., Ziejewski, M.: An assessment of primary blast injury in human brains—a numerical simulation. In: Proceedings of the 2007 ASME Summer Bioengineering Conference, pp. 349–350 (2007a)
13. Chafi, M.S., Karami, G., Ziejewski, M.: Simulation of blast–head interactions to study traumatic brain injury. In: Proceedings of the 2007 International Mechanical Engineering Congress and Exposition, pp. 211–220 (2007b)
14. Chafi, M.S., Karami, G., Ziejewski, M.: Biomechanical assessment of brain dynamic responses due to blast pressure waves. Ann Biomed Eng. Epub (2009)
15. Chakrabarty, S.P., Hanson, F.B.: Distributed parameters deterministic model for treatment of brain tumors using galerkin finite element method. Math. Biosci. **219**(2), 129–141 (2009)
16. Chan, H.S.: Mathematical model for closed head impact. In: Proceedings of the 18th Stapp Car Crash Conference, SAE 741191. Ann Arbor, Michigan, USA (1974)
17. Chan, H.S., Liu, Y.K.: The symmetric response of a fluid-filled spherical shell—a mathematical simulation of a glancing blow to the head. J. Biomech. **7**, 43–59 (1974)
18. Chen, M., Mogul, D.J.: A structurally detailed finite element human head model for simulation of transcranial magnetic stimulation. J. Neurosci. Methods **179**(1), 111–120 (2009)
19. Chen, S., Pickard, J.D., Harris, N.G.: Time course of cellular pathology after controlled cortical impact injury. Exp. Neurol. **182**(1), 87–102 (2003)
20. Chen, Y.C., Smith, D.H., Meaney, D.F.: In vitro approaches for studying blast-induced traumatic brain injury. J. Neurotrauma **26**(6), 861–876 (2009)
21. Chu, C.S., Lin, M.S., Huang, H.M., Lee, M.C.: Finite element analysis of cerebral contusion. J. Biomech. **27**(2), 187–194 (1994)
22. Claessens, M., Sauren, F., Wismans, J.: Modeling of the human head under impact conditions: a parametric study. In: Proceedings of the 41st Annual Stapp Car Crash Conference, SAE 973338. Lake Buena Vista, Florida, USA (1997)
23. Cloots, R.J., Gervaise, H.M., van Dommelen, J.A., Geers, M.G.: Biomechanics of traumatic brain injury: influences of the morphologic heterogeneities of the cerebral cortex. Ann. Biomed. Eng. **36**(7), 1203–1215 (2008)
24. Coats, B., Margulies, S.S., Ji, S.: Parametric study of head impact in the infant. Stapp. Car Crash J. **51**, 1–15 (2007)
25. Couper, Z., Albermani, F.: Infant brain subjected to oscillatory loading: material differentiation, properties, and interface conditions. Biomech. Model. Mechanobiol. **7**(2), 105–125 (2008)
26. Desantis-Klinich, K.D., Hulbert, G., Schneider, L.W.: Estimating infant head injury criteria and impact response using crash reconstruction and finite element modeling. Stapp Car Crash J. **46**, 165–194 (2002)
27. Dimasi, F., Eppinger, R.H., Gabler, H.C., Marcus, J.: Simulated head impacts with upper interior structures using rigid and anatomic brain models. In: R. Strombotne (ed.) Auto and Traffic Safety, vol. 1, no. 1. National Highway Traffic Safety Publication, Washington, D.C. USA (1991)

28. DiMasi, F.P., Eppinger, R.H., Bandak, F.A.: Computational analysis of head impact response under car crash loadings. In: Proceedings of the 39th Stapp Car Crash Conference, SAE 952718, San Diego, CA, USA (1995)
29. Engin, A.E., Liu, Y.K.: Axisymmetric response of a fluid-filled spherical shell in free vibrations. J. Biomech. **3**(1), 11–22 (1970)
30. Eppinger, R., Kleinberger, M., Morgan, R., Khaewpong, N., Bandak, F.A., Haffner, M.: Advanced injury criteria and crash test evaluation techniques. In: Proceedings of the NHTSA 14th International Technical Conference on Experimental Safety Vehicles, Paper no. 90-S1-O-11. Munich, Germany (1994)
31. Fan, W.R.S.: Internal head injury assessment. In: Proceedings of the 15th Stapp Car Crash Conference, SAE 710870. San Diego, CA, USA (1971)
32. Finkel, M.F.: The neurological consequences of explosives. J. Neurol. Sci. **249**, 63–67 (2006)
33. Finkelstein, E., Corso, P., Miller, T., and Associates.: The Incidence and Economic Burden of Injuries in the United States. Oxford University Press, New York (2006)
34. Fitch, M.T., Doller, C., Combs, C.K., Landreth, G.E., Silver, J.: Cellular and molecular mechanisms of glial scarring and progressive cavitation: in vivo and in vitro analysis of inflammation-induced secondary injury after cns trauma. J. Neurosci. **19**(19), 8182–8198 (1999)
35. Franklyn, M., Fildes, B., Zhang, L., Yang, K., Sparke, L.: Analysis of finite element models for head injury investigation: reconstruction of four real-world impacts. Stapp Car Crash J. 50 (2005)
36. Franklyn, M., Peiris, S., Huber, C., Yang, K.H.: Pediatric material properties: a review of human child and animal surrogates. Crit. Rev. Biomed. Eng. **35**(3–4), 197–342 (2007)
37. Fredriksson, R., Zhang, L., Boström, O., Yang, K.H.: Influence of Impact Speed on Head and Brain Injury Outcome in Vulnerable Road User Impacts to the Car Hood. Stapp Car Crash J. **51**, 155–167 (2007)
38. Galford, J.E., McElhaney, J.H.: A viscoelastic study of scalp, brain, and dura. J. Biomech. **3**(2), 211–221 (1970)
39. Gao, C., Tay, H., Nowinski, F., Wieslaw, L.: A finite element method based deformable brain atlas suited for surgery simulation. In: Conf Proc IEEE Eng Med Biol Soc. vol. 4, pp. 4337–4340 (2005)
40. Gao, C.P., Ang, B.T.: Biomechanical modeling of decompressive craniectomy in traumatic brain injury. Acta Neurochir. Suppl. **102**, 279–282 (2008)
41. Goldsmith, W.: The physical processes producing head injury. In: Proceedings of the Head Injury Conference, pp. 350–382. Lippincott, PA (1966)
42. Gondusky, J.S., Reiter, M.P.: Protecting military convoys in Iraq: an examination of battle injuries sustained by a mechanized battalion during Operation Iraqi Freedom II. Mil. Med. **170**(6), 546–549 (2005)
43. Greve, M.W., Zink, B.J.: Pathophysiology of traumatic brain injury. Mt Sinai J. Med. **76**(2), 97–104 (2009)
44. Hagemann, A., Rohr, K., Stiehl, H.S.: Coupling of fluid and elastic models for biomechanical simulations of brain deformations using FEM. Med. Image Anal. **6**(4), 375–388 (2002)
45. Hardy, C.H., Marcal, P.V.: Elastic Analysis of a Skull. Technical Report No. 8, Office of Naval Research, Contract No. N00014-67-A-0191–0007, Div. Eng, Brown University (1971)
46. Hardy, W.N., Foster, C.D., Mason, M.J., Yang, K.H., King, A.I., Tashman, S.: Investigation of head injury mechanisms using neutral density technology and high-speed biplanar X-ray. Stapp Car Crash J. **45**, 337–368 (2001)
47. Hardy, W.N., Mason, M.J., Foster, C.D., Shah, C.S., Kopacz, J.M., Yang, K.H., King, A.I., Bishop, J., Bey, M., Anderst, W., Tashman, S.: A study of the response of the human cadaver head to impact. Stapp Car Crash J. **51**, 17–80 (2007)

48. Ho, J., Kleiven, S.: Dynamic response of the brain with vasculature: a three-dimensional computational study. J. Biomech. **40**(13), 3006–3012 (2007)
49. Ho, J., Kleiven, S.: Can sulci protect the brain from traumatic injury? J. Biomech. **42**(13), 2074–2080 (2009)
50. Hodgson, V.R., Gurdjian, E.S., Thomas, L.M.: Development of a model for the study of head injury during impact tests. In: Proceedings of the 11th Stapp Car Crash Conference, SAE 670923. Anaheim, CA, USA (1967)
51. Hodgson, V.R., Patrick, L.M.: (1968) Dynamic response of the human cadaver head compared to a simple mathematical model. In: Proceedings of the 12th Stapp Car Crash Conference, SAE 680784. Detroit, Michigan, USA
52. Horgan, T.J., Gilchrist, M.D.: The creation of three-dimensional finite element models for simulating head impact biomechanics. Int. J. Crashworthiness **8**(4), 353–366 (2003)
53. Holmberg, P., Liljequist, S., Wagner, A.: Secondary brain injuries in thalamus and hippocampus after focal ischemia caused by mild, transient extradural compression of the somatosensori cortex in the rat. Curr. Neurovasc. Res. **6**(1), 1–11 (2009)
54. Hosey, R.R., Liu, Y.K.: A homeomorphic finite element model of the human head and neck. In: Gallagher, P.H., Simon, B.R., Johnson, T.C., Gross, J.F. (eds.) Finite Element in Biomechanics, pp. 379–401. Wiley, New York (1982)
55. Hu, J., Jin, X., Lee, J.B., Zhang, L., Chaudhary, V., Guthikonda, M., Yang, K.H., King, A.I.: Intraoperative brain shift prediction using a 3D inhomogeneous patient-specific finite element model. J. Neurosurg. **106**(1), 164–169 (2007)
56. Huang, H.M., Lee, M.C., Lee, S.Y., Chiu, W.T., Pan, L.C., Chen, C.T.: Finite element analysis of brain contusion: an indirect impact study. Med. Biol. Eng. Comput. **38**(3), 253–259 (2000)
57. Jin, X., Lee, J.B., Leung, L.Y., Zhang, L., Yang, K.H., King, A.I.: Biomechanical response of the bovine pia-arachnoid complex to tensile loading at varying strain-rates. Stapp Car Crash J. **50**, 637–649 (2006)
58. Jin, X., Ma, C., Zhang, L., Yang, K.H., King, A.I., Dong, G., Zhang, J.: Biomechanical response of the bovine pia-arachnoid complex to normal traction loading at varying strain rates. Stapp Car Crash J. **51**, 115–126 (2007)
59. Jin, X., Yang, K.H., King, A.I.: Mechanical properties of bovine pia-arachnoid complex in shear. J. Biomech (2010). [Epub ahead of print] 2010 Nov 17
60. Johnson, E.A., Young, P.G.: On the use of a patient-specific rapid-prototyped model to simulate the response of the human head to impact and comparison with analytical and finite element models. J. Biomech. **38**(1), 39–45 (2005)
61. Kang, H., Willinger, R., Diaw, R.M., Chinn, B.P.: Validation of a 3D anatomic human head model and replication of head impact in motorcycle accident by finite element modeling. In: 41st Stapp Car Crash Conference, SAE 973339. Lake Buena Vista, Florida, USA (1997)
62. Khalil, T.B., Goldsmith, W., Sackman, J.L.: Impact on a model head-helmet system. Int. J. Mech. Sci. **16**, 609–625 (1974)
63. Khalil, T.B., Hubbard, R.P.: Parametric study of head response by finite element modeling. J. Biomech. **10**(2), 119–132 (1977)
64. Khalil, T.B., Viano, D.C.: Critical issues in finite element modeling of head impact. In: Proceedings of the 26th Stapp Car Crash Conference, Ann Arbor, Michigan, USA, SAE 821150 (1982)
65. Kimpara, H., Nakahira, Y., Iwamoto, M., Miki, K., Ichihara, K., Kawano, S., Taguchi, T.: Investigation of anteroposterior head–neck responses during severe frontal impacts using a brain–spinal cord complex FE model. Stapp Car Crash J. **50**, 509–544 (2006)
66. King, A.I., Chou, C.C.: Mathematical modeling, simulation and experimental testing of biomechanical system crash response. J. Biomech. **9**, 301–317 (1976)
67. King, A.I., Yang, K.H., Hardy, W.N., Al-Bsharat, A.S., Deng, B., Begeman, P.C., Tashman, S.: Challenging problems and opportunities in impact biomechanics. In: Proceedings of the 1999 Bioengineering Conference, ASME, pp. 269–270 (1999)

68. Kleiven, S., von Holst, H.: Consequences of head size following trauma to the human head. J. Biomech. **35**(2), 153–160 (2002)
69. Kleiven, S., Hardy, W.N.: Correlation of an FE model of the human head with local brain motion—consequences for injury prediction. Stapp Car Crash J. **46**, 123–144 (2002)
70. Kleiven, S.: Influence of impact direction on the human head in prediction of subdural hematoma. J. Neurotrauma **20**(4), 365–379 (2003)
71. Kleiven, S.: Evaluation of head injury criteria using a finite element model validated against experiments on localized brain motion, intracerebral acceleration, and intracranial pressure. Int. J. Crashworthiness **11**(1), 65–79 (2006)
72. Kleiven, S.: Predictors for traumatic brain injuries evaluated through accident reconstructions. Stapp Car Crash J. **51**, 81–114 (2007)
73. Kochanek, P.M., Marion, D.W., Zhang, W., Schiding, J.K., White, M., Palmer, A.M., Clark, R.S., O'Malley, M.E., Styren, S.D., Ho, C., DeKosky, S.T.: Severe controlled cortical impact in rats: assessment of cerebral edema, blood flow, and contusion volume. J. Neurotrauma **12**(6), 1015–1025 (1995)
74. Krabbel, G., Appel, H.: Development of a finite element model of the human skull. J. Neurotrauma **12**(4), 735–742 (1995)
75. Kuijpers, A.H., Claessens, M.H., Sauren, A.A.: The influence of different boundary conditions on the response of the head to impact: a two-dimensional finite element study. J. Neurotrauma **12**(4), 715–724 (1995)
76. Kumaresan, S., Radhakrishnan, S.: Importance of partitioning membranes of the brain and the influence of the neck in head injury modelling. Med. Biol. Eng. Comput. **34**(1), 27–32 (1996)
77. Kurosawa, Y., Kato, K., Takahashi, T., Kubo, M., Uzuka, T., Fujii, Y., Takahashi, H.: 3-D finite element analysis on brain injury mechanism. In: Conf Proc IEEE Eng Med Biol Soc., 4090-3 (2008)
78. Kurtz, S.M., Thibault, K.L., et al.: Finite element analysis of the deformation of the human infant head under impact conditions. In: The Proceedings of the 8th Injury Prevention through Biomechanics Symposium. Wayne State University, Detroit, Michigan (1998)
79. Langlois, J.A., Rutland-Brown, W., Thomas, K.E.: Traumatic brain injury in the United States: Emergency Department Visits, Hospitalizations, and Deaths. Centers for Disease Control and Prevention, National Center for Injury Prevention and Control (2006)
80. Lee, M.C., Melvin, J.W., Ueno, K.: Finite element analysis of traumatic subdural hematoma. In: 31st Stapp Car Crash Conference, SAE 872201. New Orleans, LA, USA (1987)
81. Levchakov, A., Linder-Ganz, E., Raghupathi, R., Margulies, S.S., Gefen, A.: Computational studies of strain exposures in neonate and mature rat brains during closed head impact. J. Neurotrauma **23**(10), 1570–1580 (2006)
82. Li, J., Zhang, J., Yoganandan, N., Pintar, F., Gennarelli, T.: Regional brain strains and role of falx in lateral impact-induced head rotational acceleration. Biomed. Sci. Instrum. **43**, 24–29 (2007)
83. Lighthall, J.W., Melvin, J.W., Ueno, K., Toward a biomechanical criterion for functional brain injury. In: Proceedings of 12th International Technical Conference on Experimental Safety Vehicles, pp. 627–633 (1989)
84. Ling, G., Bandak, F., Armonda, R., Grant, G., Ecklund, J.: Explosive blast neurotrauma. J. Neurotrauma **2007**(26), 815–825 (2009)
85. Mao, H., Zhang, L., Yang, K.H., King, A.I.: Application of a finite element model of the brain to study traumatic brain injury mechanisms in the rat. Stapp Car Crash J. **50**, 583–600 (2006)
86. Mao, H., Jin, X., Zhang, L., Yang, K.H., Igarashi, T., Noble, L.J., King, A.I.: Finite element analysis of controlled cortical impact induced cell loss. J. Neurotrauma **27**, 877–888 (2010)
87. Mao, H., Yang, K.H., King, A.I., Yang, K.: Computational neurotrauma—design, simulation, and analysis of controlled cortical impact model. Biomech Model Mechanobiol (2010). doi:10.1007/s10237-010-0212-z

88. Margulies, S.S., Thibault, K.L.: Infant skull and suture properties: measurements and implications for mechanisms of pediatric brain injury. J. Biomech. Eng. **122**(4), 364–371 (2000)
89. Marjoux, D., Baumgartner, D., Deck, C., Willinger, R.: Head injury prediction capability of the HIC, HIP, SIMon and ULP criteria. Accid. Anal. Prev. **40**(3), 1135–1148 (2008)
90. Martinez, J.L.: Headrest and seat back proposals designed to eliminate head and neck injuries. In: Proceedings of the 12th Stapp Car Crash Conference, Detroit, Michigan, USA, SAE 680775 (1968)
91. Merchant, H.C., Crispino, A.J.: A dynamic analysis of an elastic model of the human head. J. Biomech. **7**(3), 295–301 (1974)
92. Meyer, R., Kohler, J., Homburg, A.: Explosives, 5th edition. Wiley-VCH, Weinheim, Germany (2002)
93. Miller, R.T., Margulies, S.S., Leoni, M., Nonaka, M., Chen, X., Smith, D.H., Meaney, D.F.: Finite element modeling approaches for predicting injury in an experimental model of severe diffuse axonal injury. 42nd Stapp Car Crash Conference, SAE 983154. Tempe, AZ, USA (1998)
94. Miller, R.T., Smith, D.H., Chen, X., Xu, B., Leoni, M., Nonaka, M., Meaney, D.F.: Comparing experimental data to traumatic brain injury finite element models. In: 43rd Stapp Car Crash Conference, SAE 99SC20. San Diego, CA, USA (1999)
95. Moss, W., King, M.J., Blackman, E.C.: Skull flexture form blast waves: a new mechanism for brain injury with implications for helmet design. J. Acoust. Soc. Am. **125**(4), 2650–2665 (2009)
96. Moore, D.F., Jerusalem, A., Nyein, M., Noels, L., Jaffee, M.S., Radovitzky, R.A.: Computational biology—modeling of primary blast effects on the central nervous system. Neuroimage **47**(Suppl 2), 10–20 (2009)
97. Nahum, A.M., Smith, R., Ward, C.C.: Intracranial pressure dynamics during head impact. In: Proceedings of the 21st Stapp Car Crash Conference, SAE Paper No. 770922. Society of Automotive Engineers, Warrendale, PA (1977)
98. Nahum, A.M., Smith, R.W., Raasch, F.D., Ward, C., 1979. Intracranial pressure relationships in the protected and unprotected head. In: Proceedings of the 23rd Stapp Car Crash Conference, SAE 791024. San Diego, California, USA
99. Nahum, A., Ward, C., Raasch, E., Adams, S., Schneider, D.: Experimental studies of side impact to the human head. In: Proceedings of the 24th Stapp Car Crash Conference, Troy, Michigan, USA, SAE 801301 (1980)
100. Nahum, A.M., Ward, C., Schneider, D., Raasch, F., Adams, S.: A study of impacts to the lateral protected and unprotected head. In: 25th Stapp Car Crash Conference, San Francisco, CA, SAE Paper No. 811006 (1981)
101. Newman, J.A., Shewchenko, N., Welbourne, E.: A proposed new biomechanical head injury assessment function—the maximum power index. In: 44th Stapp Car Crash Conf. SAE 2000-01-SC16 (2000)
102. Nyquist, G.W., Cavanaugh, J.M., Goldberg, S.J., King, A.I.: Facial impact tolerance and response. In: Proceedings of the 30th Stapp Car Crash Conference, SAE Paper No. 861896. Society of Automotive Engineers, Warrendale, PA (1986)
103. Okie, S.: Traumatic brain injury in the war zone. N. Engl. J. Med. **352**(20), 2043–2047 (2005)
104. Pintar, F.A., Yoganandan, N., Voo, L., Cusick, J.F., Maiman, D.J., Sances, A. Jr.: Dynamic characteristics of the human cervical spine. In: Proceedings of the 39th Stapp Car Crash Conference, San Diego, CA Nov 8–10, SAE 952722 (1995)
105. Paxinos, G., Watson, C.: The Rat Brain in Stereotaxic Coordinates. Elsevier, San Diego (2005)
106. Pena, A., Pickard, J.D., Stiller, D., Harris, N.G., Schuhmann, M.U.: Brain tissue biomechanics in cortical contusion injury: a finite element analysis. Acta Neurochir. Suppl. **95**, 333–336 (2005)

107. Prange, M.T., Kiralyfalvi, G., Margulies, S.S.: Pediatric rotational inertial brain injury: the relative influence of brain size and mechanical properties. Stapp Car Crash Conference, SAE 99SC23. San Diego, California, USA (1999)
108. Prange, M.T., Meaney, D.F., Margulies, S.S.: Defining brain mechanical properties: effects of region, direction, and species. In: Proceedings of the 44th Stapp Car Crash Conference, Atlanta, Georgia, USA, SAE 2000-01-SC15 44:205–213 (2000)
109. Prange, M.T., Margulies, S.S.: Regional, directional, and64 age-dependent properties of the brain undergoing large deformation. J. Biomech. Eng. **124**(2), 244–252 (2002)
110. Raul, J.S., Baumgartner, D., Willinger, R., Ludes, B.: Finite element modelling of human head injuries caused by a fall. Int. J. Leg. Med. **120**(4), 212–218 (2006)
111. Raul, J.S., Deck, C., Willinger, R., Ludes, B.: Finite-element models of the human head and their applications in forensic practice. Int. J. Leg. Med. **122**(5), 359–366 (2008)
112. Ross, D.T., Meaney, D.F., Sabol, M.K., Smith, D.H., Gennarelli, T.A.: Distribution of forebrain diffuse axonal injury following inertial closed head injury in miniature swine. Exp. Neurol. **126**(2), 291–299 (1994)
113. Roth, S., Raul, J.S., Ludes, B., Willinger, R.: Finite element analysis of impact and shaking inflicted to a child. Int. J. Leg. Med. **121**(3), 223–228 (2007)
114. Roth, S., Raul, J.S., Willinger, R.: Biofidelic child head FE model to simulate real world trauma. Comput. Methods Programs Biomed. **90**(3), 262–274 (2008)
115. Roth, S., Vappou, J., Raul, J.S., Willinger, R.: Child head injury criteria investigation through numerical simulation of real world trauma. Comput. Methods Programs Biomed. **93**(1), 32–45 (2009)
116. Rousseau, P., Hoshizaki, T.B., Gilchrist, M.D., Post, A.: Estimating the influence of neckform compliance on brain tissue strain during helmeted impact. Stapp Car Crash J. **54**, 37–48 (2010)
117. Ruan, J.S., Khalil, T., King, A.I.: Human head dynamic response to side impact by finite element modeling. J. Biomech. Eng. **113**(3), 276–283 (1991)
118. Ruan, J.S., Khalil, T., King, A.I.: Finite element modeling of direct head impact. 37th Stapp Conference proceedings, SAE 933114. San Antonio, TX (1993)
119. Ruan, J.S., Khalil, T., King, A.I.: Dynamic response of the human head to impact by three-dimensional finite element analysis. J. Biomech. Eng. **116**(1), 44–50 (1994)
120. Ruan, J.S., Prasad, P.: Head injury potential assessment in frontal impacts by mathematical modeling. In: Proceedings of the 38th Stapp Car Crash Conference, SAE 942212. Ft. Lauderdale, Florida, USA (1994)
121. Ruan, J.S., Prasad, P.: Coupling of a finite element human head model with a lumped parameter hybrid iii dummy model: preliminary results. J. Neurotrauma **12**(4), 725–734 (1995)
122. Ruan, J., Prasad, P.: The effects of skull thickness variations on human head dynamic impact responses. Stapp Car Crash J. **45**, 395–414 (2001)
123. Ruan, JS., Prasad, P.: Comments: on the consequences of head size following impact to the human head. J Biomech. **39**(2): 383–385; author reply 385–387 (2006)
124. Saberi, H., Seddighi, A.S., Farmanzad, F.: Finite element analysis of an elastic model of the brain: distortion due to acute epidural hematoma—the role of the intra-ventricular pressure gradient. Comput Aided Surg. **12**(2), 131–136 (2007)
125. Saczalski, K.J., Richardson, E.Q.: Nonlinear numerical prediction of human head/helmet crash impact response. Aviat. Space Environ. Med. **49**(1 Pt. 2):114–119 (1978)
126. Sarron, J.C., Caillou, J.P., Da Cunha, J., Allain, J.C., Tramecon, A.: Consequences of nonpenetrating projectile impact on a protected head: study of rear effects of protections. J. Trauma **49**(5), 923–929 (2000)
127. Scheff, S.W., Baldwin, S.A., Brown, R.W., Kraemer, P.J.: Morris water maze deficits in rats following traumatic brain injury: lateral controlled cortical impact. J. Neurotrauma **14**(9), 615–627 (1997)
128. Shafieian, M., Darvish, K.K., Stone, J.R.: Changes to the viscoelastic properties of brain tissue after traumatic axonal injury. J. Biomech. **42**(13), 2136–2142 (2009)

129. Shreiber, D.I., Bain, A.C., Meaney, D.F.: In vivo thresholds for mechanical injury to the blood–brain barrier. In: 41th Stapp Car Crash Conference, SAE 973335. Lake Buena Vista, Florida, USA (1997)
130. Shugar, T.A.: Transient structural response of the linear skull-brain system. Proceedings of the 19th Stapp Car Crash Conference, SAE 751161. San Diego, California, USA (1975)
131. Slattenschek, A., Tauffkirchen, W., Benedikter, G.: Quantification of internal head injury by means of the phantom head and the impact assessment methods. In: Proceedings of the 15th Stapp Car Crash Conference, SAE 710879. San Diego, California, USA (1971)
132. Soza, G., Grosso, R., Labsik, U., Nimsky, C., Fahlbusch, R., Greiner, G., Hastreiter, P.: Fast and adaptive finite element approach for modeling brain shift. Comput. Aided Surg. **8**(5), 241–246 (2003)
133. Sutton, R.L., Lescaudron, L., Stein, D.G.: Unilateral cortical contusion injury in the rat: vascular disruption and temporal development of cortical necrosis. J. Neurotrauma **10**(2), 135–149 (1993)
134. Takhounts, E.G., Eppinger, R.H., Campbell, J.Q., Tannous, R.E., Power, E.D., Shook, L.S.: On the development of the Simon finite element head model. Stapp Car Crash J. **47**, 107–133 (2003)
135. Takhounts, E.G., Ridella, S.A., Hasija, V., Tannous, R.E., Campbell, J.Q., Malone, D., Danelson, K., Stitzel, J., Rowson, S., Duma, S.: Investigation of traumatic brain injuries using the next generation of simulated injury monitor (simon) finite element head model. Stapp Car Crash J. **52**, 1–31 (2008)
136. Takizawa, H., Sugiura, K., Baba, M., Miller, J.D.: Analysis of intracerebral hematoma shapes by numerical computer simulation using the finite element method. Neurol Med Chir (Tokyo) **34**(2), 65–69 (1994)
137. Taylor, P.A., Ford, C.C.: Simulation of blast-induced early-time intracranial wave physics leading to traumatic brain injury. J. Biomech. Eng. **131**(6), 061007 (2009)
138. Thibault, K.T., Kurtz, S.M., Margulies, S.S.: Effect of the age-dependent properties of the braincase on the response of the infant brain to impact. BED Adv. Bioeng., In Proceedings for the Winter Annual Meeting of the ASME (1997)
139. Thurman, D.J., Alverson, C., Browne, D.: Traumatic brain injury in the United States: a report to congress. Centers for Disease Control and Prevention, National Center for Injury Prevention and Control (2000)
140. Trosseille, X., Tarriere, C., Lavaste, F., Guillon, F., Domont, A.: Development of a FEM of the human head according to a specific test protocol. In: Proceedings of 30th Stapp Car Crash Conference, SAE 922527, pp. 235–253 (1992)
141. Thunnissen, J.G.M., Wismans, J.S.H.M., Ewing, C.L., Thomas, D.J.: Human volunteer head-neck response in frontal flexion: a new analysis. In: Proceedings of the 39th Stapp Car Crash Conference, SAE 952721 (1995)
142. Turquier, F., Trosseille, X., Lavaste, F., Tarriere, C., Dômont, A., Kang, H.S., Willinger, R.: Validation study of a 3D finite element head model against experimental data. In: 40th Stapp Car Crash Conference, SAE 962431. Albuquerque, New Mexico, USA (1996)
143. Ueno, K., Melvin, J.W., Li, L., Lighthall, J.W.: Development of tissue level brain injury criteria by finite element analysis. J. Neurotrauma **12**(4), 695–706 (1995)
144. Viano, D.C., Casson, I.R., Pellman, E.J., Zhang, L., King, A.I., Yang, K.H.: Concussion in professional football: brain responses by finite element analysis: Part 9. Neurosurgery **57**(5), 891–916; discussion 891–916 (2005)
145. Ward, C.: Finite element models of the head and their use in brain injury research. In: Proceedings of the 26th Stapp Car Crash Conference, SAE 821154. Ann Arbor, Michigan, USA (1982)
146. Ward, C.C., Chan, M., Nahum, A.M.: Intracranial pressure—a brain injury criterion. In: 24th Stapp Car Crash Conference, SAE 801304, Troy, Michigan, USA (1980)
147. Ward, C.C., Nikravesh, P.E., Thompson, R.B.: Biodynamic finite element models used in brain injury research. Aviat Space Environ. Med. **49**(1 Pt. 2), 136–142 (1978)

148. Ward, C., Thompson, R.B.: The development of a detailed finite element brain model. In 19th Stapp Car Crash Conference, SAE 751163. San Diego, CA, USA (1975)
149. Willinger, C.M., Kopp, D., Cesari, D.: New concept of contrecoup lesions mechanism: modal analysis of a finite element head model, pp. 283–297. IRCOBI Verona, Italy (1992)
150. Willinger, R., Taleb, L., Kopp, C.M.: Modal and temporal analysis of head mathematical models. J. Neurotrauma **12**(4), 743–754 (1995)
151. Willinger, R., Taled, L., Pradoura, P.: Head biomechanics from the finite element model to the physical model, pp. 245–260. IRCOBI, Brunnen, Switzerland (1995)
152. Willinger, R., Kang, H.S., Diaw, B.: Three-dimensional human head finite-element model validation against two experimental impacts. Ann. Biomed. Eng. **27**(3), 403–410 (1999)
153. Wittek, A., Kikinis, R., Warfield, S.K., Miller, K.: Brain shift computation using a fully nonlinear biomechanical model. Medical Image Computing and Computer-Assisted Intervention—MICCAI **8**(Pt 2), 583–590 (2005)
154. Wittek, A., Dutta-Roy, T., Taylor, Z., Horton, A., Washio, T., Chinzei, K., Miller, K.: Subject-specific non-linear biomechanical model of needle insertion into brain. Comput. Methods Biomech. Biomed. Eng. **11**(2), 135–146 (2008)
155. Xu, W., Yang, J.: Development and validation of head finite element model for traffic injury analysis. Sheng Wu Yi Xue Gong Cheng Xue Za Zhi. **25**(3), 556–561 (2008)
156. Yoganandan, N., Li, J., Zhang, J., Pintar, F.A., Gennarelli, T.A.: Influence of angular acceleration–deceleration pulse shapes on regional brain strains. J. Biomech. **41**(10), 2253–2262 (2008)
157. Zhang, J., Yoganandan, N., Pintar, F.A., Gennarelli, T.A.: Brain strains in vehicle impact tests. Annu. Proc. Assoc. Adv. Automot. Med. **50**, 1–12 (2006)
158. Zhang, J., Yoganandan, N., Pintar, F.A., Gennarelli, T.A., Shender, B.S.: A finite element study of blast traumatic brain injury—biomed 2009. Biomed. Sci. Instrum. **45**, 119–124 (2009)
159. Zhang, L., Bae, J., Hardy, W.N., Monson, K.L., Manley, G.T., Goldsmith, W., Yang, K.H., King, A.I.: Computational study of the contribution of the vasculature on the dynamic response of the brain. Stapp Car Crash J. **46**, 145–164 (2002)
160. Zhang, L., Yang, K.H., King, A.I.: Comparison of brain responses between frontal and lateral impacts by finite element modeling. J. Neurotrauma **18**(1), 21–30 (2001a)
161. Zhang, L., Yang, K.H., Dwarampudi, R., Omori, K., Li, T., Chang, K., Hardy, W.N., Khalil, T.B., King, A.I.: Recent advances in brain injury research: a new human head model development and validation. Stapp Car Crash J. **45**, 369–394 (2001b)
162. Zhou, C., Khalil, T.B., King, A.I.: Shear stress distribution in the porcine brain due to rotational impact. In: 38th Stapp Car Crash Conference, SAE 942214. Ft. Lauderdale, Florida, USA (1994)
163. Zhou, C., Khalil, T.B., King, A.I.: A new model comparing impact responses of the homogeneous and inhomogeneous human brain. In: Proceedings of the 39th Stapp Car Crash Conference, SAE Paper No. 952714. Society of Automotive Engineers, Warrendale, PA (1995)
164. Zhu, Q., Prange, M., Margulies, S.: Predicting unconsciousness from a pediatric brain injury threshold. Dev. Neurosci. **28**(4–5), 388–395 (2006)
165. Zhu, F., Mao, H., Dal Cengio Leonardi, A., Wagner, C., Chou, C., Jin, X., Bir, C., VandeVord, P., Yang, K.H., King, A.I.: Development of an FE model of the rat head subjected to air shock loading. Stapp Car Crash J. **54**, 211–225 (2010)
166. Zong, Z., Lee, H.P., Lu, C.: A three-dimensional human head finite element model and power flow in a human head subject to impact loading. J. Biomech. **39**(2), 284–292 (2006)
167. Zoroya, G.: Key Iraq wound: Brain Trauma, in USA Today (2005)

Neural Tissue Biomechanics: Biomechanics and Models of Structural Neurological Disorders

Shaokoon Cheng

Abstract Research interests in biomechanical modeling of the intracranial system and structural neurological disorders have increased in the last two decades. The lack of clarity on the physiology of the intracranial system has resulted in some disparities in these models. The aim of this chapter is to provide a thorough physiological background of the intracranial system, the mechanics of several types of structural neurological disorders that arise when the system is disturbed and outline the benefits in modeling these disorders. This chapter concludes by discussing some of the impending issues that need to be resolved in order to improve our understanding of a diseased intracranial system using computational models.

1 Introduction

The biomechanical modeling of structural neurological disorders is challenging because the physiology of a healthy intracranial system and how it changes in structural neurological disorders is not yet fully understood. From the biomechanical standpoint, accurate understanding of intracranial system physiology such as cerebrospinal fluid (CSF) flow, motion of the soft tissues (brain and spinal cord) and their biomechanical properties are very important. This is because they are crucial boundary conditions that determine the accuracy of the computational models created using finite element analysis or computational fluid dynamics.

Despite the difficulties, focus on these computational models has increased in recent years as they are powerful tools to study intracranial dynamics

S. Cheng (✉)
Neuroscience Research Australia, University of New South Wales, Randwick, Australia
e-mail: s.cheng@neura.edu.au

Stud Mechanobiol Tissue Eng Biomater (2011) 3: 121–134
DOI: 10.1007/8415_2010_14

Published Online: 3 November 2010

non-invasively. For example, computational fluid dynamic models of the cerebral ventricular system can reveal the flow and pressure field and their detail temporal changes that are associated with the CSF flow of a specific subject. They also allow the simulation of drug distribution injected into the CNS, thus reducing the amount of human testing that is necessary.

In this chapter, a detail physiological background of the intracranial system will be discussed and a review of the computational models of structural neurological diseases in the literature will be provided.

2 Physiology of the Intracranial System

2.1 Cerebrospinal Fluid (CSF) Physiology

The brain, blood and CSF are the three main constituents of the intracranial system. The circulation of blood and CSF within the rigid cranium is a dynamic process. As the brain is incompressible, the volumetric balance between the cerebral blood and CSF is important to regulate the intracranial pressure. The specific gravity of CSF is lower than that of the brain (1.007 vs. 1.040) and the biological fluid creates a buoyancy phenomenon which protects the brain parenchyma from injury by reducing its weight by approximately 96%. CSF is an important transport medium for the circulation and distribution of a variety of ions, proteins, micronutrients, peptides and growth factors. The colorless fluid has mechanical properties similar to that of water [7] and approximately 650 ml is produced daily with a circadian variation of 3.5-fold difference in a day [51].

Although choroid plexus is widely known as the exclusive site of CSF production, there are also convincing evidences which support the extrachoroidal production of CSF [46, 47]. In fact, extrachoroidal production of CSF was estimated to represent 40% of the total CSF daily production and a large portion of it is likely to originate from the brain capillaries. Interstitial fluid originating from the brain capillaries adequately substitutes CSF in the subarachnoid space as these fluids have similar chemical composition [15]. In neural tissues, the extracellular spaces serve as essential conduits for the interstitial fluid to carry nutrients and oxygen. The direction of interstitial fluid diffusion reflects the ultrastructural arrangements of the brain tissue and this can be studied using magnetic resonance diffusion tensor imaging.

The absorption of CSF is generally known to occur at the arachnoid villi/pacchionian granulation although no experimental evidences have shown that the biological fluid is transported across them. By injecting a dye tracer intrathecally and measuring its excretion in urine, Dandy and Blackfan [16] demonstrated that CSF absorption is a process that should occur in the capillaries of the subarachnoid space. This was further confirmed by Greitz [26, 28] who demonstrated the absorption of CSF by brain capillaries using MR imaging and radionuclide cisternography. Thus, similar to CSF production, CSF absorption is unlikely a single or anatomically localized process and is likely to occur over the entire brain. In addition, CSF production and absorption are dynamics processes that may be affected by factors such as CSF pressure and cerebral perfusion pressure [17].

2.2 CSF Flow Measured with Functional MRI

The belief that choroid plexus is the exclusive site of CSF production has led to the assumption that it is also responsible for CSF circulation. In addition, conventional view on CSF flow is that the biological fluid is driven in bulk and is unidirectional. For many years, the mechanics of CSF circulation was unclear and technical limitations in measuring CSF flow prevented a complete understanding on CSF flow physiology. There were different views on the mechanisms of CSF circulation and they were hypothesized as the pulsation of the choroid plexus [2], the pulsation of large arteries at the base of the brain [52] and the compressive squeezing action of the third ventricle [19]. The advent of imaging techniques using MRI (e.g. phase contrast MRI) has been instrumental in delineating the true mechanisms of CSF circulation and its flow characteristics from the cerebral ventricular spaces into the spinal canal and around the subarachnoid spaces.

CSF flow waveforms measured using MRI show distinct temporal patterns which are reliable and reproducible [38]. These studies showed that CSF flow in the cerebral ventricles and subarachnoid space are pulsatile. In fact, cardiac-gated MRI is able to demonstrate the distinct relationship between intracranial vascular pulsation and CSF flow. CSF flows in the craniocaudal direction during systole and caudocranial direction during diastole. During systole, CSF flow in the cranium begins with its rapid displacement from the intracranial subarachnoid space into the cervical spinal canal. Approximately 50–100 ms later, CSF flows from the lateral to the third ventricle [57]. The delay in CSF flow in the ventricular system suggest that CSF circulation is unlikely caused by a single mechanism such as the pulsation of a single cerebral artery. The magnitude of CSF flow rate is highest at the cerebral aqueduct as this is the narrowest section of the ventricular pathway. In normal subjects, this flow rate is ~ 12 mm/s [21] and maybe affected by respiration [39]. CSF flow in the anteroposterior direction in the lateral ventricle is negligible. In the spinal subarachnoid space, CSF flow is also pulsatile and resembles a meandering river crossing back and forth around the spinal canal. CSF flow characteristics in the cervical spinal canal are similar to that in the cerebral ventricles. Spinal pulsations are involved in circulating CSF flow in the spinal subarachnoid space in early systole, just before the brain acts as the main CSF pump.

Based on reports from the MRI studies, it is now clear that CSF is driven by the expansion of brain parenchyma and the mechanism involves the transmission of arterial pulses into the cerebrovascular system which leads to the systolic expansion of brain volume [22, 23, 29]. In fact, neurosurgeons are familiar with viewing the "pulsating brain" during surgery. As the surface of the brain is densely vascularized, brain fluid is first displaced over the subarachnoid space.

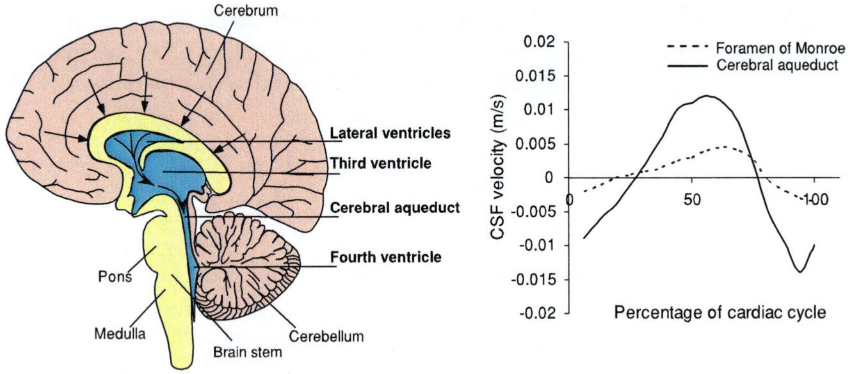

Fig. 1 A picture of the brain in the mid-sagittal plane. The *arrows* on top of the lateral ventricles represent the compressional effects of brain perfusion during systole and *arrows* in the ventricles show CSF flow. Figure on the *right* shows CSF flow velocity in the foramen of Monroe and cerebral aqueduct [22] in a healthy human

The increase in brain volume results in forces directed towards the lateral ventricles which increases CSF pressure and displaces the fluid from the ventricular system (see Fig. 1). If choroid plexus is the source of CSF circulation, then the caudal displacement of CSF should be accompanied by the expansion of the lateral ventricles. However, by measuring brain tissue motion, Feinberg and Mark [23] showed that the volume of the lateral ventricle decreases during systole. Also, if the squeezing action of the thalamus is responsible for pulsating CSF around the cranium, then a decrease in the anteroposterior dimension of the third ventricle and CSF flow in this direction should be observed. However, none of these observations were made in brain motion MRI studies. On the contrary, elongation of the third ventricle's floor during systole was reported.

3 Structural Neurological Disorders

3.1 Hydrocephalus

Clearly, the dynamic buffering of CSF to accommodate the increase in intracranial blood volume is an important mechanism to maintain a healthy intracranial system. An increase in resistance of CSF flow, given a constant CSF production rate will cause hydrocephalus. Hydrocephalus is a structural neurological disorder that is marked by enlarged cerebral ventricles caused by the accumulation of CSF (see Fig. 2). The pathophysiology of hydrocephalus is likely to involve more than one mechanism and is therefore a complex medical entity.

Although brain is widely known to be incompressible, MR images of a hydrocephalic brain clearly reveals that the ventricles are enlarged at the expense

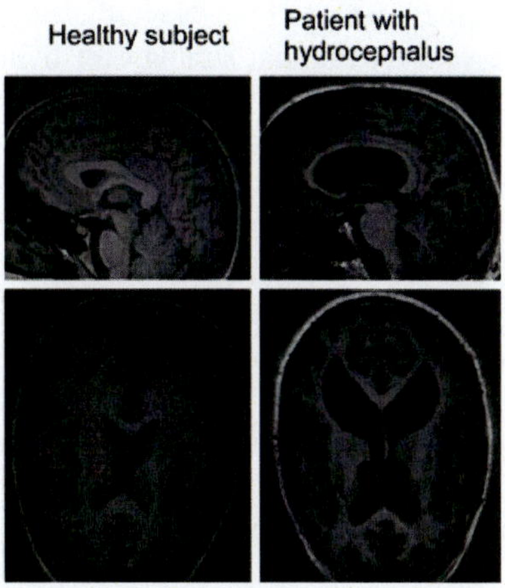

Fig. 2 MRI images of a healthy human and a patient with hydrocephalus. The first and second row of images show the brain in the mid-saggital plane and axial plane respectively

of a compressed brain. The fact that the brain is able to reconstitute back to its original geometry after treatment suggests that hydrocephalus is not related to the loss of brain tissue and that a unique mechanism is involved to protect the brain cells from injury during the disorder. Hydrocephalus is generally classified as three different types and they are non-communicating, communicating and normal pressure hydrocephalus. The term "communicating" was derived from radiological classification as to whether an obstruction is visible on a computer tomography (CT) or magnetic resonance image. These definitions are illusive because in most cases of communicating hydrocephalus, CSF flow is obstructed at the basal cisterns or in the subarachnoid space [34, 37]. Moreover, it is not uncommon to observe short periods of raised intracranial pressure in patients with normal pressure hydrocephalus [30], thus contradicting its definition [36].

The large ventricles and compressed brain tissue appears to show that a large pressure gradient exist between the ventricles and subarachnoid space. However, this pressure gradient (also known as transmantle pressure gradient) was not found with mechanical pressure sensors [61]. When small pressure pulses were simulated in the lateral ventricles of a lamb, Di Rocco et al. [18] were able to produce communicating hydrocephalus even though the mean ventricular CSF pressure was normal. Recent theories on hydrocephalus suggest that the disorder is caused by the increase in CSF pulsatile forces [27]. According to MRI studies, the higher CSF pulsatile forces in hydrocephalic patients are manifested as a three- to fivefold increase in CSF flow rate at the cerebral aqueduct [9]. This finding initiated a series of studies that aim to investigate the reliability of using CSF flow rate at the cerebral aqueduct as an effective biomarker to differentiate patients with normal pressure hydrocephalus and cerebral atrophy [8]. Current treatments for hydrocephalus involve the diversion of CSF to other regions of the body through a mechanical shunt implant or the creation of a natural bypass in the third ventricle using an endoscope [endoscopic third ventriculostomy (ETV)]. Shunts are commonly associated with problems such as mechanical malfunction and their initial placement, revisions and removal cost billions of dollars in healthcare expenditure in the US alone [54]. Nevertheless, these mechanical devices have remained as the primary treatment strategy for hydrocephalus.

3.2 Slit Ventricle Syndrome and Intracranial Hypertension

The careful selection of patients with enlarged ventricles suitable for shunt treatment is required as not all of them respond well to this procedure. For example, patients with cerebral atrophy were also presented with large cerebral ventricles but shunting these patients would be useless and may be fatal. In addition, in some hydrocephalic patients, over drainage may result from shunt treatment and this may result in the collapsation of the cerebral ventricles (slit ventricle syndrome). While some other patients show similar signs of hydrocephalus such as raised intracranial pressure, their ventricles may not be necessarily enlarged. This type of intracranial disorder is known as idiopathic intracranial hypertension. The causes of slit ventricle syndrome and idiopathic intracranial hypertension are not known but increase in mechanical properties of the brain tissues is believed to play an important role in the pathogenesis of these disorders.

3.3 Syringomyelia

Syringomyelia is a disorder of the spinal cord that is characterized by a fluid filled cavity known as cyst/syrinx that elongates and expands over time. The syrinx compresses the spinal cord and causes deterioration of the spinal cord function. Syrinx could form in the spinal cord central canal (canalicular syringomyelia) or within the spinal cord parenchyma (extracanalicular syringomyelia). As the fluid pressure in the syrinx is higher than the CSF pressure at the spinal subarachnoid space, it is unclear how fluid can flow against the pressure gradient and into the syrinx although animal experiments suggest that such flow does occur [10]. An MR study has revealed that CSF flow pattern in the spinal subarachnoid space of patients with syrinomelia is complex and mean peak caudal CSF velocity is significantly higher ($\sim$2 times) compared to normal subjects [62].

4 Conventional Research on Brain Biomechanics

Extensive research has been conducted in an effort to understand the biomechanical responses of the intracranial system. Traditional methods involve studies where intracranial pressure and volume relationship are used to infer brain biomechanics. By infusing artificial CSF, brain tissue elastic properties (termed elastance) or the ability of the brain to respond to raised intracranial pressure (termed compliance) are estimated from the relationship between the change in intracranial pressure and CSF volume (P–V relation). The mathematical nature of the P–V relation is exponential [25, 42, 45, 60, 64] and could be described by several exponential functions [58]. Although P–V relation depicts the response of the brain, it is also important to know that the nature of these curves is likely to be multifactorial. Thus, P–V relation does not represent the elasticity of the brain tissues exclusively but rather, the elasticity of the intracranial system [59] which consists of the cerebral vasculature, dura and skull. In addition, the mechanisms for spatial compensation in the cranial system, such as the alteration of CSF dynamics and displacement of blood from the cerebrovascular bed, are all time dependent processes. The P–V relation in the intracranial system is therefore, a dynamic response and does not necessarily give direct information about a patient's remaining compensatory abilities at the point of measurement. Hence, the P–V studies are limited in elucidating the biomechanics of brain deformation in structural neurological disorders. In addition, the stress and strain field distributions of the geometrically complex brain and how they relate to the functional deficits of the patient cannot be understood using P–V studies and require a more sophisticated class of mathematical models.

5 Computational Models of Brain Biomechanics

5.1 Finite Element Analysis Models of Hydrocephalus

Finite element analysis (FEA) is useful to study structural neurological disorders as it show stress–strain distribution in the neural tissues during those disorders. The first FEA model of hydrocephalus was created by Nagashima et al. [49]. In that study, the brain was modeled as an isotropic biphasic material—porous material which consists of elastic/viscoelastic solid matrix saturated with fluid. The concept of modeling brain as a biphasic material was initiated by Hakim et al. [31]. This theory proposes that under the circumstance of an increasing mass lesion or hydrocephalus, brain tissue is compressed at the expense of collapsing extracellular spaces and the expulsion of interstitial fluid. Hakim's hypothesis is a plausible explanation as to why brain can be severely compressed in hydrocephalus. In fact, a study using CT scan has showed that fluid content in brain tissues is different in a normal and hydrocephalus brain [56]. Furthermore, the effectiveness of the biphasic models to characterize the deformation of other biological tissues such as articular cartilage [48] under a range of loading conditions has also been demonstrated.

Modeling neural tissues as biphasic materials provide information that conventional single phase material model is unable to give. For example, information on how interstitial fluid pressure changes as the brain deforms can be

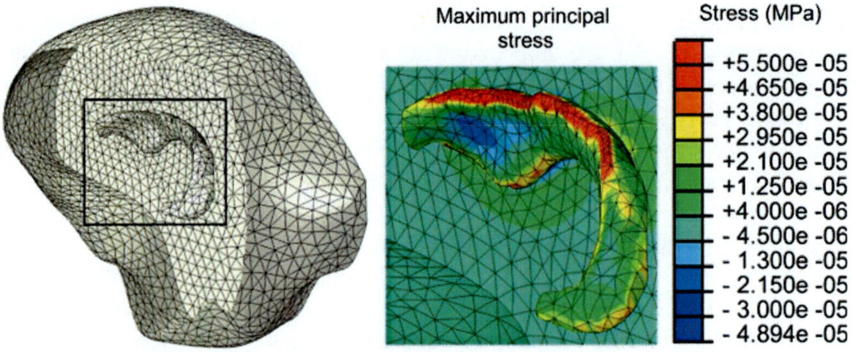

Fig. 3 Picture on the *left* shows a three-dimensional model of the brain and cerebral ventricles. The biomechanical stresses (in MPa) along the periventricular region in hydrocephalus is shown on the *right*. Positive stresses imply that the tissues are in tension while negative stresses imply that the tissues are in compression. Pictures are adopted with permission from Cheng and Bilston [12]

investigated. This is an important aspect when studying the pathophysiology of edema in neural tissues. Periventricular lucency (PVL) is a known feature in hydrocephalus and is characterized as a blurring of the lateral ventricular margins on a CT or MR scan [50]. The etiology of PVL was unclear and was debated as the extravasation of ventricular CSF due to the disruption of the ependyma. By using a two-dimensional FEA model of a brain and incorporating biphasic material properties, Pena et al. [55] were able to demonstrate the regional expansion of extracellular spaces in brain tissues around the ventricles and how this was caused by the combination of tensile and compressive stress around the convex and concave ventricular geometry. The study demonstrated that PVL is not solely associated with the increased in intraventricular pressure but is also related to the geometry of the ventricles.

The primary concern when modeling the brain as a two-dimensional structure is that it over-simplifies its complex geometry and sacrifices information in the missing dimension. Three-dimensional biphasic model of the cerebral ventricles ([12]; Fig. 3) shows that stress magnitude at the ventricles is four times higher than what is predicted by a two-dimensional model reconstructed from the MR images of a similar subject. In general, the loading condition in the FEA models is such that pore pressure at the brain surface is assumed as null and pressure is applied at the surface of the ventricles. In reality, this is a misrepresentation of the actual boundary condition in hydrocephalus and modelers should not interpret the stress field as the actual mechanical stress that is associated with the disorder. The pattern of stress distributions however, is useful to understand which region of the brain is most severely affected as it deforms mechanically. This information can then be correlated with histological studies and symptoms of hydrocephalus to understand the pathogenesis of the disorder.

5.2 Computational Fluid Dynamic Models of Hydrocephalus

Apart from using FEA to study brain deformation in hydrocephalus, computational fluid dynamics (CFD) has also been used to study CSF flow in both healthy and patients with hydrocephalus [13, 14, 24, 33, 40, 41, 43, 44]. As the Reynolds number is low, CSF flow is assumed laminar and is modeled as a Newtonian fluid with physical properties similar to water (viscosity = 1 mPas, density = 10^3 kg/m^3). By simulating CSF flow in a model of the cerebral aqueduct of a healthy individual, Jacobson et al. [35] shows that only a pressure of 1.1 Pa is required to push CSF across the structure. This finding explains why previous measurements of transmantle pressure difference were unfruitful [61] as the mechanical pressure transducers used were not sensitive enough to measure the small pressure differences. The study performed by Jacobson et al. [35] suggests that large transmantle pressure gradient is unlikely involved in the development of hydrocephalus.

The first three-dimensional model of CSF flow was performed by Fin and Grebe [24]. The significance of that study is that fluid structure interaction (FSI) of CSF and brain structures are modeled and results show that the pressure drop for a given flow decreased with the implementation of FSI. As mentioned, CSF circulation is closely related to brain motion and an accurate CFD model of CSF flow should incorporate FSI. Modeling the FSI of CSF flow in the cerebral aqueduct is straightforward as the effects of brain perfusion can be ignored. This is because the brain structures that form the cerebral aqueduct is passive and is not a component of driving force for CSF circulation. However, modeling the FSI of CSF flow at the lateral ventricles would involve modeling the systolic expansion of the brain parenchyma and the dynamic motion of the ventricle walls. While this type of model may be achieved by simplifying and representing the brain structures around the lateral ventricles as a membrane, it is not sensible to assign the mechanical properties of the brain tissues to the membrane structure. FSI of CSF flow for the full ventricular system was achieved by using a two-dimensional model [13] and the authors acknowledged that the model was only a proof of concept and the primary motivation of the study was to propose mechanisms for the development of hydrocephalus. Thus, although a complete three-dimensional FSI model of the entire ventricular system is ideal, the complex boundary conditions at the lateral ventricles and geometry of the brain structure makes this type of model technically difficult to create.

5.3 Computational Fluid Dynamic Models of Syringomyelia

One hypothesis for the development of syringomyelia is that the disorder is associated with increase in CSF pressure at the SAS as a result of coughing and sneezing. By using a one-dimensional co-axial tube model, Carpenter et al. [11] and Berkouk et al. [3] show that shock wave formation (due to cough or sneeze) and their reflection at sites of blockages in the spinal subarachnoid spaces (SAS) could generate transient regions of high pressure within the spinal cord and causes syrinx formation. Nevertheless, by using a two-dimensional FSI model of the spinal column, Bertram et al. [4] shows that the degree of wave propagation that could proceed to shock formation in spinal cord at the region where syrinx is usually formed is quantitatively infeasible. Despite the debate, computational fluid dynamic models of syringomyelia have generally highlighted the role of increased CSF pressure in the SAS and that they are important in the development of syrinx. For example, it is common that syrinx is associated with scarring around the spinal cord from prior infection or haemorrhage (arachnoiditis) and by modeling arachnoiditis as a porous blockage across the SAS, increased in peak CSF pressure (above the arachnoiditis) by 20-fold [6] was demonstrated. Other computational models of syringomyelia have also help to demonstrate that the pulsation of arteries in the cord could drive CSF along the perivascular spaces and into the spinal cord [5].

6 Future Work

Understanding the physiology of the intracranial system is the first step towards the sensible modeling of brain deformation and CSF flow in structural neurological disorders. The accuracy of the computational models are dependent on realistic boundary conditions, accurate biomechanical properties of the brain and geometry of the brain or CSF space representing the specific subject of interest. Although existing models have revealed interesting and useful information on the biomechanics of hydrocephalus and syringomyelia, they still lack a realistic representation of the true intracranial dynamics associated with the disorders. As mentioned, the extrachoroidal production of CSF, absorption of CSF, cerebral perfusion and the dynamic interaction of the perfused brain structure with the CSF are all important factors that should be modeled.

The accuracy of mechanical properties of the brain and spinal cord is important for modeling their deformation. As the mechanical properties of neural tissues is both strain and strain-rate dependent, technical constraints (e.g. post-mortem changes) have made it difficult to measure these properties at a slow strain rate that is similar to the deformation of the neural tissues in hydrocephalus or syringomyelia. Table 1 shows the diversity in mechanical properties used in FEA models of hydrocephalus and signify that further research on the mechanical properties of the brain and spinal cord tissues are warranted.

Modeling neural tissues as biphasic material in structural neurological disorders is important as the production, diffusion and absorption of interstitial fluid is likely to play an important role when brain deforms slowly. Interestingly, the mechanism of interstitial fluid pressurization plays an important role in the load bearing capability of cartilage [1] and perhaps, similar mechanism maybe involved in protecting the neuronal structures during hydrocephalus. Biphasic material may

Table 1 Summary of brain mechanical properties used in biphasic models of hydrocephalus

References	Gray matter (Pa)	White matter (Pa)	Poisson's ratio	Extracellular spaces	Details
[49]	3	0.3	0.4999	0.2	2D model
[55]	10,000	10,000	0.3	0.2	2D model
[63]	584.4	584.4	0.35	0.2	2D model
[12]	350	350	0.35	0.2	3D model

also explain the change in neural tissue properties (increased in tissue stiffness) in slit ventricle syndrome and intracranial hypertension as this material is not compressible without the loss of interstitial fluid. Although the biphasic properties of the spinal cord are known to play a pivotal role in the pathogenesis of syringomelia, they have not been modeled [32, 53]. There are very little work done to characterize the biphasic properties of neural tissues and further work in this area is warranted. This type of work should also incorporate the anisotropicity of the white matter as the direction of tissue fibres affect interstitial fluid diffusion and tissue stiffness.

Although it seems that we are far from an ideal model, understanding structural neurological disorders with computational models is not presently hopeless. Rigorous validation of computer models with MR studies and the careful interpretation of simulated results will still render these models as powerful tools to both test hypotheses and elucidate the mechanisms of structural neurological disorders. These models also remain as potential tools to predict the efficacy of different treatment strategies in patients with structural neurological disorders.

References

1. Ateshian, G.A.: The role of interstitial fluid pressurization in articular cartilage lubrication. J. Biomech. **42**(9), 1163–1176 (2009)
2. Bering, E.A.: Choroid plexus and arterial pulsation of cerebrospinal fluid; demonstration of the choroid plexuses as a cerebrospinal fluid pump. Am. Med. Assoc.: Arch. Neurol. Psychiatry **73**(2), 165–172 (1955)
3. Berkouk, K., Carpenter, P.W., Lucey, A.D.: Pressure wave propagation in fluid-filled co-axial elastic tubes. Part 1: Basic theory. J. Biomech. Eng. **125**(6), 852–856 (2003)
4. Bertram, C.D., Brodbelt, A.R., Stoodley, M.A.: The origins of syringomyelia: numerical models of fluid/structure interactions in the spinal cord. J. Biomech. Eng. **127**(7), 1099–1109 (2005)
5. Bilston, L.E., Fletcher, D.F., Brodbelt, A.R., et al. Arterial pulsation-driven cerebrospinal fluid flow in the perivascular space: a computational model. Comput. Methods Biomech. Biomed. Eng. **6**(4), 235–241 (2003)
6. Bilston, L.E., Fletcher, D.F., Stoodley, M.A.: Focal spinal arachnoiditis increases subarachnoid space pressure: a computational study. Clin. Biomech. (Bristol, Avon) **21**(6), 579–584 (2006)
7. Bloomfield, I.G., Johnston, I.H., Bilston, L.E.: Effects of proteins, blood cells and glucose on the viscosity of cerebrospinal fluid. Pediatr. Neurosurg. **28**(5), 246–251 (1998)

8. Bradley, W., Whittemore, A., Kortman, K., et al. Marked cerebrospinal fluid void: indicator of successful shunt in patients with suspected normal pressure hydrocephalus. Radiology **178**, 459–466 (1991)
9. Bradley, W.G., Kortman, K.E., Burgoyne, B.: Flowing cerebrospinal fluid in normal and hydrocephalic states: appearance on MR images. Radiology **159**(3), 611–616 (1986)
10. Brodbelt, A.R., Stoodley, M.A., Watling, A.M., et al.: Altered subarachnoid space compliance and fluid flow in an animal model of posttraumatic syringomyelia. Spine **28**(20), E413–E419 (2003)
11. Carpenter, P.W., Berkouk, K., Lucey, A.D.: Pressure wave propagation in fluid-filled co-axial elastic tubes. Part 2: Mechanisms for the pathogenesis of syringomyelia. J. Biomech. Eng. **125**(6), 857–863 (2003)
12. Cheng, S., Bilston, L.: Computational model of the cerebral ventricles in hydrocephalus. J. Biomech. Eng. **132**(5), 054501 (2010)
13. Cheng, S., Jacobson, E., Bilston, L.E.: Models of the pulsatile hydrodynamics of cerebrospinal fluid flow in the normal and abnormal intracranial system. Comp. Methods Biomech. Biomed. Eng. **10**(2), 151–157 (2007)
14. Cheng, S., Tan, K., Bilston, L.E.: The effects of the interthalamic adhesion position on cerebrospinal fluid dynamics in the cerebral ventricles. J. Biomech. **43**(3), 579–582 (2009)
15. Cserr, H.F.: Convection of Brain Interstitial Fluid. Raven Press, New York (1984)
16. Dandy, W.E., Blackfan, K.D.: Internal hydrocephalus. An experimental, clinical and pathological study. Am. J. Dis. Child. **8**, 406–481 (1914)
17. Davson, H., Segal, M.B.: Physiology of the CSF and Blood-Brain Barrier. CRC Press, Boca Raton (1996)
18. Di Rocco, C., Pettorossi, V.E., Caldarelli, M., et al.: Communicating hydrocephalus induced by mechanically increased amplitude of the intraventricular cerebrospinal fluid pressure: experimental studies. Exp. Neurol. **59**(1), 40–52 (1978)
19. DuBoulay, G., O'Connell, J., Currie, J., et al.: Further investigations on pulsatile movements in the cerebrospinal fluid pathways. Acta Radiol.: Diagn. **13**, 496–523 (1972)
20. Dutta-Roy, T., Wittek, A., Miller, K.: Biomechanical modeling of normal pressure hydrocephalus. J. Biomech. **41**(10), 2263–2271 (2008)
21. Enzmann, D.R., Pelc, N.J.: Normal flow patterns of intracranial and spinal cerebrospinal fluid defined with phase-contrast cine MR imaging. Radiology **178**, 467–474 (1991)
22. Enzmann, D.R., Pelc, N.J.: Brain motion: measurement with phase-contrast MR imaging. Radiology **185**, 653–660 (1992)
23. Feinberg, D.A., Mark, A.S.: Human brain motion and cerebrospinal fluid circulation demonstrated with MR velocity imaging. Radiology **163**(3), 793–799 (1987)
24. Fin, L., Grebe, R.: Three dimensional modeling of the cerebrospinal fluid dynamics and brain interactions in the aqueduct of sylvius. Comp. Methods Biomech. Biomed. Eng. **6**(3), 163–170 (2003)
25. Friden, H.G., Ekstedt, J.: Volume/pressure relationship of the cerebrospinal space in humans. Neurosurgery **13**(4), 351–366 (1983)
26. Greitz, D.: Cerebrospinal fluid circulation and associated intracranial dynamics. A radiological investigation using MR imaging and radionuclide cisternography. Acta Radiol. Suppl. **386**, 1–23 (1993)
27. Greitz, D.: Radiological assessment of hydrocephalus: new theories and implications of therapy. Neurosurg. Rev. **27**(3), 145–165 (2004)
28. Greitz, D., Hannerz, J.: A proposed model of cerebrospinal fluid circulation: observations with radionuclide cisternography. Am. J. Neuroradiol. **17**(3), 431–438 (1996)
29. Greitz, D., Wirestam, R., Franck, A., et al.: Pulsatile brain movement and associated hydrodynamics studied by magnetic resonance phase imaging. The Monro-Kellie doctrine revisited. Neuroradiology **34**(5), 370–380 (1992)
30. Hakim, S., Adams, R.D.: The special clinical problem of symptomatic hydrocephalus with normal cerebrospinal fluid pressure: observations on cerebrospinal fluid hydrodynamics. J. Neurol. Sci. **2**, 307–327 (1965)

31. Hakim, S., Venegas, J., Burton, J.: The physics of the cranial cavity, hydrocephalus and normal pressure hydrocephalus: mathematical interpretations and mathematical models. Surg. Neurol. **5**, 187–210 (1976)
32. Heiss, J.D., Patronas, N., DeVroom, H.L., et al.: Elucidating the pathophysiology of syringomyelia. J. Neurosurg. **91**(4), 553–562 (1999)
33. Howden, L., Giddings, D., Power, H., et al.: Three-dimensional cerebrospinal fluid flow within the human ventricular system. Comp. Methods Biomech. Biomed. Eng. **11**(2), 123–133 (2008)
34. Ishii, M., Suzuki, S., Julow, J.: Subarachnoid haemorrhage and communicating hydrocephalus. Scanning electron microscope observations. Acta Neurochir. (Wien) **50**, 265–272 (1979)
35. Jacobson, E.E., Fletcher, D.F., Morgan, M.K., et al.: Fluid dynamics of the cerebral aqueduct. Pediatr. Neurosurg. **24**, 229–236 (1996)
36. Jagdish, C., Hulme, A., Cooper, R.: Intracranial pressure in patients with dementia and communicating hydrocephalus. J. Neurosurg. **40**, 376–380 (1974)
37. Julow, J., Ishii, M., Iwabuchi, T.: Scanning electron microscope of subarachnoid macrophages and subarachnoid haemorrhage and their possible role in the formation of subarachnoid fibrosis. Acta Neurochir. (Wien) **50**, 273–279 (1979)
38. Kim, D.S., Choi, J.U., Huh, R., et al.: Quantitative assessment of cerebrospinal fluid hydrodynamics using a phase-contrast cine MR image in hydrocephalus. Childs Nerv. Syst. **15**(9), 461–467 (1999)
39. Klose, U., Strik, C., Kiefer, C., et al.: Detection of a relation between respiration and CSF pulsation with an echoplanar technique. J. Magn. Reson. Imaging **11**(7), 438–444 (2000)
40. Kurtcuoglu, V., Poulikakos, D., Ventikos, Y.: Computational modeling of the mechanical behavior of the cerebrospinal fluid system. J. Biomech. Eng. **127**(2), 264–269 (2005)
41. Kurtcuoglu, V., Soellinger, M., Summers, P., et al.: Computational investigation of subject-specific cerebrospinal fluid flow in the third ventricle and aqueduct of Sylvius. J. Biomech. **40**(6), 1235–1245 (2007)
42. Leech, P., Miller, J.D.: Intracranial volume-pressure relationships during experimental brain compression in primates. 1. Pressure response to changes in ventricular volume. J. Neurol. Neurosurg. Psychiatry **37**, 1093–1098 (1974)
43. Linninger, A.A., Tsakiris, C., Zhu, D.C., et al.: Pulsatile cerebrospinal fluid dynamics in the human brain. IEEE Trans. Biomed. Eng. **52**(4), 557–565 (2005)
44. Linninger, A.A., Xenos, M., Zhu, D.C., et al. Cerebrospinal fluid flow in the normal and hydrocephalic human brain. IEEE Trans. Biomed. Eng. **54**(2), 291–302 (2007)
45. Lofgren, J., Essen, C.V., Zwetnow, N.: The pressure volume curve of the cerebrospinal fluid space in dogs. Acta Neurol. Scand. **49**, 557–574 (1973)
46. Milhorat, T.H.: Failure of choroid plexectomy as treatment of hydrocephalus. Surg. Gynecol. Obstet. **139**, 505–508 (1974)
47. Milhorat, T.H., Hammock, M.K., Fenstermacher, J.D., et al.: Cerebrospinal fluid production by the choroid plexus and brain. Science **173**(994), 330–332 (1971)
48. Mow, V.C., Kuei, S.C., Lai, W.M., et al.: Biphasic creep and stress relaxation of articular cartilage in compression: theory and experiments. J. Biomech. Eng. **102**, 73–83 (1980)
49. Nagashima, T., Tamaki, N., Matsumoto, S., et al.: Biomechanics of hydrocephalus: a new theoretical model. Neurosurgery **21**, 898–904 (1987)
50. Naidich, T.P., Epstein, F., Lin, J.P., et al.: Evaluation of pediatric hydrocephalus by computed tomography. Radiology **119**(2), 337–345 (1976)
51. Nilsson, C., Stahlberg, F., Thomsen, C., et al.: Circadian variation in human cerebrospinal fluid production measured by magnetic resonance imaging. Am. J. Physiol. **262**, R20–R24 (1992)
52. O'Connell, J.E.: Cerebrospinal fluid mechanics. Proc. R. Soc. Med. **63**(5), 507–518 (1970)
53. Oldfield, E.H., Muraszko, K., Shawker, T.H., et al.: Pathophysiology of syringomyelia associated with Chiari I malformation of the cerebellar tonsils. Implications for diagnosis and treatment. J. Neurosurg. **80**(1), 3–15 (1994)

54. Patwardhan, R.V., Nanda, A.: Implanted ventricular shunts in the united states: the billion-dollar-a year cost of hydrocephalus treatment. Neurosurgery **56**(1), 139–144 (2005)
55. Pena, A., Bolton, M.D., Whitehouse, H., et al.: Effects of brain ventricular shape on periventricular biomechanics: a finite-element analysis. Neurosurgery **45**(1), 107–118 (1999)
56. Penn, R.D., Bacus, J.W.: The brain as a sponge: a computed tomographic look at Hakim's hypothesis. Neurosurgery **14**(6), 670–675 (1984)
57. Schroth, G., Klose, U.: Cerebrospinal fluid flow. I. Physiology of cardiac-related pulsation. Neuroradiology **35**(1), 1–9 (1992)
58. Sivaloganathan, S., Tenti, G., Drake, J.M.: Mathematical pressure-volume models of the cerebrospinal fluid. Appl. Math. Comput. **94**, 243–266 (1998)
59. Sklar, F.H., Diehl, J.T., Beyer, C.W.: Brain elasticity changes with ventriculomegaly. J. Neurosurg. **53**, 173–179 (1980)
60. Sklar, F.H., Elashvili, I.: The pressure-volume function of brain elasticity. Physiological considerations and clinical applications. J. Neurosurg. **47**(5), 670–679 (1977)
61. Stephensen, H., Tisell, M., Wikkelso, C.: There is no transmantle pressure gradient in communicating or noncommunicating hydrocephalus. Neurosurgery **50**(4), 763–771 (2002)
62. Struck, A.F., Haughton, V.M.: Idiopathic syringomyelia: phase-contrast MR of cerebrospinal fluid flow dynamics at level of foramen magnum. Radiology **253**(1), 184–190 (2009)
63. Taylor, Z., Miller, K.: Reassessment of brain elasticity for analysis of biomechanisms of hydrocephalus. J. Biomech. **37**(8), 1263–1269 (2004)
64. Tenti, G., Drake, J.M., Sivaloganathan, S.: Brain biomechanics: mathematical modeling of hydrocephalus. Neurol. Res. **22**(1), 19–24 (2000)

Computational Biomechanics of the Brain; Application to Neuroimage Registration

Karol Miller, Adam Wittek, Grand Joldes, Jiajie Ma and Ben Jamin Zwick

Abstract We present selected topics in the area of mathematical and numerical modelling of the brain biomechanics for brain image registration. We show how to describe registration in purely mechanical terms, such as displacements, strains and stresses and perform it using established methods of continuum mechanics. We advocate the use of fully non-linear theory of continuum mechanics. We discuss in some detail modelling geometry, boundary conditions, loading and material properties. We consider numerical problems such as the use of hexahedral and mixed hexahedral-tetrahedral meshes as well as meshless spatial discretisation schemes as well as the effects of model complexity on accuracy of brain deformation computation. We advocate the use of Total Lagrangian Formulation of both finite element and meshless methods together with explicit time-stepping procedures. We support our recommendations and conclusions with an example of the computation of the brain shift for intra-operative image registration.

1 Introduction

Mathematical modelling and computer simulation are standard tools commonly used in engineering. Computational mechanics has had a profound impact on science and technology. It allows simulation of complex systems that would be very difficult or impossible to treat using analytical methods. A challenging task for the

K. Miller (✉) · A. Wittek · G. Joldes · J. Ma · B. J. Zwick
Intelligent Systems for Medicine Laboratory, School of Mechanical Engineering,
The University of Western Australia, 35 Stirling Highway,
Crawley, WA 6009, Australia
e-mail: kmiller@mech.uwa.edu.au

Stud Mechanobiol Tissue Eng Biomater (2011) 3: 135–157
DOI: 10.1007/8415_2011_80

Published Online: 28 April 2011

future is to extend the success of computational mechanics to fields outside traditional engineering, in particular to biology, biomedical sciences, and medicine [1].

In computational sciences, the selection of the physical and mathematical model of the phenomenon to be investigated has a major influence on the accuracy of the simulation results. Model selection is often a very subjective process; different modellers may choose different models to describe the same phenomenon. Nevertheless, valid computer simulations of a physical reality cannot be obtained without a proper model selection [1].

In this Chapter we present how various aspects of image-guided neurosurgery can benefit from the application of methods of computational mechanics. We discuss issues related to the model selection and numerical algorithms used for obtaining the solution. We use an example from computational radiology as a vehicle to demonstrate the appropriateness and effectiveness of our methods. Other important applications of biomechanics of the brain e.g. to traumatic injury modelling are discussed elsewhere in this book as well as in the recent literature [2, 3].

1.1 Computational Radiology

Nakaji and Speltzer [4] list the "accurate localisation of the target" as the first principle in modern neurosurgical approaches. Neurosurgical interventions have extremely localised areas of therapeutic effect. As a result, they have to be applied precisely in relation to the patient's current (i.e. intra-operative) anatomy, directly over the specific location of anatomic or functional abnormality [5].

As only pre-operative anatomy of the patient is known precisely from medical images (usually Magnetic Resonance Images–MRI), it is now recognised that the ability to predict soft organ deformation (and therefore intra-operative anatomy) during the operation is the main problem in performing reliable surgery on soft organs. We are particularly interested in problems arising in image-guided neurosurgery (Fig. 1). In this context it is very important to be able to predict the effect of procedures on the position of pathologies and critical healthy areas in the brain. If displacements within the brain can be computed during the operation, they can be used to warp pre-operative high-quality MR images so that they represent the current, intra-operative configuration of the brain. Such "warped" images would then facilitate accurate neuronavigation.

The neuroimage registration problem involves large deformations, non-linear material properties and non-linear boundary conditions as well as the difficult issue of generating patient-specific computational models. Moreover, the computations must be conducted during the operation and seamlessly incorporated in the clinical workflow. We estimate that for this to happen the calculation of brain displacements and warping of pre-operative images should not take more than approximately 40 s. This forms a stringent requirement for computational efficiency of methods used.

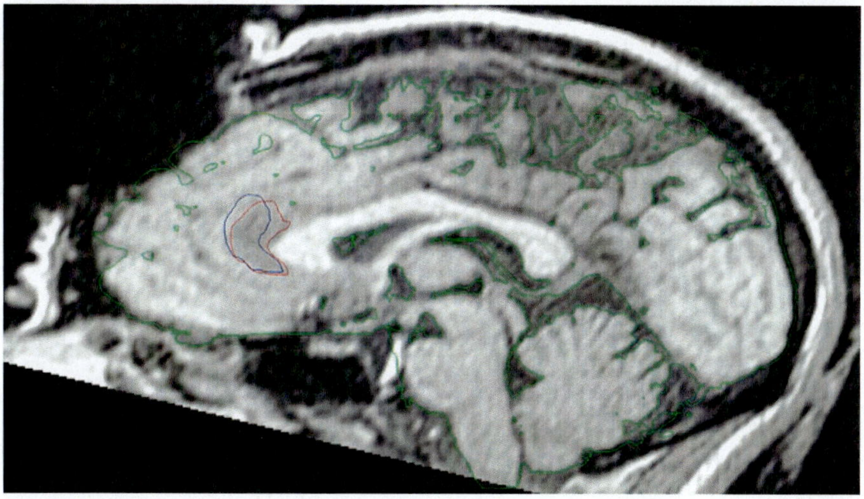

Fig. 1 Comparison of the brain surface determined from images acquired pre-operatively (*green*) with the intra-operative images acquired after craniotomy. The pre-operative position of the tumor (*blue*) is shown against the intra-operative tumor segmentation (*red*). MRI images were provided by Department of Surgery, Brigham and Women's Hospital (Harvard Medical School, Boston, Massachusetts, USA)

Following the introduction (Sect. 1), in Sect. 2 we discuss issues related to modelling geometry, boundary conditions, loading and material properties of the brain, and numerical algorithms devised to efficiently solve brain deformation behaviour models. In Sect. 3 we consider an example application in the area of computational radiology—brain shift computation for neuroimage registration. We conclude with some reflections about the state of the field.

2 What is and What is Not Important in Modelling the Brain Biomechanics?

Modelling the behaviour of the brain remains a key issue to provide *a priori* knowledge for image-guided surgery. The biomechanical property experiments significantly contribute to the understanding of the physics of brain tissue [6–11].

Miga, Paulsen and collaborators [12–21] have developed a sophisticated model of brain tissue undergoing surgery, incorporating simulations of forces associated with tumour tissue, and simulations of retraction and resection forces. Careful validation experiments indicate their model is able to closely match observed deformations. They indicate further improvements in accuracy will be possible by incorporating sparse data from inexpensive intra-operative imaging devices. This work has demonstrated that computer aided updating of pre-operative brain images can restore close correspondence between the pre-operative data and the intra-operative configuration of the subject. A practical difficulty of these models is the extensive time necessary to mesh the brain and solve the problem which is too long for intra-operative purposes.

Simple biomechanical models have been used to interpolate the full brain deformation based on sparse measured displacements. Miga et al. [22] measured the visible intra-operative cortex shift using a laser range scanner. The displacement of deep brain structures was then obtained applying these displacements as boundary conditions to the brain mesh. A similar surface based approach was proposed by e.g. Skrinjar et al. [23] who imaged the brain surface with a stereo vision system.

The approximation-based registration formulates the problem as a functional minimisation. This functional can generally be decomposed into a similarity energy and a regularisation energy. Computation of the similarity energy often relies on a block (or feature) matching algorithm. In 1998, Yeung showed impressive registration results on a phantom using an approximation formulation combining ultrasound speckle tracking with a mechanical finite element model [24]. Hagemann et al. [25] improved the basic block matching algorithm by selecting relevant anatomical landmarks in the image and taking into account the anisotropic matching error in the global functional. Shen and Davatzikos [26] investigated this idea of anatomical landmarks and proposed an attribute vector for each voxel reflecting the underlying anatomy at different scales.

A number of groups have investigated the use of intra-operative ultrasound in order to provide data that could be used to update pre-operative models to account for brain shift. In references [27, 28] a tracked, free-hand ultrasound probe have been utilised. Pennec et al. [29] demonstrated a system that utilises full-volume, intensity-based registration and 3D ultrasound, rather than the landmark-based methods discussed above.

These studies show that several intra-operative imaging modalities have the potential to accurately measure brain deformation, but that further study is needed. It is worth noting, that all but one [30] studies used linear formulations (elastic or poroelastic) that assume infinitesimality of deformation. As accurate computation of finite deformations of the brain is attempted, this assumption is not appropriate. Our modelling methods, described in detail in the following sections, are fully non-linear and allow accurate computations of large deformations.

2.1 Geometry Discretisation

Detailed geometric information is needed to define the domain in which the deformation field needs to be computed. Image registration systems require patient-specific data. Such data are available from radiological images (for an example see Fig. 2), however, they are significantly inferior in quality to the data available from anatomical atlases. The main components to consider in the brain

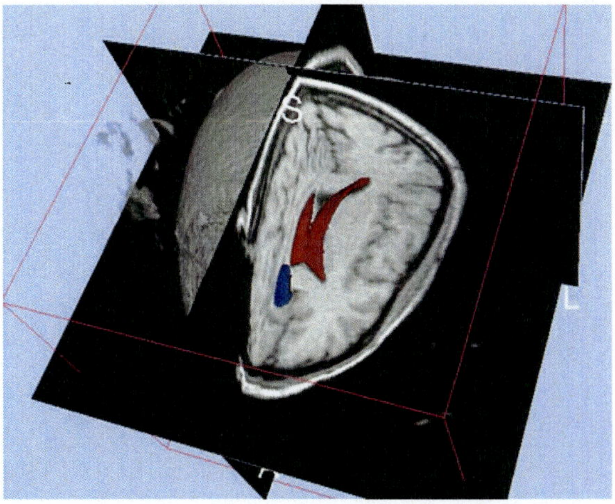

Fig. 2 3D magnetic resonance image presented as a tri-planar cross-section. Parts of the tumor (*red*) and ventricles (*blue*) are clearly visible. Public domain software Slicer (www.slicer.org) developed by our collaborators from Surgical Planning Laboratory, Harvard Medical School, was used to generate the image

model are the brain parenchyma, ventricles and tumour (if present). They need to be identified (segmented) in radiological images (in practice magnetic resonance images). As the segmentation and meshing of ventricles and especially the tumour still presents itself as a formidable challenge one needs to consider whether even simpler models that do not explicitly contain brain structures of complicated shapes might be sufficient. This question is considered in some detail in Sect. 3.

The accuracy of neurosurgery is not better than 1 mm [5]. Voxel size in high quality pre-operative MR images is usually of similar magnitude. Therefore, we can conclude that patient-specific models of the brain geometry can be constructed with approximately 1 mm accuracy, and that higher accuracy is probably not required.

A necessary step in the development of the numerical model of the brain is the creation of a computational grid which in most practical cases is a finite element mesh or a cloud of points required by a meshless method. Because of the stringent computation time requirements, the mesh must be constructed using low order elements that are not computationally intensive. The linear under-integrated hexahedron is the preferred choice.

Many algorithms are now available for fast and accurate automatic mesh generation using tetrahedral elements, but not for automatic hexahedral mesh generation [31–33]. Template based meshing algorithms can be used for meshing different organs using hexahedrons [34–36], but these types of algorithms only work for healthy organs. In the case of severe pathologies (such as a brain tumour), such algorithms cannot be used, as the shape, size and position of the pathology is unpredictable. This is one reason why many authors proposed the use of tetrahedral meshes for their models [37–40]. In order to automate the simulation process, mixed meshes having both hexahedral and linear tetrahedral elements are the most convenient (see Sect. 3 for examples of such meshes).

The under-integrated hexahedral elements require the use of an hourglass control algorithm in order to eliminate the instabilities, known as zero energy modes, which arise from the single-point integration. One of the most popular and powerful

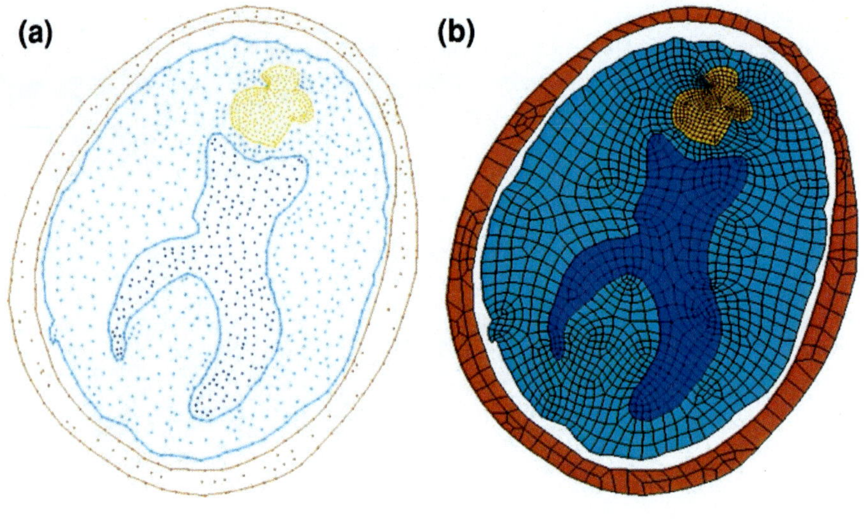

Fig. 3 A 2D slice of the brain discretised by **a** nodes of MTLED [49] method; and **b** quadrilateral finite elements. Development of a good-quality finite element mesh is time-consuming. Generation of the meshless grid is almost instantaneous

hourglass control algorithms, that is currently available in many commercial software finite element packages, is the one proposed in [41]. This method is applicable for hexahedral and quadrilateral elements with arbitrary geometry undergoing large deformations. We adapted this method to the Total Lagrangian formulation so that many quantities involved can be pre-computed [42], making the hourglass control mechanism very efficient from the computational point of view.

In modelling of incompressible continua, artificial stiffening (often referred to as volumetric locking) affects many standard elements including the linear tetrahedral element, see e.g. [43]. This phenomenon occurs also for nearly incompressible materials and therefore introducing slight compressibility does not solve the problem. A number of improved linear tetrahedral elements with anti-locking features have been proposed by different authors [44–47]. The average nodal pressure (ANP) tetrahedral element proposed in [44] is computationally inexpensive and provides much better results for nearly incompressible materials compared to the standard tetrahedral element. Nevertheless, one shortcoming of the ANP element and its implementation in a finite element code is the handling of interfaces between different materials. We extended the formulation of the ANP element so that all elements in a mesh are treated in the same way, requiring no special handling of the interface elements [48].

An alternative to using the finite element method is to use our recently developed Meshless Total Lagrangian Explicit Dynamics algorithm (MTLED) [49]. The problem of computational grid generation disappears as one needs only to drop a cloud of points into the volume defined by a 3D medical image [50–55], see Fig. 3.

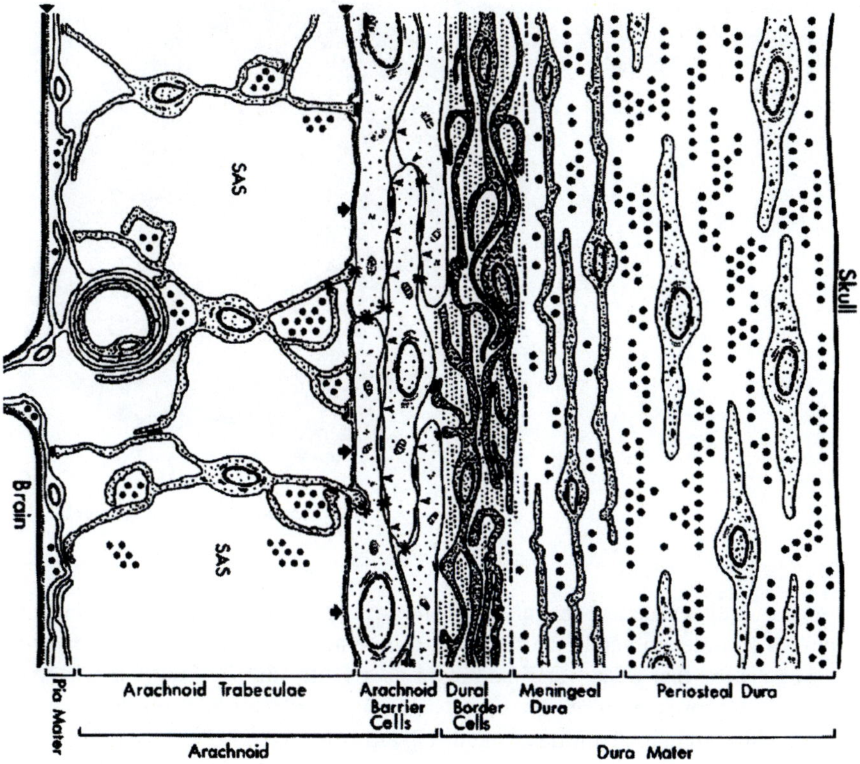

Fig. 4 Structure of the brain-skull interface, adapted from [56]

The use of meshless methods is motivated by simple, automatic computational grid generation for patient-specific simulations. We use a modified Element-Free Galerkin method [49] that is meshless in the sense that deformation is calculated at nodes that are not part of an element mesh. Node placement is almost arbitrary. Volumetric integration is performed over a regular background grid that does not conform to the simulation geometry.

2.2 Boundary Conditions

The formulation of appropriate boundary conditions for computation of brain deformation constitutes a significant problem because of complexity of the brain–skull interface, see Fig. 4.

A number of researchers fix the brain surface to the skull [57, 58]. We do not recommend this approach. Our experience [59–63] suggests the existence of a small gap between the brain and the skull that allows for the motion of the brain within the cranial cavity. Therefore, a simple and effective model of the brain–skull interface is a frictionless contact that allows separation.

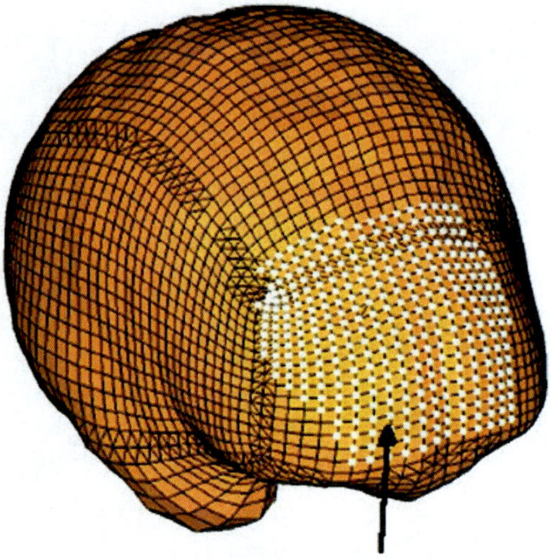

Fig. 5 Model loading through prescribed nodal displacements at the exposed brain surface

As the skull is orders of magnitude stiffer than the brain tissue its rigidity can be assumed. In order to handle the brain-skull interaction we developed a very efficient algorithm that treats this interaction as a finite sliding, frictionless contact between a deformable object (the brain) and a rigid surface (the skull) [64]. Unlike contacts in commercial finite element solvers (e.g. ABAQUS, LS-DYNA), our contact algorithm has no configuration parameters (as it only imposes kinematic restrictions on the movement of the brain surface nodes) and is very fast, with the speed almost independent of the mesh density of the skull surface.

2.3 Loading

We advocate loading the models through imposed displacements on the model surface [61, 62, 65], see Fig. 5. In the case of neurosurgical simulation, this loading will be imposed by known motion of a surgical tool. In the case of intra-operative image registration the current (intra-operative) position of the exposed part of the brain surface can be measured using a variety of techniques [66]. This information can then be used to define model loading.

As suggested in papers [65, 67–69] for problems where loading is prescribed as forced motion of boundaries, the unknown deformation field within the domain depends very weakly on the mechanical properties of the continuum. This feature is of great importance in biomechanical modelling where there are always uncertainties in patient-specific properties of tissues. It allows us to use a simple (but suitable for finite deformation solution procedures) Neo-Hookean constitutive model.

2.4 Solution Algorithms

The algorithms implemented in the great majority of commercial finite element programs use the Updated Lagrangian formulation, where all variables are referred to the current (i.e. from the end of the previous time step) configuration of the system (ANSYS [70], ABAQUS [71], ADINA [72], LS-DYNA [73], etc.). The advantage of this approach is the simplicity of incremental strain description and low internal memory requirements. The disadvantage is that all derivatives with respect to spatial coordinates must be recomputed in each time step, because the reference configuration is changing.

We use the Total Lagrangian formulation, where all variables are referred to the original configuration of the system. The decisive advantage of this formulation is that all derivatives with respect to spatial coordinates are calculated with respect to the original configuration and therefore can be pre-computed—this is particularly important for time-critical applications such as surgical simulation and intra-operative image registration.

Because biological tissue behaviour can be described in general using hyper-elastic or hyper-viscoelastic models, such as that given in Equations 1 and 2, the use of the Total Lagrangian formulation also leads to a simplification of material law implementation as these material models can be easily described using the deformation gradient.

The integration of equilibrium equations in the time domain can be done using either implicit or explicit methods [74–76]. The most commonly used implicit integration methods, such as Newmark's constant acceleration method, are unconditionally stable. This implies that their time step is limited only by the accuracy considerations. However, the implicit methods require solution of a set of non-linear algebraic equations at each time step. Furthermore, iterations need to be performed for each time step of implicit integration to control the error and prevent divergence. Therefore, the number of numerical operations per each time step can be three orders of magnitude larger than for explicit integration [74].

On the other hand, in explicit methods, such as the central difference method, treatment of non-linearities is very straightforward and no iterations are required. By using a lumped (diagonal) mass matrix [74], the equations of motion can be decoupled and no system of equations must be solved. Computations are done at the element or support domain level eliminating the need for assembling the stiffness matrix of the entire model. Thus, the computational cost of each time step and internal memory requirements are substantially smaller for explicit than for implicit integration. There is no need for iterations anywhere in the algorithm. These features make explicit integration suitable for real time applications.

However, the explicit methods are only conditionally stable. Normally a severe restriction on the time step size has to be included in order to receive satisfactory simulation results. Stiffness of soft tissues is very low [6, 77–79], e.g. stiffness of the brain is about eight orders of magnitude lower than that of common engineering materials such as steel. Since the maximum time step allowed for stability is (roughly speaking) inversely proportional to the square root of Young's modulus divided by the mass density [73], it is possible to conduct simulations of brain deformation with much longer time steps than in typical dynamic simulations in engineering. This was confirmed in our previous simulations of brain shift using the commercial finite element solver LS-DYNA [60, 61]. Therefore, when developing the suite of finite element algorithms for computation of brain tissue deformation, we combined Total Lagrange formulation with explicit time integration.

A detailed description of the Total Lagrange Explicit Dynamics (TLED) algorithm is presented in [80]. The main benefits of the TLED algorithm are:

- allows pre-computing of many variables involved (e.g. derivatives with respect to spatial coordinates and hourglass control parameters),
- no accumulation of errors—increase stability for quasi-static solutions,
- easy implementation of the material law for hyper-elastic materials using the deformation gradient,
- straightforward treatment of non-linearities,
- no iterations required for a time step,
- no system of equations needs to be solved,
- low computational cost for each time step.

3 Application Example: Computer Simulation of the Brain Shift

3.1 Medical Context

A particularly exciting application of non-rigid image registration is in intra-operative image-guided procedures, where high resolution pre-operative scans are warped onto sparse intra-operative ones [81, 82]. We are especially interested in registering high-resolution pre-operative MRI with lower quality intra-operative imaging modalities, such as multi-planar MRI and intra-operative ultrasound. To achieve accurate matching of these modalities accurate and fast algorithms to compute tissue deformations are fundamental.

Here we present examples of computational results of the brain shift. To account for various types of situations that occur in neurosurgery, we analyzed five cases of craniotomy-induced brain shift. These cases represent different situations that may occur in neurosurgery as characterised by tumours located in different parts of the brain: anteriorly (Cases 1, 2), laterally (Case 3) and posteriorly (Cases 4 and 5), Fig. 6. Detailed information about craniotomy size, tumour and brain volume is given in Table 1.

For the studied cases the maximum craniotomy-induced displacement of the cortical surface, as observed on intra-operative MR images, was about 7.7 mm.

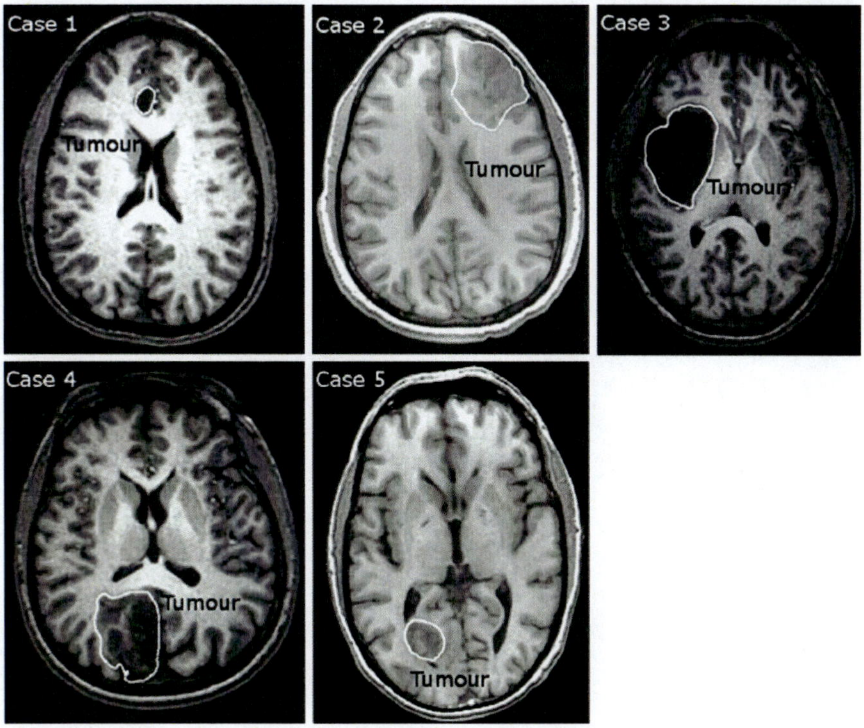

Fig. 6 Preoperative MRIs (*inferior view*) showing tumour location in the cases analysed in this study. Tumours are labelled and highlighted with white line

Table 1 Volumes of tumour and ventricles for the craniotomy cases analysed in the study. The volumes were obtained from the segmented pre-operative magnetic resonance images

	Parenchyma volume [m^3]	Tumour volume [m^3]	Tumour to parenchyma volume ratio
Case 1	1.17×10^{-3}	0.0027×10^{-3}	0.002
Case 2	1.44×10^{-3}	0.0625×10^{-3}	0.043
Case 3	1.38×10^{-3}	0.0745×10^{-3}	0.054
Case 4	1.13×10^{-3}	0.0141×10^{-3}	0.012
Case 5	1.43×10^{-3}	0.0073×10^{-3}	0.005

3.2 Finite Element Models of The Brain

Three-dimensional patient-specific brain meshes (Fig. 7) were constructed from the segmented preoperative magnetic resonance images (MRIs). The segmentation was done using seed growing algorithm followed, in some cases, by manual corrections. A detailed presentation of the meshes properties is given in Table 2.

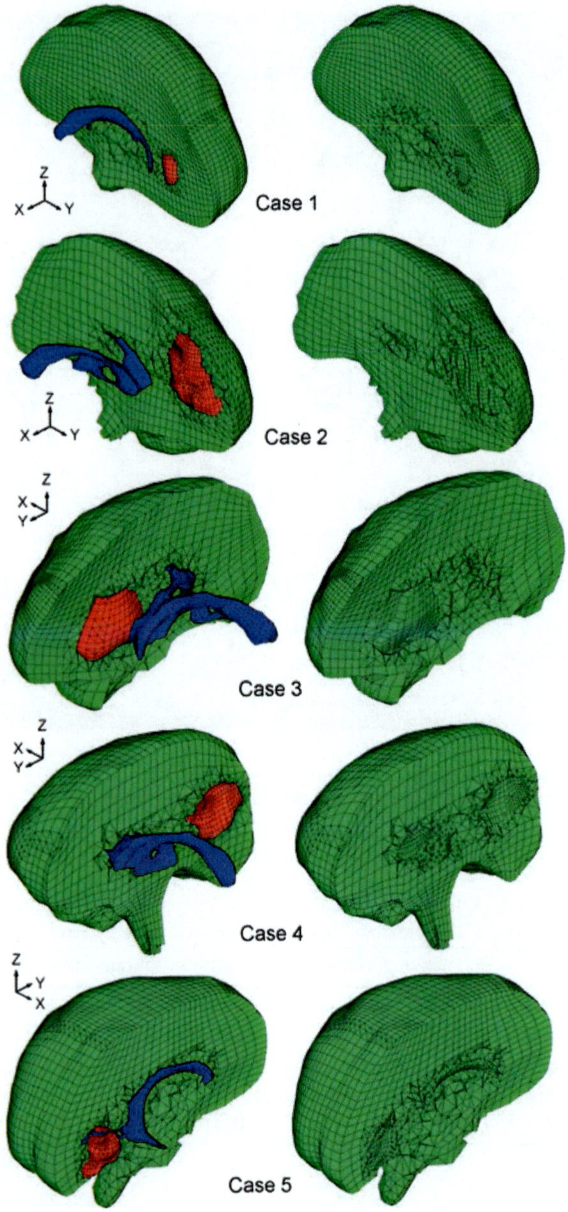

Fig. 7 The complete model (**A**, *left*) and homogeneous one (**C**, *right*) for five cases of craniotomy induced brain shift analysed here. In the complete model, the parenchyma (*Green*), Tumour (*Red*), and Ventricles (*Blue*) were simulated. In the homogeneous model, the tumour and ventricles were included in the parenchyma

To evaluate how the predicted brain shift deformation is affected by model complexity, patient-specific finite element models of five cases of craniotomy-induced brain shift originally developed and validated by Joldes et al. [48] were used as the starting point.

Table 2 Summary of the patient-specific brain meshes built in this study (models (**A**), see below)

	Case1	Case2	Case3	Case4	Case5
Number of hexahedral elements	14447	10258	10127	9032	8944
Number of tetrahedral elements	13563	20316	23275	23688	21160
Number of nodes	18806	15433	15804	14732	14069
Number of degrees of freedom	55452	45315	46896	43794	42018

The models by Joldes et al. [48], referred to as the complete model or Model (**A**), are geometrically accurate representation of five clinical cases of craniotomy and include the parenchyma, ventricles, tumour and skull (Fig. 7). The brain shift deformations predicted by these models were used as a reference point. When conducting analysis of the model complexity, the complete model was simplified to create the following models with decreasing order of complexity:

Model (**B**): Simplified model consisting of the parenchyma, tumour, ventricles and skull. The ventricles were modelled as an empty cavity. This simplification removes the need for meshing of the ventricles.

Model (**C**): Homogenous model consisting of parenchyma and skull (Fig. 7). The tumour and ventricles were not modelled and were included in the parenchyma. This simplification further removes the need to segment the ventricles. For this model, segmentation of the parenchyma and skull is sufficient.

3.3 Boundary Conditions, Loading and Brain Tissue Constitutive Model

Following (Wittek et al. [61]; Joldes et al. [48]; Miller et al. [83]), all the models in this study were loaded by prescribing nodal displacements to a set of nodes under the craniotomy determined from the segmented pre- and intra-operative cortical surfaces. At the brain surface nodes where the displacements were not prescribed, the boundary conditions were defined as a frictionless contact interface with the rigid skull. This interface prevents the brain surface from penetrating the skull while allowing for sliding and separation between the brain and the skull.

The neo-Hookean hyperelastic constitutive model (Yeoh [84]) with Poisson's ratio of 0.49 was used for tumour and parenchyma. In all the models, the Young's modulus of the parenchyma was designated a value of 3,000 Pa. In the complete models, a Young's modulus of 6,000 Pa was used for the tumour, and, following Wittek et al. [61], the ventricles were modelled as a very soft compressible elastic solid with Young's modulus of 10 Pa and Poisson's ratio of 0.1 to allow ventricular volume decrease during brain shift.

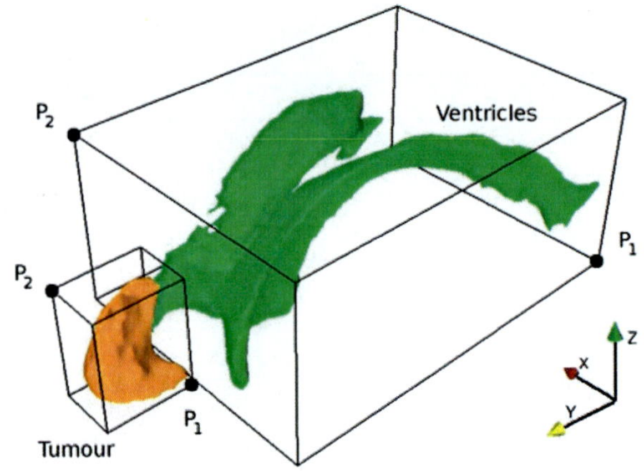

Fig. 8 Definition of ventricles' bounds. Vertices P_1 and P_2 define a cuboidal box that bounds the ventricles. The box faces are formed by planes perpendicular to X, Y and Z axes

For all models, the specialised nonlinear finite element solver (Joldes et al. [42, 48, 64, 85, 86]; Miller et al. [80, 83]) for real-time computation of soft organ deformation was used to calculate the craniotomy-induced deformation field within the brain.

3.4 Evaluation of Effects of Model Complexity

We compare the intra-operative geometries of the tumour and ventricles obtained by registration of the pre-operative data using the deformation fields predicted by different models. When conducting the registration, we determined to which element of the finite element mesh a given point of the tumour or ventricles belongs to and calculated its intra-operative position from the nodal displacements using the element shape functions for interpolation.

Universally accepted "gold standard" for evaluating the accuracy of prediction of intra-operative deformation field within an organ (e.g. brain) for image-guided surgery has not been developed yet [87]. Objective metrics of the images alignment/overlap can be provided by automated methods, such as e.g. Dice coefficient (Dice [88]), Mutual Information [89, 90] and Normalised Cross-Correlation [91]. From the perspective of biomechanical modelling for computing the deformation field within the brain, one of the key deficiencies of such methods is that they quantify the differences/similarities between two images using metrics that do not have straightforward geometrical (in Euclidean distance sense) interpretation.

In this study, following Joldes et al. [48], the intra-operative Euclidean bounds of tumour and ventricles are used to provide quantitative information about the deformed shape and position of the tumour and ventricles. The bounds can be interpreted as the X, Y, Z coordinates of vertices P_1 and P_2 defining cuboidal boxes bounding the tumour and ventricles (Fig. 8). The coordinates of vertices P_1 and P_2

Table 3 Differences in the X, Y, Z coordinates of vertices P_1 and P_2 defining the bounds of the tumour and ventricles (Fig. 9) between the complete models (**A**) and the homogeneous ones (**C**). The numbers in bold font indicate the maximum differences in the X, Y, and Z directions

	Tumour						Ventricles					
Unit: mm	X		Y		Z		X		Y		Z	
	P_1	P_2	P_1	P_2	P_1	P_2	P_1	P_2	P_1	P_2	P_1	P_2
Model (**C**): Homogenous mesh, tumour and ventricles modelled as parenchyma												
Case 1	0.01	0.10	0.16	0.10	0.02	0.01	**0.16**	0.07	0.06	0.05	0.05	**0.14**
Case 2	0.09	0.01	0.25	**0.85**	**0.15**	0.09	0.02	0.02	0.01	**0.09**	0.12	0.06
Case 3	0.00	**0.14**	0.01	0.02	0.02	0.02	0.04	0.05	0.02	0.03	0.01	0.01
Case 4	0.00	0.00	0.00	0.04	0.03	0.03	0.01	0.01	0.01	0.01	0.02	0.00
Case 5	0.00	0.02	0.01	0.00	0.01	0.02	0.01	0.00	0.01	0.01	0.00	0.00

can be determined automatically, which makes them less prone to subjective errors than the measures based on anatomical landmarks manually selected by experts.

To provide a comparison of the brain shift deformation prediction from the finite element modelling perspective, displacements of the nodes defining the tumour and ventricles are compared one by one, between the complete model (**A**) and model (**C**) in which homogenous constitutive properties were used for the entire brain. The homogenous model was selected because the largest differences (compared to the complete model **A**) in the tumour and ventricular bounds (see Table 3) were observed for this model.

A qualitative comparison of the results obtained in the present study with the intra-operative images is also presented. For selected sections through the brain where tumour and ventricles are clearly visible, the registered (i.e. warped using deformation fields predicted by means of the models we used here) contours of tumour and ventricles are super-imposed on the intra-operative MR images. This comparison was done only for model (**C**) in which the largest differences (compared with the complete model **A**) in the intra-operative bounds of tumour and ventricles were observed. The intra-operative contours of tumour and ventricles obtained using models (**B**) are not presented here as the intra-operative bounds predicted by these models are not visually distinguishable from those predicted by Model (**A**).

4 Results

The differences in X, Y, Z coordinates of the bounds of the tumour and ventricles (Fig. 8) between the complete model (**A**) and model (**B**) (i.e. the one with the ventricles simplified as an empty cavity) were less than 0.01 mm. Given that the resolution of intra-operative magnetic resonance images MRIs used in this study was $0.86 \times 0.86 \times 2.5$ mm^3, these differences are negligible. This suggests that treating the ventricles as an empty cavity has negligible effect on the predicted

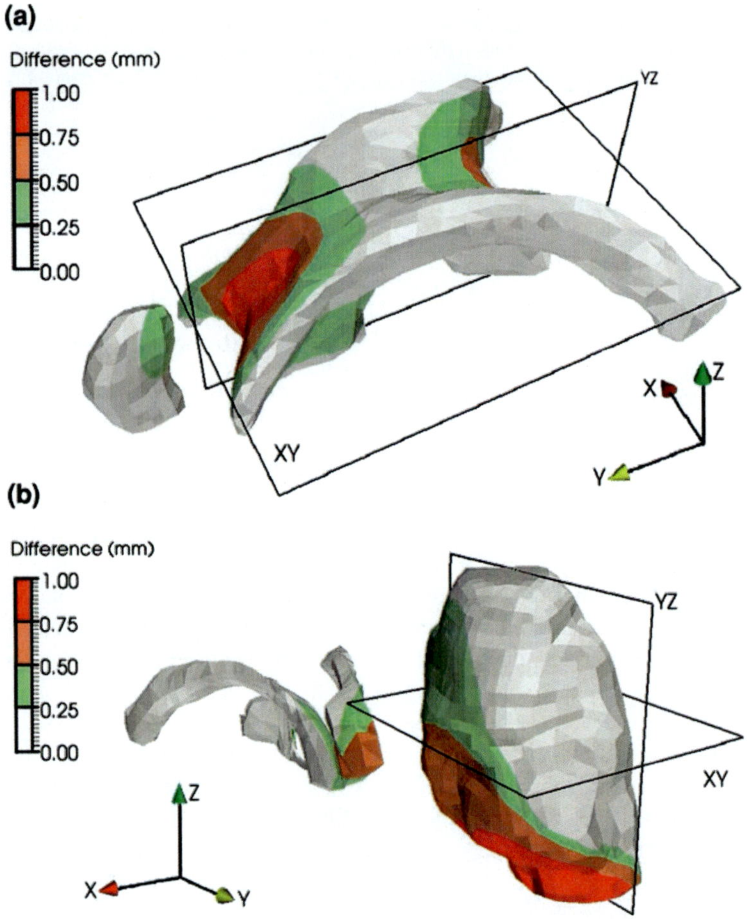

Fig. 9 The resultant differences between nodal displacements predicted by the complete model (A) and homogenous model (C) for tumour and ventricles. **a** Case 1 and **b** Case 2. The differences are indicated by a *colour bar*

brain shift deformations. Consequently, meshing of the ventricles can be eliminated when building the finite element models for predicting craniotomy-induced brain shift deformations.

The differences between the intra-operative bounds of ventricles and tumour predicted by the complete models (**A**) and simplified ones (**C**) are listed in Table 3. The differences were within the resolution of the intra-operative images ($0.86 \times 0.86 \times 2.5$ mm^3) for all the analysed cases. The largest observed differences were 0.85 mm in the Y (anterior-posterior, Fig. 9) direction for the tumour (Case 2) and 0.16 mm in the X (lateral) direction for the ventricles (Case 1).

The results regarding the effects of varying the model complexity on prediction of the intra-operative bounds of tumour and ventricles are confirmed by analysis of

Table 4 Maximum and median differences in the X, Y and Z displacements of nodes defining the tumour and ventricles between the complete models (**A**) and the homogenous ones (**C**). The numbers in bold font indicate single largest difference and highest median difference in tumour and ventricles. Please note that the maximum differences in the X, Y, Z directions may not have been observed for the same node

	Tumour						Ventricles					
Unit: mm	X		Y		Z		X		Y		Z	
	Max	Mdn	Max	Mdn	Max	Mdn	Max	Mdn	Max	Mdn	Max	Mdn
Model (**C**): Homogenous mesh, tumour and ventricles modelled as parenchyma												
Case 1	0.16	0.05	0.40	0.17	0.18	0.05	0.50	0.12	**0.90**	**0.15**	0.56	0.15
Case 2	0.28	0.08	**0.95**	**0.26**	0.31	0.10	0.61	0.14	0.62	0.12	0.44	0.09
Case 3	0.24	0.06	0.17	0.04	0.15	0.02	0.12	0.04	0.07	0.02	0.07	0.01
Case 4	0.07	0.02	0.23	0.04	0.10	0.02	0.03	0.01	0.04	0.02	0.03	0.01
Case 5	0.04	0.01	0.04	0.01	0.03	0.01	0.05	0.01	0.03	0.01	0.04	0.00

the effects of model complexity on the calculated nodal displacements. The comparison of the nodal displacements is presented only between the complete models (**A**) and homogenous ones (**C**), as the differences in nodal displacements for models (**A**) and (**B**) were very small. For tumour, the largest differences in the nodal displacements were observed for Case 2 in the Y (anterior-posterior) direction: maximum difference of 0.95 mm and median difference of 0.26 mm (Table 4, Fig. 9). For the ventricles, the largest differences were observed for Case 1, also in the anterior-posterior direction: maximum difference of 0.90 mm and median difference of 0.15 mm (Table 3, Fig. 9). The maximum differences in the calculated nodal displacements, between the complete models (**A**) and homogenous ones (**C**), are close to the resolution of the intra-operative MRIs ($0.86 \times 0.86 \times 2.5\ mm^3$) (Table 3). As indicated in Fig. 9, for vast majority of the nodes, the differences are well below this resolution.

Comparison of the intra-operative contours of tumour and ventricles predicted using models of different complexity yields results consistent with those obtained when analysing the effects of model complexity on nodal displacements and intra-operative bounds of tumour and ventricles. The predicted contours overlap and are very close to those in the intra-operative images, except for minor local mis-registration. This was observed not only for the complete model (**A**) but also for homogenous one (**C**) in which no distinction was made between the constitutive properties of the parenchyma, tumour and ventricles (Figs. 10 and 11).

5 Conclusions

Computational mechanics has led to greater understanding and advances in modern science and technology [1]. It is now in a position to make a similar impact in medicine. We have discussed modelling approaches to two applications

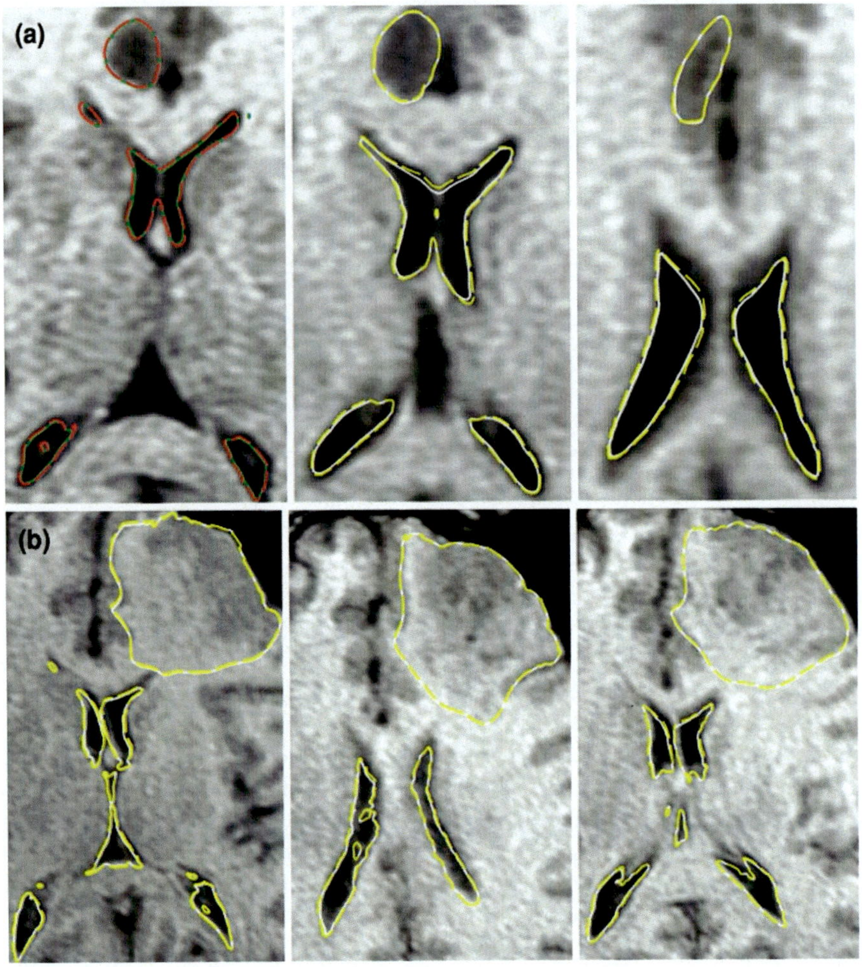

Fig. 10 Transverse contour sections of the tumour and ventricles in Case 1(**a**) and 2(**b**), computed using the complete model (Model A, *White*) and homogenous one (Model C, *Yellow dash line*) are imposed on the intra-operative MR images. The sections are enlarged and cropped to highlight the pathology. To ensure that the sections are not subject to the choice of the authors, we used the following approach. The first sections are manually selected so both the tumour and ventricles are clearly visible; then we show sections of the 10th and 20th slices after the first one

of clinical relevance: surgical simulation and neuroimage registration. Mechanical terms such as displacements and forces can be used to characterise these problems, and therefore the standard methods of continuum mechanics can be applied. Moreover similar methods may find applications in modelling the development of structural diseases of the brain [63, 92–94].

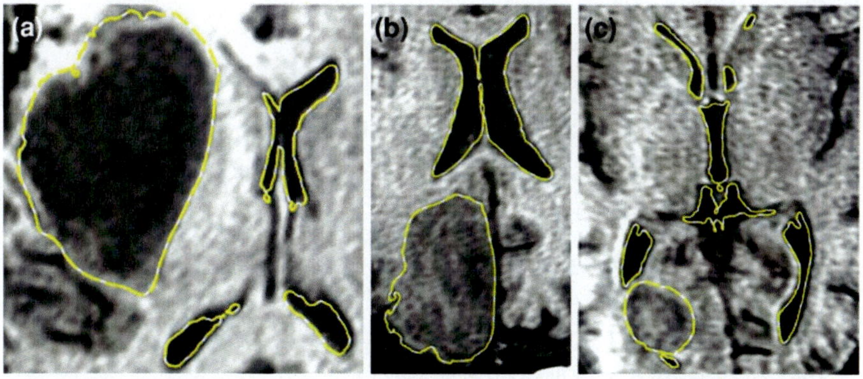

Fig. 11 Transverse contours of the tumour and ventricles in Case 3(**a**), 4(**b**), and 5(**c**), computed using the complete model (Model **A**, *White*) and homogenous one (Model **C**, *Yellow dash line*) are imposed on the intra-operative MR images. We selected the sections in which both the tumour and ventricles are clearly visible. The contour overlaps and there is no observable difference for the scale shown

Because of the large displacements involved ($\sim$10–20 mm in the case of a brain shift) and the strongly non-linear mechanical response of tissue to external loading, we use non-linear finite element procedures for the numerical solution of the proposed models.

A number of challenges still prevent the wide acceptance of Computer-Integrated Surgery systems based on computational biomechanical models. As we deal with individual patients, methods to produce patient-specific computational grids quickly and reliably must be improved. Substantial progress in automatic meshing methods is required, while meshless methods may provide an alternative solution. Computational efficiency is an important issue, as intra-operative applications, requiring reliable results within approximately 40 s, are most appealing. The use of the Total Lagrangian Formulation of the finite element method [75, 80], where all field variables are related to the original (known) configuration of the system and therefore most spatial derivatives can be calculated before the simulation commences, during the pre-processing stage, offers such a possibility. Implementation of these algorithms on graphics processing units leads to computation times well within the limits required for intra-operative applications.

Acknowledgments The financial support of the Australian Research Council (Grants No. DP0343112, DP0664534 and LX0560460) and NIH (Grant No. 1-RO3-CA126466-01A1) is gratefully acknowledged. We thank our collaborators Dr Ron Kikinis and Dr Simon Warfield from Harvard Medical School, and Dr Kiyoyuki Chinzei and Dr Toshikatsu Washio from Surgical Assist Technology Group of AIST, Japan for help in various aspects of our work.The medical images used in the present study (provided by Dr Simon Warfield) were obtained in the investigation supported by a research grant from the Whitaker Foundation and by NIH grants R21 MH67054, R01 LM007861, P41 RR13218 and P01 CA67165.

References

1. Oden, J.T., et al.: Research directions in computational mechanics. Comput. Methods Appl. Mech. Eng. **192**, 913–922 (2003)
2. Cloots, R.J.H., et al.: Biomechanics of traumatic brain injury: influences of the morphologic heterogeneities of the cerebral cortex. Ann. Biomed. Eng. **36**(7), 1203–1215 (2008)
3. Kleiven, S., von Holst, H.: Consequences of head size following trauma to the human head. J. Biomech. **35**(2), 153–160 (2002)
4. Nakaji, P., Speltzer, R.F.: The marriage of technique, technology, and judgement. Innov. Surg. Approach **51**, 177–185 (2004)
5. Bucholz, R., MacNeil, W., McDurmont, L.: The operating room of the future. Clin. Neurosurg. **51**, 228–237 (2004)
6. Miller, K.: Biomechanics of brain for computer integrated surgery. Publishing House of Warsaw University of Technology, Warsaw (2002)
7. Miller, K., Chinzei, K.: Constitutive modelling of brain tissue: experiment and theory. J. Biomech. **30**(11–12), 1115–1121 (1997)
8. Miller, K., et al.: Mechanical properties of brain tissue in vivo: experiment and computer simulation. J. Biomech. **33**(11), 1369–1376 (2000)
9. Miller, K., Chinzei, K.: New UWA robot–possible application to robotic surgery. Biomed. Sci. Instrum. **36**, 135–140 (2000)
10. Chinzei, K., Miller, K.: Towards MRI guided surgical manipulator. Med. Sci. Monit. **7**(1), 153–163 (2001)
11. Miller, K., Chinzei, K.: Mechanical properties of brain tissue in tension. J. Biomech. **35**(4), 483–490 (2002)
12. Miga, M.I., et al.: Model-updated image guidance: initial clinical experiences with gravity-induced brain deformation. IEEE Trans. Med. Imaging. **18**(10), 866–874 (1999)
13. Miga, M.I., et al.: Updated neuroimaging using intraoperative brain modeling and sparse data. Stereotact. Funct. Neurosurg. **72**(2–4), 103–106 (1999)
14. Miga, M.I., et al.: In vivo quantification of a homogeneous brain deformation model for updating preoperative images during surgery. IEEE Trans. Biomed. Eng. **47**(2), 266–273 (2000)
15. Miga, M.I., et al.: In vivo modeling of interstitial pressure in the brain under surgical load using finite elements. J. Biomech. Eng. **122**(4), 354–363 (2000)
16. Miga, M.I., et al.: In vivo analysis of heterogeneous brain deformation computations for model-updated image guidance. Comput. Methods Biomech. Biomed. Eng. **3**(2), 129–146 (2000)
17. Miga, M.I., et al.: Modeling of retraction and resection for intraoperative updating of images. Neurosurgery **49**(1), 75–84 (2001). discussion 84–85
18. Paulsen, K.D., et al.: A computational model for tracking subsurface tissue deformation during stereotactic neurosurgery. IEEE Trans. Biomed. Eng. **46**(2), 213–225 (1999)
19. Roberts, D.W., et al.: Intraoperative brain shift and deformation: a quantitative analysis of cortical displacement in 28 cases. Neurosurgery **43**(4), 749–758 (1998). discussion 758–760
20. Roberts, D.W., et al.: Intraoperatively updated neuroimaging using brain modeling and sparse data. Neurosurgery **45**(5), 1199–1206 (1999). discussion 1206–1207
21. Roberts, D.W., et al.: Intra-operative image updating. Stereotact. Funct. Neurosurg. **76**(3–4), 148–150 (2001)
22. Miga, M.I., et al.: Cortical surface registration for image-guided neurosurgery using laser-range scanning. IEEE Trans. Med. Imaging. **22**(8), 973–985 (2003)
23. Skrinjar, O., Nabavi, A., Duncan, J.: Model-driven brain shift compensation. Med. Image Anal. **6**(4), 361–373 (2002)
24. Yeung, F., et al.: Feature-adaptive motion tracking of ultrasound image sequences using a deformable mesh. IEEE Trans. Med. Imaging. **17**(6), 945–956 (1998)

25. Hagemann, A., Rohr, K., Stiehl, H.S.: Coupling of fluid and elastic models for biomechanical simulations of brain deformations using FEM. Med. Image Anal. **6**(4), 375–388 (2002)
26. Shen, D., Davatzikos, C., ER, H.A.M.M.: Hierarchical attribute matching mechanism for elastic registration. IEEE Trans. Med. Imaging **21**(11), 1421–1439 (2002)
27. Dey, D., et al.: Automatic fusion of freehand endoscopic brain images to three-dimensional surfaces: creating stereoscopic panoramas. IEEE Trans. Med. Imaging **21**(1), 23–30 (2002)
28. Gobbi, D.G., Peters, T.M.: Generalized 3D nonlinear transformations for medical imaging: an object-oriented implementation in VTK. Comput. Med. Imaging Graph. **27**(4), 255–265 (2003)
29. Pennec, X., Cachier, P., Ayache, N.: Tracking brain deformations in time sequences of 3D US images. Pattern Recognit. Lett. **24**, 810–813 (2003)
30. Xu, M., Nowinski, W.L.: Talairach-Tournoux brain atlas registration using a metalforming principle-based finite element method. Med. Image Anal. **5**(4), 271–279 (2001)
31. Owen, S.J.: A survey of unstructured mesh generation technology. In: 7th International Meshing Roundtable. Dearborn, Michigan (1998)
32. Viceconti, M., Taddei, F.: Automatic generation of finite element meshes from computed tomography data. Crit. Rev. Biomed. Eng. **31**(1), 27–72 (2003)
33. Owen, S.J.: Hex-dominant mesh generation using 3D constrained triangulation. Comput. Aided. Des. **33**, 211–220 (2001)
34. Castellano-Smith, A.D., et al.: Constructing patient specific models for correcting intraoperative brain deformation. In: 4th International Conference on Medical Image Computing and Computer Assisted Intervention MICCAI, Lecture Notes in Computer Science 2208. Utrecht, The Netherlands (2001)
35. Couteau, B., Payan, Y., Lavallée, S.: The mesh-matching algorithm: an automatic 3D mesh generator fo finite element structures. J. Biomech. **33**, 1005–1009 (2000)
36. Luboz, V., et al.: Orbital and maxillofacial computer aided surgery: patient-specific finite element models to predict surgical outcomes. Comput. Methods Biomech. Biomed. Eng. **8**(4), 259–265 (2005)
37. Ferrant, M., et al.: Deformable modeling for characterizing biomedical shape changes. In: Discrete geometry for computer imagery: 9th International Conference. Uppsala, Springer-Verlag GmbH, Sweden (2000)
38. Ferrant, M., et al.: Serial registration of intraoperative MR images of the brain. Med. Image Anal. **6**(4), 337–359 (2002)
39. Clatz, O., et al.: Patient specific biomechanical model of the brain: application to Parkinson's disease procedure. In: International Symposium on Surgery Simulation and Soft Tissue Modeling (IS4TM'03). Springer-Verlag, France (2003)
40. Warfield, S.K., et al.: Real-time registration of volumetric brain MRI by biomechanical simulation of deformation during image guided surgery. Comput. Vis. Sci. **5**, 3–11 (2002)
41. Flanagan, D.P., Belytschko, T.: A uniform strain hexahedron and quadrilateral with orthogonal hourglass control. Int. J. Numer. Method. Eng. **17**, 679–706 (1981)
42. Joldes, G.R., Wittek, A., Miller, K.: An efficient hourglass control implementation for the uniform strain hexahedron using the total Lagrangian formulation. Commun. Numer. Method. Eng. **24**, 1315–1323 (2008)
43. Hughes, T.J.R.: The finite element method: linear static and dynamic finite element analysis, p. 682. Dover Publications, Mineola (2000)
44. Bonet, J., Burton, A.J.: A simple averaged nodal pressure tetrahedral element for incompressible and nearly incompressible dynamic explicit applications. Commun. Numer. Method. Eng. **14**, 437–449 (1998)
45. Bonet, J., Marriott, H., Hassan, O.: An averaged nodal deformation gradient linear tetrahedral element for large strain explicit dynamic applications. Commun. Numer. Method. Eng. **17**, 551–561 (2001)
46. Zienkiewicz, O.C., et al.: Triangles and tetrahedra in explicit dynamic codes for solids. Int. J. Numer. Method. Eng. **43**, 565–583 (1998)

47. Dohrmann, C.R., et al.: Node-based uniform strain elements for three-node triangular and four-node tetrahedral meshes. Int. J. Numer. Method. Eng. **47**, 1549–1568 (2000)
48. Joldes, G.R., Wittek, A., Miller, K.: Non-locking tetrahedral finite element for surgical simulation. Commun. Numer. Method. Eng. **25**(7), 827–836 (2009)
49. Horton, A, et al.: A meshless total lagrangian explicit dynamics algorithm for surgical simulation. Commun. Numer. Method. Eng. (2010). doi:10.1002/cnm.1374.
50. Horton, A., Wittek, A., Miller, K.: Computer simulation of brain shift using an element free Galerkin method. In: 7th International Symposium on Computer Methods in Biomechanics and Biomedical Engineering CMBEE 2006. Antibes, France (2006)
51. Horton, A., Wittek, A., Miller, K.: Towards meshless methods for surgical simulation. In: Computational Biomechanics for Medicine Workshop, Medical Image Computing and Computer-Assisted Intervention MICCAI 2006. Copenhagen, Denmark (2006)
52. Horton, A., Wittek, A., Miller. K.: Subject-specific biomechanical simulation of brain indentation using a meshless method. In: International Conference on Medical Image Computing and Computer-Assisted Intervention MICCAI 2007, Springer, Brisbane (2007)
53. Belytschko, T., Lu, Y.Y., Gu, L.: Element-free Galerkin methods. Int. J. Numer. Method. Eng. **37**, 229–256 (1994)
54. Liu, G.R.: Mesh free methods: moving beyond the finite element method. CRC Press, Boca Raton (2003)
55. Li, S., Liu, W.K.: Meshfree particle methods. Springer-Verlag, Berlin (2004)
56. Haines, D.E., Harkey, H.L., Al-Mefty, O.: The "subdural" space: a new look at an outdated concept. Neurosurg. **32**, 111–120 (1993)
57. Hagemann, A., et al.: Biomechanical Modeling of the Human Head for Physically Based, Nonrigid Image Registration. IEEE Transactions on Medical Imaging - Special Issue on Model-Based Analysis of Medical Images, **18**(10), 875–884 (1999)
58. Miga, M.I., et al.: In vivo quantification of a homogenous brain deformation model for updating preoperative images during surgery. IEEE Trans. Biomed. Eng. **47**(2), 266–273 (2000)
59. Wittek, A., Omori, K.: Parametric study of effects of brain-skull boundary conditions and brain material properties on responses of simplified finite element brain model under angular acceleration in sagittal plane. JSME Int. J. **46**(4), 1388–1398 (2003)
60. Wittek, A. et al.: Brain shift computation using a fully nonlinear biomechanical model. In: 8th International Conference on Medical Image Computing and Computer Assisted Surgery MICCAI 2005. Palm Springs, California (2005)
61. Wittek, A., et al.: Patient-specific model of brain deformation: application to medical image registration. J. Biomech. **40**, 919–929 (2007)
62. Wittek, A., Hawkins, T., and Miller, K., On the unimportance of constitutive models in computing brain deformation for image-guided surgery. Biomech. model. mechanobiol., (2008). doi:10.1007/s10237-008-0118-1
63. Dutta-Roy, T., Wittek, A., Miller, K.: Biomechanical modelling of normal pressure hydrocephalus. J. Biomech. **41**(10), 2263–2271 (2008)
64. Joldes, G.R., et al.: Realistic and efficient brain-skull interaction model for brain shift computation. In: Computational Biomechanics for Medicine III Workshop, MICCAI, New-York (2008)
65. Miller, K., Wittek, A.: Neuroimage registration as displacement-zero traction problem of solid mechanics. In: Computational Biomechanics for medicine MICCAI-associated Workshop. MICCAI, Copenhagen (2006)
66. Miga, M.I., et al.: Cortical surface registration for image-guided neurosurgery using laser-range scanning. IEEE Trans. Med. Imaging **22**(8), 973–985 (2003)
67. Miller, K.: Method of testing very soft biological tissues in compression. J. Biomech. **38**, 153–158 (2005)
68. Miller, K.: Biomechanics without mechanics: calculating soft tissue deformation without differential equations of equilibrium. In: 5th Symposium on Computer Methods in Biomechanics and Biomedical Engineering CMBBE2004. Madrid, Spain: First Numerics (2005)

69. Miller, K.: How to test very soft biological tissues in extension. J. Biomech. **34**(5), 651–657 (2001)
70. ANSYS.: ANSYS home page. 28/03/2008]; Available from: www.ansys.com
71. ABAQUS, I.: ABAQUS Online Documentation: version 6.5-1. 2004
72. ADINA, R., ADINA, D.: home page. 28/03/2008]; Available from: www.adina.com
73. Hallquist, J.O.: LS-DYNA theory manual. Livermore software technology corporation, Livermore (2005)
74. Belytschko, T.: A survey of numerical methods and computer programs for dynamic structural analysis. Nuclear Eng. Des. **37**, 23–34 (1976)
75. Bathe, K.-J.: Finite element procedures. Prentice-Hall, New Jersey (1996)
76. Crisfield, M.A.: Non-linear dynamics, in non-linear finite element analysis of solids and structures, pp. 447–489. John Wiley and Sons, Chichester (1998)
77. Miller, K., Chinzei, K.: Constitutive modelling of brain tissue; experiment and theory. J. Biomech. **30**(11/12), 1115–1121 (1997)
78. Miller, K., Chinzei, K.: Mechanical properties of brain tissue in tension. J. Biomech. **35**, 483–490 (2002)
79. Miller, K., et al.: Mechanical properties of brain tissue in vivo: experiment and computer simulation. J. Biomech. **33**, 1369–1376 (2000)
80. Miller, K., et al.: Total Lagrangian explicit dynamics finite element algorithm for computing soft tissue deformation. Commun. Numer. Methods Eng. **23**, 121–134 (2007)
81. Ferrant, M., et al.: Registration of 3-D intraoperative MR images of the brain using a finite-element biomechanical model. IEEE Trans. Med. Imaging **20**, 1384–1397 (2001)
82. Warfield, S.K., et al.: Capturing intraoperative deformations: research experience at Brigham and Womens's hospital. Med. Image Anal. **9**(2), 145–162 (2005)
83. Miller, K., Wittek, A., Horton, A., Dutta Roy, T., Berger, J. and Morriss, L.: Modelling brain deformation for computer-integrated neurosurgery (invited review). Int. J. Numer. Method. Biomed. Eng. **26**(1), 117–138 (2010)
84. Yeoh, O.: Some Forms of Strain-Energy Function for Rubber. Rubber. Chem. Technol. 66, 754–771 (1993)
85. Joldes, G., Wittek, A., Miller, K.: Suite of finite element algorithms for accurate computationof soft tissue deformation for surgical simulation. Med. Image Anal. **13**(6), 912–919 (2009a)
86. Joldes, G., Wittek, A., Miller, K.: Computation of intra-operative brain shift using dynamic relaxation. Comput. Methods. Appl. Mech. Eng. **198**, 3313–3320 (2009b)
87. Chakravarty, M.M., et al.: Towards a validation of atlas warping techniques. Med. Image Anal. **12**(6), 713–726 (2008)
88. Dice, L.R.: Measures of the amount of ecologic association between species. Ecology **26**, 297–302 (1945)
89. Viola, P., Wells III, W.M.: Alignment by maximization of mutual information. Int. J. Comput. Vis. **24**(2), 137–154 (1997)
90. Wells III, W.M., et al.: Multi-modal volume registration by maximization of mutual information. Med. Image Anal. **1**(1), 35–51 (1996)
91. Rexilius, J., et al.: A novel nonrigid registration algorithm and applications, in medical image computing and computer-assisted intervention – MICCAI 2001. pp. 923–931. (2001)
92. Taylor, Z., Miller, K.: Reassessment of brain elasticity for analysis of biomechanisms of hydrocephalus. J. Biomech. **37**, 1263–1269 (2004)
93. Miller, K., Taylor, Z., Nowinski, W.L.: Towards computing brain deformations for diagnosis prognosis and nerosurgical simulation. J. Mech. Med. Biol. **5**(1), 105–121 (2005)
94. Berger, J., et al.: Coupling finite element and mesh-free methods for modelling brain defromations in response to tumour growth. In: Computational Biomechanics for Medicine III MICCAI-Associated Workshop. MICCAI, New York (2008)

Biomechanical Aspects of Spinal Cord Injury

Thomas R. Oxland, Timothy Bhatnagar, Anthony M. Choo, Marcel F. Dvorak, Wolfram Tetzlaff and Peter A. Cripton

Abstract Research into the biomechanics of spinal cord injury has progressed rapidly over the past decade via a number of research approaches. Cadaveric experimentation, animal model development and computational simulations continue to contribute much insight into the relation between spinal column injury (i.e. vertebrae, discs, ligaments) and spinal cord damage. Efforts in this research field are directed towards providing clinicians information on how to design optimal intervention strategies, and establishing preventative strategies that may be

---

T. R. Oxland (✉), T. Bhatnagar and P. A. Cripton
Department of Orthopaedics and Mechanical Engineering, University of British Columbia, 5460 Blusson Pavilion, 818 West 10th Avenue, Vancouver, BC V5Z 1M9, Canada
e-mail: toxland@interchange.ubc.ca

T. R. Oxland, T. Bhatnagar, M. F. Dvorak, W. Tetzlaff and P. A. Cripton
International Collaboration On Repair Discoveries (ICORD), Vancouver, Canada

A. M. Choo
Department of Bioengineering, University of Pennsylvania, Philadelphia, PA, USA

M. F. Dvorak
Combined Neurosurgical and Orthopaedic Spine Program (CNOSP), Department of Orthopaedics, University of British Columbia, Vancouver, Canada

W. Tetzlaff
Department of Zoology, University of British Columbia, Vancouver, Canada

W. Tetzlaff
Department of Surgery, University of British Columbia, Vancouver, Canada

Stud Mechanobiol Tissue Eng Biomater (2011) 3: 159–180
DOI: 10.1007/8415_2010_37

Published Online: 30 November 2010

targeted to specific injury mechanisms, thus reducing the severity and extent of permanent paralysis and improving the health related quality of life of individuals with spinal cord injury. While there has been advancement in our understanding of the complex mechanical response of the cord to a variety of distinctly different injury mechanisms, there is still much to discover regarding the behavior of the spinal cord during the traumatic event and the ensuing pathophysiological degradation of the spinal cord.

1 Introduction

Spinal cord injuries (SCIs) are devastating to the injured individuals and their families. Approximately 15–40 per million new cases of SCI occur worldwide each year and although the patient demographic and mechanisms of injury may be changing, the overall incidence of these injuries has not changed appreciably over the past several decades [1–4]. SCI most commonly affects young males in high-energy injuries from motor vehicle accidents [5], but recent demographic data indicates a rising incidence in elderly individuals who sustain a SCI after a low energy fall [6, 7]. The economic costs associated with the medical care and lost earnings for SCIs represent a significant burden to society, estimated in the billions of dollars per year in North America [1].

The spinal cord is encased and protected by the vertebrae, intervertebral discs, and ligaments that make up the surrounding spinal column. During the dynamic events that result in a SCI (e.g. a motor vehicle collision or fall), various combinations of forces and moments are applied to the spine. If applied with sufficient magnitude, these forces will disrupt the spinal column, as manifested by fracturing of the vertebrae and/or rupturing of the ligaments. These spinal column injuries occur in specific patterns, often described as injury types, with SCIs being most commonly associated with higher energy fracture-dislocations and burst fractures [1, 7]. These two injury types account for about 80% of the spinal column failures resulting in SCI, and they occur with similar frequency, see Fig. 1 [7, 8]. In a burst fracture, the posterior aspect of the vertebral body is fractured off as the body "bursts", and this bony fragment is propelled into the spinal cord (Fig. 2a). In a fracture-dislocation, adjacent vertebral bodies are broken, ligaments are torn, and adjacent vertebral segments are translated causing compression and shear of the spinal cord between the cephalad and caudal spinal elements (Fig. 2b). In a specific pattern of injury that can occur without fracture, termed flexion-distraction, posterior distractive forces rupture ligaments and thereby stretch the spinal cord (Fig. 2c). It was recognized over twenty-five years ago that the pattern of spinal column injury (the "injury type") influenced the degree of neurological dysfunction [8, 9].

Despite known differences in neurological outcomes, clinical treatment for SCI does not differentiate between spinal column injury types other than to guide the

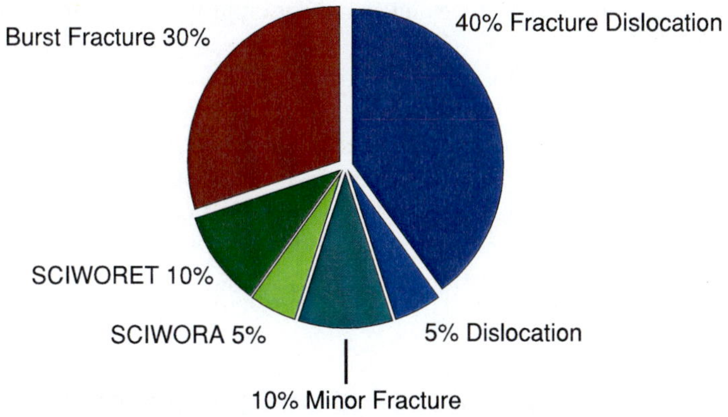

Fig. 1 The relative frequency of spinal column injuries that are associated with spinal cord injuries. SCIWORET—spinal cord injury without obvious radiological evidence of trauma, SCIWORA—spinal cord injury without obvious radiological abnormality [1]

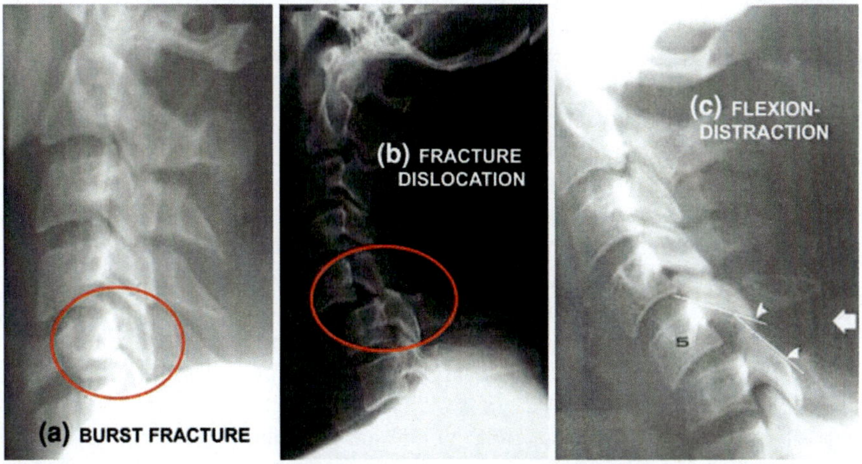

Fig. 2 Sagittal radiological images of cervical spinal column injuries. **a** Burst fracture—the indicated vertebra shows loss of vertebral height and the posterior wall of the vertebral body has moved posteriorly into the spinal canal, thereby impinging on the spinal cord. **b** Fracture dislocation—an apparent dislocation of the spinal column resulting in spinal canal occlusion. **c** Flexion-distraction—adjacent vertebrae are subluxed with significant relative rotation resulting in a tensile load on the spinal cord

technique of surgical reduction and stabilization. By not distinguishing SCIs by their injury mechanism, we may be ignoring potentially important differences in the actual neuropathology of the injury, a point made recently by Charles Tator [10], where he noted that 'disregard for the heterogeneity of human SCI may be an important reason for the failure to find clinically effective therapies'. With respect to injury mechanism, there is a call for a better understanding of the relation between spinal column injury type and spinal cord injury—or the biomechanics of spinal cord injury.

The objective of this chapter is to provide an overview of current research on the biomechanical aspects of spinal cord injury. Also, existing SCI research models are categorized into cadaveric-based, animal, and mathematical/simulation models and briefly summarized. There are inherent limitations with each model type that will be noted, but they all do contribute to our understanding of the problem and will hopefully lead to enhanced treatment methods in the future and the development of approaches for injury prevention.

2 Clinically Relevant Spinal Cord Injury Mechanisms

Preclinical SCI models have traditionally included spinal cord transection, contusion, and compression [11–13]. In humans, however, other injury mechanisms such as cord shearing from vertebral dislocation, and cord stretching from spinal distraction, occur frequently. This creates a potentially important disparity between experimental paradigms and clinical injuries.

Similar to SCI in the clinical population, animal models yield different temporal stages of spinal cord injury. To differentiate between spinal cord damage from the initial physical insult and damage resulting from the ensuing cascade of biochemical processes, post-trauma injury has been popularly divided into primary (immediately following trauma) and secondary injury (develops in minutes to days post-trauma). Primary injury can manifest itself in vascular damage leading to hemorrhage and membrane disruptions causing uncontrolled permeability of neurons and axons (Fig. 3). This catastrophic failure to maintain homeostasis triggers extensive necrosis at the lesion epicenter and axonal interruption in the adjacent white matter. The mechanical injury is regarded to be irreversible and hence SCI research has concentrated on intervening in secondary degenerative processes [10, 14].

In the wake of primary injury, a wave of secondary injury is triggered that is characterized by energy depletion, failure of ion pumps, calcium overload, excitotoxicity, free radical formation, lipid peroxydation, microglial activation (Fig. 4), pro-inflammatory cytokine release and infiltration with inflammatory cells. These inflammatory cells further generate oxygen free radicals that exacerbate lipid peroxidation and protein nitrosylation. The net result is a period of prolonged tissue damage extending over weeks with apoptotic cell death of oligodendrocytes and neurons [15].

The biomechanical response of the spinal cord to trauma is governed by the structure of the cord—the softer, more delicate grey and white matter tissues are surrounded by the stiffer and more durable meninges (pia, arachnoid and dura mater). While the spinal column will interact directly with the meninges, the ensuing injury typically manifests itself in the grey and white matter. Additionally, the grey and white matters consist of different cellular components and are structurally organized in distinctly different ways. The grey matter of the spinal cord

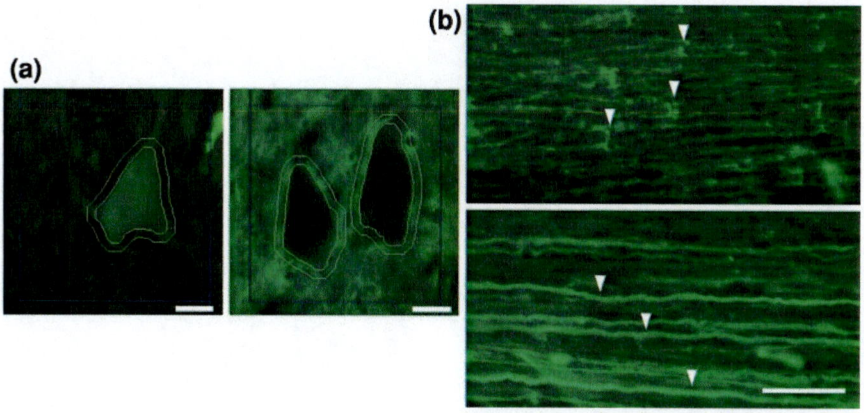

Fig. 3 Dextran (a fluorescent marker) is introduced into the extra-cellular fluid within the spinal cord and uptake into cellular bodies post-trauma indicates increased permeability in cell membranes. **a** Cell bodies that do (*left*) and do not (*right*) exhibit membrane compromise due to injury. **b** Axons that do (*bottom*) and do not (*top*) exhibit increased membrane permeability due to injury. *White arrows* indicate regions of dextran uptake

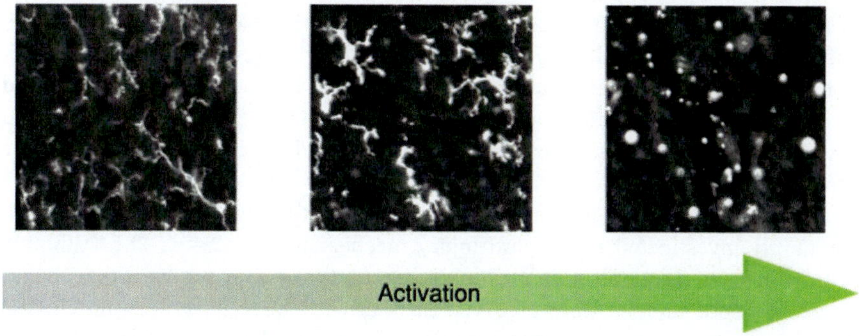

Fig. 4 Microglial activation due to traumatic spinal cord injury. As microglia activate to initiate inflammation, there are distinct morphology changes over time and quantification of activated bodies is possible

is composed mostly of neuronal cell bodies which are arranged in a largely isotropic manner. In contrast, the majority of white matter consists of axons that originate from the neuronal cell bodies in the brain/brainstem or sensory ganglia or from propriospinal neurons. The white matter is highly anisotropic since the axons predominately course rostro-caudally along the long axis of the spinal cord [16–18]. In addition, these axons are myelinated, i.e. wrapped by many layers of compact cell membrane produced by the oligodendrocytes. These differences are highlighted when observing the cord tissue response to the different injury mechanism that we understand can impose anisotropic loading schemes on the cord [19].

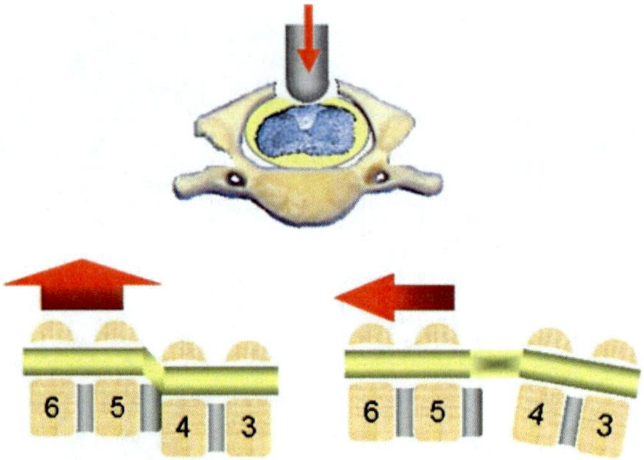

Fig. 5 Different injury mechanisms of SCI. *Top*: contusion—axial view of the spinal cord under posterior compression. Clinically, note that the contusion would impact the anterior, rather than the posterior aspect of the cord. *Left*: dislocation—sagittal view of a posterior translation of two vertebral bodies (C5-6) with respect to adjacent vertebrae (C3-4). *Right*: distraction—sagittal view of an inferior translation of two vertebral bodies (C5-6) with respect to adjacent vertebrae (C3-4)

Contusion injuries are a localized compression of the cord due to bone fragments intruding at high speed into the spinal canal during a vertebral burst fracture. Dislocation injuries result from spinal column shearing (either laterally, or anterior/posteriorly) which exposes the spinal column to compressive, tensile and shearing loads. Distraction injuries occur during axial extension of the spine beyond physiologically safe thresholds [20–22], resulting in a tensile load on the spinal cord (Fig. 5).

Spinal cord contusion injury (associated with burst-fracture injuries) is the most commonly investigated spinal cord injury mechanism [11, 23–26, 28]. Although distinctions between the spinal column injury types that produce the above-mentioned cord injuries were observed qualitatively, there have been no clinical data to directly compare different resulting spinal cord damage patterns. This has led to the development of numerous non-clinical models to facilitate an experimental approach to understanding the biomechanics of spinal cord injury.

3 Models of Spinal Cord Injury

3.1 Cadaveric Models of Spinal Cord Injury

To gain a better understanding of gross spinal cord behavior during clinically relevant spine injuries, research groups have used surrogate spinal cords [27] in cadaveric models. Saari et al. studied the spinal column-cord interactions during trauma caused by head-first impact. A previously developed radio-opaque

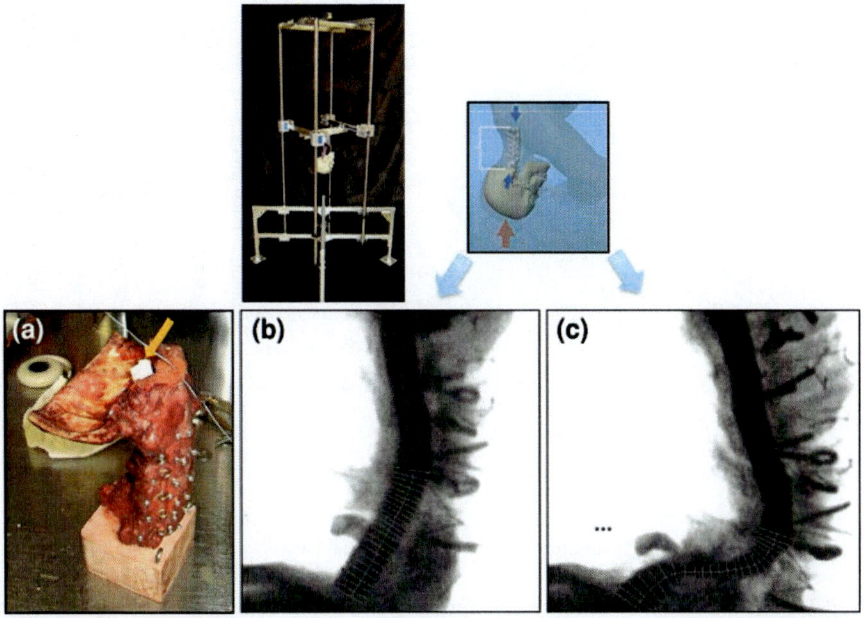

Fig. 6 **a** Cadaveric cervical spine specimen including a section of the occipital bone. A section of the radio-opaque surrogate spinal cord [28] has been inserted in the spinal canal and it is visible protruding from the foramen magnum (*Arrow*). **b**, **c** Example of two high-speed x-ray images from the same specimen during a head-first impact test in the drop tower shown at the *top left*. The x-ray images are taken at 1,000 frames per second and 'b' is prior to impact and 'c', at maximum deformation is associated with a traumatic atlanto-axial dislocation. Both transverse and longitudinal measurements of cord deformation can be made using these images. [27]

surrogate cord [28] was used with a high-speed X-ray system to monitor the cord morphology throughout impact (Fig. 6). The anterior-posterior diameter of the cord was measured at 1.5 mm increments along its length for each frame of the radiographic footage. Measurements of interest from these trials focused on the column-cord interactions during trauma. The surrogate cord model has improved our understanding of the spinal cord injuries that occur when the column experiences fracture-dislocations and hyperextension injuries due to clinically relevant head-first impacts. The resulting cord deformations were used to determine the theoretical neurological outcome of the impact based on in vivo studies carried out previously by Kearney et al. [29]. Probability of recovery ranged between 8% for the atlantoaxial dislocation injury and 95% for mild hyperextension injuries. Experimental approaches such as this enable one to address the effects of intrinsic (such as specimen age, gender and bone quality) and extrinsic factors (such as muscle activation and impact surface friction and angles) on the potential for spinal cord injury in an impact situation.

Nightingale et al. noted the importance of the boundary conditions in cervical spine injury [30] and Camacho et al. suggested that decreased friction at the head-

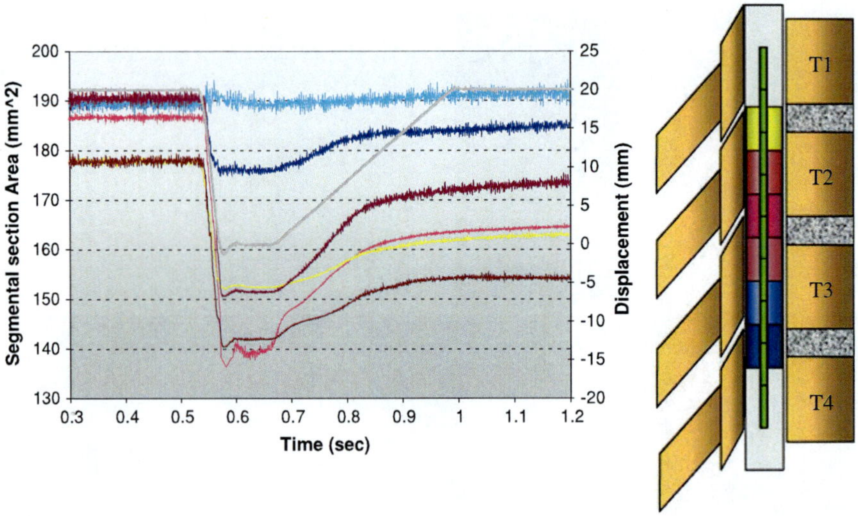

Fig. 7 A spinal canal occlusion transducer measures spinal canal occlusion during injury of the thoracic spine at different levels (color-based). The spinal canal, from T1 to T4, experienced different amounts of encroachment during the transient imposed trauma. The residual occlusion was always less than during the injury [32]

impact surface interface would reduce the likelihood of neck injury [31]. A study by Zhu et al. investigated the effect of translational constraint conditions on the amount of spinal canal occlusion in a cadaveric model of the thoracic spine [32]. In that study, a human cadaveric upper thoracic spine model was used that included a surrogate spinal cord capable of measuring the spinal canal occlusion in real-time. This surrogate cord produced electrical signals that were proportional to the degree of canal occlusion at several discrete points along the spine (Fig. 7). Under flexion-compression loading, the cranial aspect of the spine was fixed in some specimens and was permitted to translate anteriorly in other specimens. The hypothesis of this work was that sagittal-plane translation of the head during impact could reduce the potential for spinal cord occlusion. The tests indicated that free anterior translation (i.e. unconstrained) of the cranial aspect of the spine resulted in less occlusion of the spinal canal during the injury process (Fig. 8). This result suggests that the severity of a head-first impact for spinal cord injury may be mitigated by safety equipment that facilitates motion of the head during impact.

Bilston et al. also conducted a study that utilized a synthetic head and cervical spine model to investigate the effects of fast, transient motions causing hyper-flexion and hyper-extension on the axial strain magnitudes in a surrogate silicone gel spinal cord [33]. The study produced head and neck kinematics to match previous live and cadaveric studies and determined that based on injurious strain thresholds in animal models, the simulated motions in the study could result in cervical SCI with neurologic deficit. However, the authors state that more verification with clinical data is required before the results of the study can be validated.

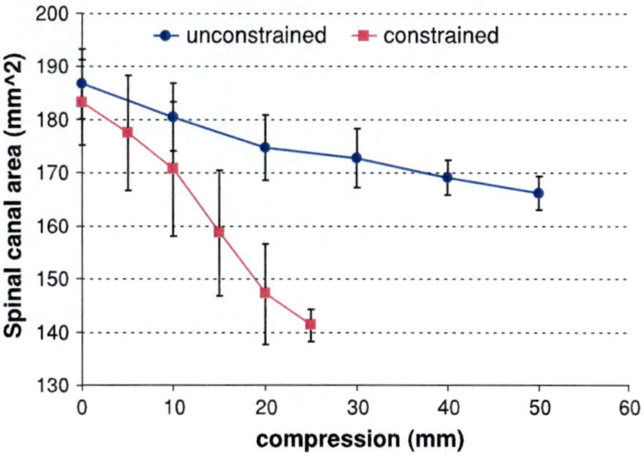

Fig. 8 Spinal canal occlusion during axial compression of cadaveric thoracic spine (T1–T4). When the cranial aspect of the spine is permitted to translate anterior/posteriorly, the severity of spinal canal occlusion is significantly reduced [32]

Studies using human cadaveric spines with surrogate spinal cords can provide important data from a normally inaccessible area of the body. While there are limitations associated with recreating the true structure of the heterogeneous spinal cord, the data do contribute significantly to our understanding of the spinal column-cord interaction during injury.

3.2 Mathematical Models of Spinal Cord Injury

Computational modeling is possibly the fastest growing category of research models for investigating spine biomechanics. These models can predict kinematic and kinetic changes in the spine due to surgical interventions such as anatomical resection or the implantation of devices. They may be especially valuable in spinal cord research due to the delicate mechanical structure of the spinal cord and the highly-sensitive physiological tolerance to injury which makes direct mechanical experimentation difficult and complex.

A widely used method of mathematical simulation is finite element modeling (FEM), which is designed to determine internal mechanical response (e.g. stresses, strains, pressures) throughout the entire model based on pre-determined external stimuli. When used in SCI research, these computer-generated models can describe what the spinal cord is locally experiencing physically, during prescribed injury simulations.

These modeling tools have been used to simulate burst-fracture contusion, fracture-dislocation and flexion-distraction injuries to the human spinal cord (Fig. 9) [19]. With continually increasing software and computation capabilities, the complexity of these models has been consistently on the rise. A sophisticated model of contusion in a rat model has been reported recently [34]. Further enhancements to these models may be created to mimic the highly complex nature of the spinal cord by including cerebrospinal fluid (CSF) behavior, interactions between the meninges and parenchyma of the spinal cord and other soft-tissue

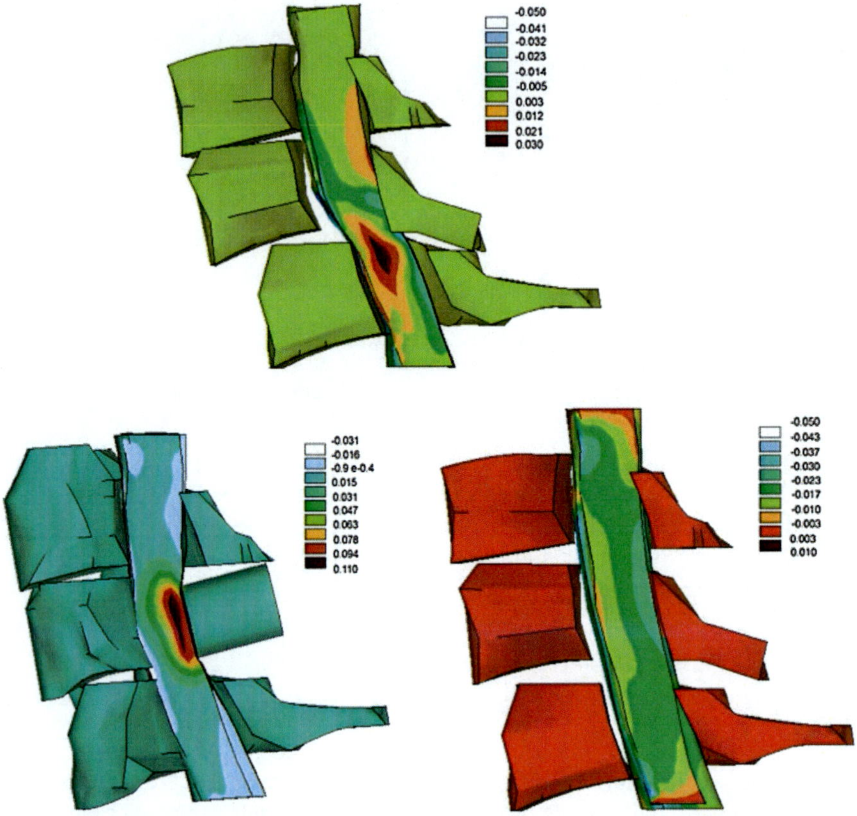

Fig. 9 Finite element analysis of a three-vertebrae model of the human cervical spine [19]. This model used linear elastic properties for all structures to simplify the task of observing how the spinal cord experiences different loading during different injury mechanisms. The color plots indicate areas of larger strain in the cord. *Bottom-left*: contusion (burst-fracture). *Top*: Fracture-dislocation. *Bottom-right*: Flexion-distraction

effects during injury [35]. As seen in the figures, these computational models enable one to observe the distribution of tissue stresses and strains during an injury event, which is very difficult to visualize with physical model experimentation.

Finite element models have several limitations. These computational models are largely governed by user-defined material properties that are traditionally sourced from existing literature, but there can be disparities. For example, Ichihara et al. reported that the gray matter in the bovine spinal cord was stiffer than the white matter [16]. In that same year, Ozawa et al. found no significant differences between the mechanical characteristics of gray and white matter obtained from rabbit spinal cords [17]. These differences in results can be attributed to differences in the experimental approach such as the direction of measurement [18] as well as the presence of regional heterogeneity in the mechanical properties of the tissue [36]. These inherent technical difficulties in determining the mechanical characteristics of extremely soft tissues continue to be addressed and validation of these models continues to progress in parallel with physical model-based research. (Further information regarding Finite-Element Models is presented in another chapter in this monograph.)

3.3 *Animal Models of Spinal Cord Injury*

To investigate in vivo spinal cord behavior, animal models must be used. It is hoped that mathematical models may be useful future adjuncts to these in vivo studies, but the current understanding of the relationship between mechanical trauma and the ensuing neurobiological responses is not sufficiently understood. While cadaveric studies are very useful for studying spinal column and canal behavior during injury, the spinal cord parenchyma (i.e. grey and white matter) experiences rapid degradation post-mortem, resulting in a mechanical behavior that no longer represents what would occur in a normal in vivo environment [37, 38] and a physiological response that is not relevant from a clinical perspective. Hence, in vivo animal models are currently the best setting in which to evaluate both biomechanical as well as neurophysiological responses to trauma.

Rodent models have largely replaced cat and dog models in the study of the neurobiology of SCI. The spinal cord contusion model, which transiently compresses the spinal cord in a dorsal-to-ventral direction, has gained widespread application as a clinically relevant model of human SCI [39–43]. The contused rat spinal cord exhibits a central cavitating lesion that is surrounded by a glial scar at chronic time-points post trauma—similar to the lesions observed in human SCIs [44, 45].

4 Biomechanical Parameters in Animal Models

All animal models control one or more mechanical parameters in order to produce a consistent SCI lesion. The weight-drop method was pioneered nearly a century ago by Allen in 1911 and it is still used today [46]. In this injury model, a mass is dropped from a prescribed height onto the spinal cord which has been exposed via a laminectomy [39]. This dorsal-ventral contusion of the animal's spinal cord resembles the anterior-posterior burst fracture mechanism that frequently occurs in humans [1], although clinically the fragment would impact the anterior aspect of the spinal cord. The initial potential energy of the drop-mass, commonly expressed as grams-centimeter, is used to biomechanically characterize the injury. As would be expected, an increase in either the mass or drop-height produces a more severe injury. A recent study illustrated that increasing the g-cm product increases the extent of primary mechanical disruption to the blood–brain barrier as visualized by extravasation of different molecular weight markers from the blood stream into the spinal cord parenchyma [46].

A limitation of the weight-drop model is that the g-cm product is an indeterminant quantity, where differing combinations of the drop-mass and the drop-height that yield equivalent g-cm products produce different injury severities [47]. In an attempt to circumvent this ambiguity, in rats, a constant drop mass (e.g. 10 g) is used in combination with different drop heights (e.g. 12.5, 25, and 50 mm) to achieve different known impact velocities (e.g. 48.9, 69.0, and 97.4 cm/s) respectively [46].

Alternatively, the SCI could be parameterized by controlling the contusion displacement [48–51] or the impact force [42]. The choice of controlling contusion injuries using force or displacement feedback remains controversial [42, 43]. Both parameters have correlated with experimental SCI severity [23, 42, 43, 49, 52]. Typically, a 1.1 mm contusion injury in the rat spinal cord results in a peak impact force of approximately 2.6 N [23]. However, there can be substantial variability. In a behavioural analysis of rats subjected to displacement controlled contusions, a force window of ± 1.3 standard deviations was necessary to isolate distinct biomechanical groups resulting in an exclusion rate of 30% [53]. Analogously, in force controlled injuries, an inclusion/exclusion window around the mean resultant displacement has also been advocated in order to reduce the variability within different severity groups [43]. These inclusion/exclusion criteria illustrate the inherent biomechanical variability of the in vivo spinal cord.

In the past, uncontrolled deflection of the animal's vertebral column, which supports the spinal cord from below, introduced substantial errors into the actual impact delivered to the spinal cord. Measurements in a cat model showed 4.6 mm of cord/thorax deflection resulted in only 1.1 mm of actual spinal cord compression [47]. The current weight-drop device explicitly measures the deflection of the vertebral column in order to correct the impact data [13], however, it is also possible to mitigate the uncontrolled deflection of the vertebral column with rigid vertebral clamping during impact [54]. Nevertheless, other factors such as the cerebrospinal fluid [55] and spinal cord vasculature will still introduce some biomechanical variability between animals. In a study of cervical contusion injuries, we were able to control the injury displacement to within 2% (1.1 mm s.d. 0.02 mm), but still observed a primary hemorrhage variation of 60% between animals. However, this variation in hemorrhage was significantly correlated with our impact force measurements [54].

Another biomechanical parameter that is often overlooked is the impact velocity to the spinal cord. The impact speed in the weight drop model varies between 48.9 cm/s in mild injuries to 97.4 cm/s in severe injuries [46]. The contusion model developed at Ohio State University has a peak velocity of 30 cm/s [56] while the force controlled model developed at the University of Kentucky reaches speeds of 13 cm/s [42]. Anderson used a pneumatic impactor to produce SCI at 6 m/s and showed that the extent of hemorrhage was largely controlled by the impact speed while the magnitude of the contusion displacement correlated strongly with post-traumatic spinal cord function [52]. Kearney showed that the product of the magnitude of spinal cord compression and the velocity of injury could be used to predict residual spinal cord function. Some spinal cord injury models such as the aneurysm clip compression [57, 58] and forceps compression [59] do not standardize the injury velocity. It was recently reported that the impact speed does not influence SCI severity when tested over a range of 10–40 cm/s [60]. Research from our group suggests that a wider range in the velocity magnitude is required in order to clearly see the influence of the impact speed [56]. Similarly, several in vitro and ex vivo studies have also emphasized the need to model trauma at high rates [61–64]. Injuries on the order of meters per second are believed to occur in human SCI [65–67].

Each of the injury models has strengths and limitations. An important feature of both the controlled displacement and controlled force models is the measurement of the complementary biomechanical response variable (resultant force or resultant displacement) thereby providing an a priori method to exclude outliers. Irrespective of which mechanical parameter is controlled in the injury model, it is perhaps more important to be aware of the model's limitations [68]. Although a highly repeatable model is essential for statistically detecting treatment effects, a single repeatable injury model may not be able to test the robustness of an experimental therapy.

In addition to the biomechanical parameters discussed above, recent work from our group suggests that distinct biomechanical mechanisms of injury lead to differing SCI pathology thereby suggesting the need for a broader spectrum of clinically relevant injury models.

5 Clinical Relevance in Animal Models

Due to the heterogeneous layout of the spinal cord and the links between biomechanical parameters of injury and the severity of the ensuing neurological deficit [41], outcomes from rodent-model SCI may be important in understanding how the human spinal cord responds to injury. Recently, Choo et al. performed contusion, dislocation and distraction injuries with an in vivo rat model [26]. The histological analyses utilize markers that are able to isolate and highlight the primary and secondary injury phenomena within the damaged cord, such that they may be quantified for evaluation and comparison.

In the contusion mechanism, the ensuing damage is localized to the spinal level at which the contusion occurred, with limited rostral or caudal extension, since the mechanism of injury consists of a localized compressive load. There is typically a lesion centre filled with hemorrhagic, and necrotic tissue that will ultimately become cleared to give rise to a cystic cavity surrounded by a glial scar (Fig. 10a). An increase in local membrane permeability, both in the cell body as well as the axon, is characteristic for the early stage after the injury event (Figs. 11a and 12a).

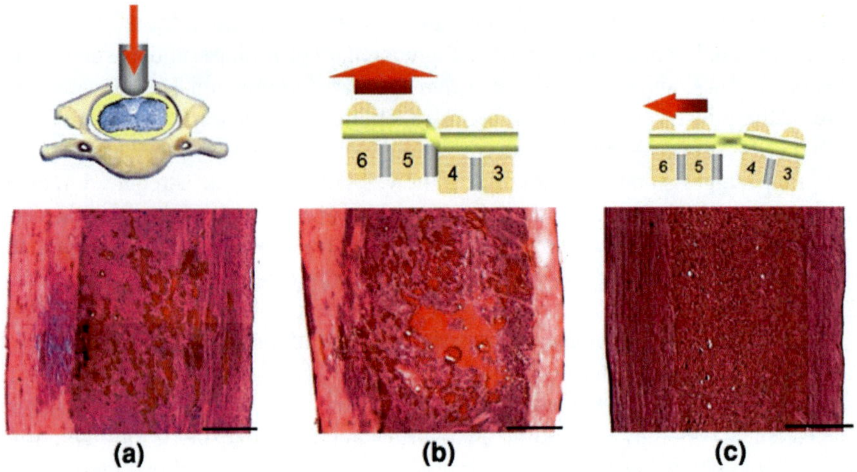

Fig. 10 Primary injury hemorrhage. The pattern of hemorrhage in the spinal cord is dependent on the injury mechanism [26]; **a** Contusion injury produces a localized area of hemorrhage. **b** Dislocation injury will also yield a central hemorrhage area. **c** Distraction injuries do not produce any central hemorrhage area

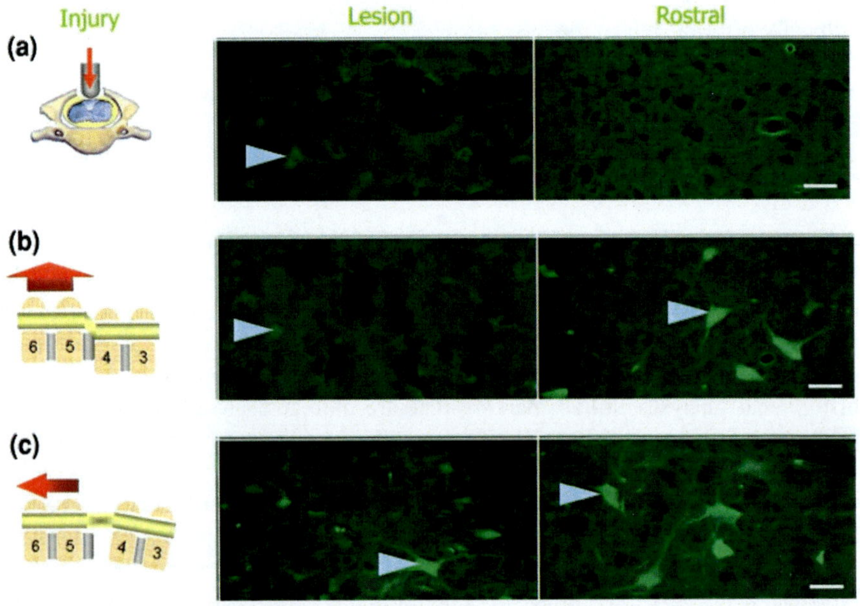

Fig. 11 Neuronal cell body membrane compromise (*grey matter*). Histological analysis at the level of injury (lesion) and rostral to the injury level show uptake of Dextran into cell bodies based on the type of cord injury administered [26]. **a** Contusion injury, **b** Dislocation injury and **c** Distraction injury. *Arrows* indicate cell bodies where membrane permeability is evident

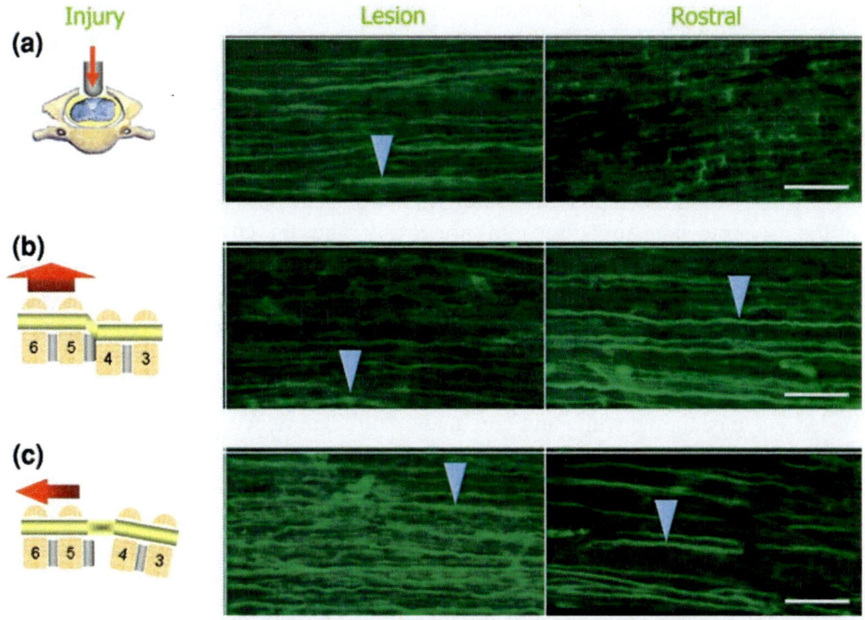

Fig. 12 Axon membrane compromise (*white matter*). Histological analysis at the level of injury (lesion) and rostral to the injury level show uptake of Dextran into cell bodies based on the type of cord injury administered [26]. **a** Contusion injury, **b** Dislocation injury and **c** Distraction injury. *Arrows* indicate axonal bodies where membrane permeability is evident

During a spinal column dislocation, the injury epicenter is the point at which the spinal cord is pinched between the two adjacent vertebrae. At this point, the loading is similar to the contusion mechanism, in that a localized compressive load is suddenly imposed. However, the gross motion of the spinal column imposes loads on other parts of the cord close to the injury epicenter as well; the spinal cord segments on the rostral and caudal side of the epicenter experience a tensile load, which causes different types of primary and secondary damage in the grey and white matter than compressive loading. The dislocation mechanism similarly produces central hemorrhaging in the grey matter (Fig. 10b) at the injury level as well as increased membrane permeability, but diffuse damage also extends rostrally and caudally in both the grey and white matter (Figs. 11b and 12b). As seen from the figure, tissue that is injured through tension and compression will result in a more diffuse damage pattern than with localized compression alone. A study by Fiford et al. yielded similar results by finding a larger spread of damage within the spinal cord due to a dislocation injury in the rat [69], when compared with contusion mechanism injury results. Although the injury mechanism investigated was a lateral dislocation rather than the anterior-posterior dislocation studied by Choo et al., both results indicated damage to the white and grey matter at the injury site (determined by quantification of hemorrhage and axonal transport dysfunction, respectively), with additional signs of damage extending rostrally and caudally from the level of injury. Furthermore, a study by Clarke et al. directly investigated the differences between the lateral and anterior-posterior dislocation mechanisms in a thoracolumbar spinal cord injury in rats, finding that a posterior dislocation resulted in significantly greater damage to the cord, represented by hemorrhaging and degeneration of neurons in the grey matter and more axonal transport dysfunction within the axons of the white matter [70].

The distraction mechanism does not yield any local compression of the cord as in the two mechanisms discussed previously and no ensuing hemorrhage is observed (Fig. 10c). Axial extension of the spinal column results in an excessive tensile load borne by the spinal cord. Rather than being anchored at a pinch point as in the dislocation mechanism, the anchor points during this injury are the nerve roots and possibly the denticulate ligaments. While it is possible that the meninges may bear more load in this mechanism since the entire cross-section of the cord is loaded via tension, the white and grey matter still experience diffuse injury similar to the rostral and caudal damage in the dislocation mechanism which is indicated by the increased permeability in cell bodies and axons, rostral and caudal to the injury epicenter (Figs. 11c and 12c).

Choo et al. quantified differences in damage patterns throughout the caudal, epicenter and rostral regions of the spinal cord due to the various injury mechanisms (Fig. 13). All indications of primary and secondary damage from this study were summarized in this format and the data strongly indicate that the tissues of the spinal cord have different responses to different imposed mechanical stimuli. This observation is also evident from the previously mentioned studies from Fiford et al. [69] and Clarke et al. [70], which indicated distinctly different biological responses from various injury mechanisms.

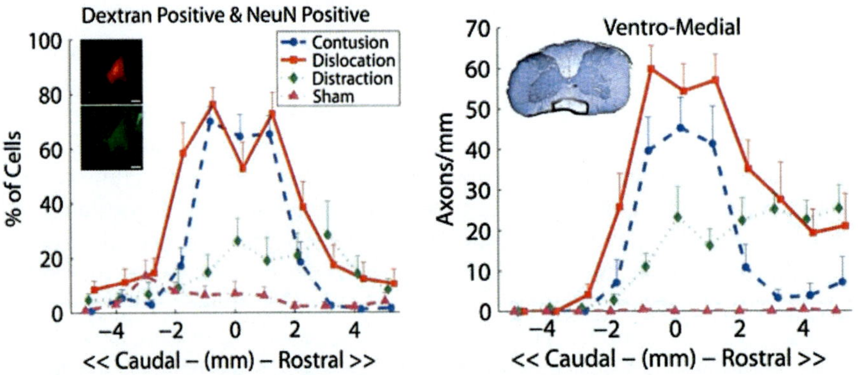

Fig. 13 Distribution of spinal cord cells that experienced increased membrane permeability categorized by injury mechanism [26]. Data presented from 5 mm caudal to 5 mm rostral (0 represents injury level). *Left*: Percentage of identified cell bodies in the ventral lobes of the grey matter. *Right*: Number of axons per mm (axial length) in the ventral column of white matter

6 Novel Injury Prevention Devices

The field of SCI biomechanics has also lead to innovations in the area of injury prevention, in an effort to prevent, mitigate or aid recovery from these injuries. One typical mechanism of cervical spinal cord injury that has been studied extensively is head-first impact. These injuries occur during diving into shallow water where the head strikes the pool bottom. They also occur during sports such as football or hockey where head-first collisions against other players or the boards (in hockey) or ground (in football) can occur. The type of injury that will result and the determinants of whether any spinal injury will result from these impacts are dependent on the spinal alignment, the angle of approach of the head with respect to the impact surface and other characteristics of the interface between the head and the surface it is striking. In an early investigation using full body cadavers wearing football helmets, Hodgson and Thomas [71] found it necessary to artificially create a fusion at the atlanto-occipital joint in order to constrain the spine to an aligned posture and thereby transmit sufficient force through the cervical spine to result in injury. Nightingale et al. showed that the cadaveric osseoligamentous cervical spine's axial stiffness increased with increasing constraint on the head [72]. In cadaveric head-first impact experiments in the same lab, some specimens avoided spine injury when the head translated and rotated along a low friction and inclined surface. Many specimens that impacted flat surfaces or padded surfaces in the same experiment developed a wide array of unstable fractures [65]. These results are consistent with the study of spinal cord occlusion as a function of constraint conducted by Zhu et al. as described previously.

A prevention strategy for compressive neck injury might therefore involve minimizing the friction of the impact surface thereby facilitating the bending of the spine away from a rigidly aligned posture. Halldin used a finite element model of a human head and cervical spine to evaluate this concept [73]. He simulated a head-first impact with an experimental automotive roof during rollover. The roof structure had an asymmetrical spring that deflected towards the front of the vehicle during head impact causing an occupant's head to translate anteriorly thereby bending the spine away from an aligned posture. Twenty-seven percent and 44% reductions in cervical spine axial force for perpendicular and -15° oblique impacts respectively were reported in this study.

Cripton and Nelson have used the concept of facilitating movement of the head during head-first impacts to develop a unique helmet that will passively guide the head in an anterior or posterior translation that will ideally reduce the likelihood of spinal cord injury (www.pronecktor.com, [74]). The helmet has been named PRO-NECK-TOR™ and it is a double-shell assembly that is functional as a rigid helmet during regular use, but it incorporates a guide mechanism that will deploy during head-first impacts above a pre-determined axial force injury threshold (inset of Fig. 14).

The helmet has been tested experimentally under head-first impact by fitting a version of the helmet that facilitates head rotation in the sagittal plane to a

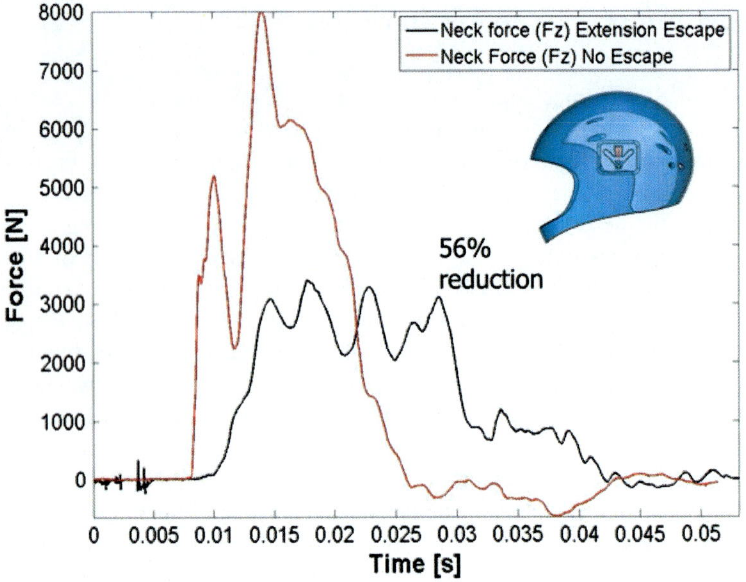

Fig. 14 Loading experienced at the neck during a head-first impact. When using the novel helmet design (*inset*) to facilitate extension of the head during impact, the axial load at the neck responsible for injury was dramatically decreased. [74]

surrogate head and neck assembly. Load data was measured at the lower neck and indicated that the maximum axial load on the neck during impact was 56% less than during impact without the custom helmet for one typical impact against a low-friction 15° angled impact surface (Fig. 14).

This work has large application to the sports industry in which helmets are already being used, especially those activities such as football and ice hockey where head-first impact occurs commonly.

Other devices have been proposed to prevent cervical spine and spinal cord injuries during head first impacts in sports. Many of these devices provide a mechanical linkage between the helmet and the shoulders themselves or to shoulder pads. The linkage has been created using stiffened collars between the helmet and shoulders, hydraulic mechanical struts, or with an overall solid structure attached to the shoulders. In all cases, this linkage is designed to act as a secondary load path for the force that is experienced during the head first impact. If the force is transmitted to the shoulders via the linkage then the cervical spine would experience a decreased force relative to what would be experienced without the linkage. There are several patents describing such devices but there is almost no published biomechanical test data available about these devices.

To our knowledge only one such device is being sold commercially and used in sports. That is the Leatt Brace which is gaining acceptance in motocross racing and which, according to the company's website, has been tested at testing facilities in Germany and South Africa. The Leatt Brace is a composite neck brace system which provides a mechanical stop to helmet motion when the head experiences an impact. The Leatt brace attempts to prevent neck injuries by limiting helmet motion in this way.

One challenge of the helmet-shoulder linkage or brace mechanisms described above is that they may limit the head motion that is necessary for the wearer to participate in the sport. Another challenge is the transmission of the helmet force to the shoulders since the soft tissue on the shoulder girdle and the shoulder girdle itself is compliant and will deflect when the force is transmitted from the helmet.

7 Summary

There is much effort in the field of spinal cord research to further our understanding of the biomechanical aspects of traumatic acute injury. Experimental models continue to be introduced and the methods of observation of spinal column/cord interactions and the ensuing pathophysiology are continually updated and expanded upon. Sub-categorization of acute SCI by injury mechanism may be important in our future ability to direct treatment to specific subsets of the heterogeneous SCI patient population. As clinically relevant injury mechanisms are further investigated, it is apparent that biomechanics and neurophysiology of SCI are intimately linked and must be simultaneously addressed to eventually take experimental results from research to clinical trials. Advancement in preventative strategies for SCI is clearly an area for further research and development.

Acknowledgments Funding from the Canada Research Chairs program, the Canadian Institutes for Health Research, the National Sciences and Engineering Research Council of Canada and the Rick Hansen Foundation for this research is gratefully acknowledged.

References

1. Sekhon, L.H., Fehlings, M.G.: Epidemiology, demographics, and pathophysiology of acute spinal cord injury. Spine (Phila. Pa 1976) **26**(24 Suppl), S2–S12 (2001)
2. Kattail, D., et al.: Epidemiology and clinical outcomes of acute spine trauma and spinal cord injury: experience from a specialized spine trauma center in Canada in comparison with a large national registry. J. Trauma **67**(5), 936–943 (2009)
3. Jackson, A.B., et al.: A demographic profile of new traumatic spinal cord injuries: change and stability over 30 years. Arch. Phys. Med. Rehabil. **85**(11), 1740–1748 (2004)
4. O'Connor, P.J.: Trends in spinal cord injury. Accid. Anal. Prev. **38**(1), 71–77 (2006)
5. Tator, C.H.: Epidemiology and general characteristics of the spinal cordinjured patient. In: Tator, C.H., Benzel, E.C. (eds.) Contemporary Management of Spinal Cord Injury: From Impact to Rehabilitation, 3rd ed edn, pp. 15–19. American Association of Neurological Surgeons, Park Ridge, IL (2000)
6. O'Connor, P.: Incidence and patterns of spinal cord injury in Australia. Accid. Anal. Prev. **34**(4), 405–415 (2002)

7. Pickett, G.E., et al.: Epidemiology of traumatic spinal cord injury in Canada. Spine (Phila. Pa 1976) **31**(7), 799–805 (2006)
8. Tator, C.H.: Spine-spinal cord relationships in spinal cord trauma. Clin. Neurosurg. **30**, 479–494 (1983)
9. Marar, B.C.: The pattern of neurological damage as an aid to the diagnosis of the mechanism in cervical-spine injuries. J. Bone Joint Surg. Am. **56**(8), 1648–1654 (1974)
10. Tator, C.H.: Review of treatment trials in human spinal cord injury: issues, difficulties, and recommendations. Neurosurgery **59**(5), 957–982 (2006). discussion 982-7
11. Kwon, B.K., et al.: Animal models used in spinal cord regeneration research. Spine (Phila. Pa 1976) **27**(14), 1504–1510 (2002)
12. Rosenzweig, E.S., McDonald, J.W.: Rodent models for treatment of spinal cord injury: research trends and progress towards useful repair. Curr. Opin. Neurol. **17**, 121–131 (2004)
13. Young, W.: Spinal cord contusion models. Prog. Brain Res. **137**, 231–255 (2002)
14. Hawryluk, G.W., et al.: Protection and repair of the injured spinal cord: a review of completed, ongoing, and planned clinical trials for acute spinal cord injury. Neurosurg. Focus **25**(5), E14 (2008)
15. Crowe, M.J., et al.: Apoptosis and delayed degeneration after spinal cord injury in rats and monkeys. Nat. Med. **3**(1), 73–76 (1997)
16. Ichihara, K., et al.: Gray matter of the bovine cervical spinal cord is mechanically more rigid and fragile than the white matter. J. Neurotrauma **18**(3), 361–367 (2001)
17. Ozawa, H., et al.: Comparison of spinal cord gray matter and white matter softness: measurement by pipette aspiration method. J. Neurosurg. **95**(2 Suppl), 221–224 (2001)
18. Arbogast, K.B., Margulies, S.S.: Material characterization of the brainstem from oscillatory shear tests. J. Biomech. **31**(9), 801–807 (1998)
19. Greaves, C.Y., et al.: A three-dimensional finite element model of the cervical spine with spinal cord: an investigation of three injury mechanisms. Ann. Biomed. Eng. **36**(3), 396–405 (2008)
20. Breig, A., el-Nadi, A.F.: Biomechanics of the cervical spinal cord. Relief of contact pressure on and overstretching of the spinal cord. Acta. Radiol. Diagn. (Stockh.) **4**(6), 602–624 (1966)
21. Breig, A.: Overstretching of and circumscribed pathological tension in the spinal cord–a basic cause of symptoms in cord disorders. J. Biomech. **3**(1), 7–9 (1970)
22. Silberstein, M., McLean, K.: Non-contiguous spinal injury: clinical and imaging features, and postulated mechanism. Paraplegia **32**(12), 817–823 (1994)
23. Behrmann, D.L., et al.: Spinal cord injury produced by consistent mechanical displacement of the cord in rats: behavioral and histologic analysis. J. Neurotrauma **9**(3), 197–217 (1992)
24. Wrathall, J.R.: Spinal cord injury models. J. Neurotrauma **9**(Suppl 1), S129–S134 (1992)
25. Basso, D.M., et al.: Graded histological and locomotor outcomes after spinal cord contusion using the NYU weight-drop device versus transection. Exp. Neurol. **139**(2), 244–256 (1996)
26. Choo, A.M., et al.: Contusion, dislocation, and distraction: primary hemorrhage and membrane permeability in distinct mechanisms of spinal cord injury. J. Neurosurg. Spine **6**(3), 255–266 (2007)
27. Saari, A., Itshayek, E., Nelson, T.S., Morley, P.L., Cripton P.A.: Spinal cord deformation during injury of the cervical spine in head-first impact. International Research Council on the Biomechanics of Impacts—2006 Annual Meeting, September 20–22 2006, Madrid, Spain
28. Kroeker, S.G., et al.: The development of an improved physical surrogate model of the human spinal cord–tension and transverse compression. J. Biomech. **42**(7), 878–883 (2009)
29. Kearney, P.A., et al.: Interaction of contact velocity and cord compression in determining the severity of spinal cord injury. J. Neurotrauma **5**(3), 187–208 (1988)
30. Nightingale, R.W., et al.: The influence of end condition on human cervical spine injury mechanisms. Society of Automotive Engineers Transactions, Paper 912915, 391–399 (1991)
31. Camacho, D.L., et al.: Surface friction in near-vertex head and neck impact increases risk of injury. J. Biomech. **32**(3), 293–301 (1999)
32. Zhu, Q., et al.: Translational constraint influences dynamic spinal canal occlusion of the thoracic spine: an in vitro experimental study. J. Biomech. **41**(1), 171–179 (2008)

33. Bilston, L.E., Thibault, L.E.: Biomechanics of cervical spinal cord injury in flexion and extension: A physical model to estimate spinal cord deformations. IJ Crash **2**(2), 207–218 (1997)
34. Maikos, J.T., et al.: Finite element analysis of spinal cord injury in the rat. J. Neurotrauma **25**(7), 795–816 (2008)
35. Russell, C., et al.: Creation of a finite element model of the rat cervical spine from magnetic resonance images. J. Biomech. **41**(S1), S516 (2008)
36. Coats, B., Margulies, S.S.: Material properties of porcine parietal cortex. J. Biomech. **39**(13), 2521–2525 (2006)
37. Oakland, R.J., et al.: The biomechanical response of spinal cord tissue to uniaxial loading. Proc. Inst. Mech. Eng. H. **220**(4), 489–492 (2006)
38. Metz, H., et al.: A comparison of the elasticity of live, dead, and fixed brain tissue. J. Biomech. **3**(4), 453–458 (1970)
39. Gruner, J.A.: A monitored contusion model of spinal cord injury in the rat. J. Neurotrauma **9**(2), 123–126 (1992). discussion 126-8
40. Jakeman, L.B., et al.: Traumatic spinal cord injury produced by controlled contusion in mouse. J. Neurotrauma **17**(4), 299–319 (2000)
41. Stokes, B.T., Jakeman, L.B.: Experimental modelling of human spinal cord injury: a model that crosses the species barrier and mimics the spectrum of human cytopathology. Spinal Cord **40**(3), 101–109 (2002)
42. Scheff, S.W., et al.: Experimental modeling of spinal cord injury: characterization of a force-defined injury device. J. Neurotrauma **20**(2), 179–193 (2003)
43. Ghasemlou, N., et al.: Tissue displacement and impact force are important contributors to outcome after spinal cord contusion injury. Exp. Neurol. **196**(1), 9–17 (2005)
44. Metz, G.A., et al.: Validation of the weight-drop contusion model in rats: a comparative study of human spinal cord injury. J. Neurotrauma **17**(1), 1–17 (2000)
45. Fleming, J.C., et al.: The cellular inflammatory response in human spinal cords after injury. Brain **129**(Pt 12), 3249–3269 (2006)
46. Maikos, J.T., Shreiber, D.I.: Immediate damage to the blood-spinal cord barrier due to mechanical trauma. J. Neurotrauma **24**(3), 492–507 (2007)
47. Dohrmann, G.J., et al.: Biomechanics of experimental spinal cord trauma. J. Neurosurg. **48**(6), 993–1001 (1978)
48. Bresnahan, J.C., et al.: A behavioral and anatomical analysis of spinal cord injury produced by a feedback-controlled impaction device. Exp. Neurol. **95**(3), 548–570 (1987)
49. Noyes, D.H.: Electromechanical impactor for producing experimental spinal cord injury in animals. Med. Biol. Eng. Comput. **25**(3), 335–340 (1987)
50. Noyes, D.H.: Correlation between parameters of spinal cord impact and resultant injury. Exp. Neurol. **95**(3), 535–547 (1987)
51. Somerson, S.K., Stokes, B.T.: Functional analysis of an electromechanical spinal cord injury device. Exp. Neurol. **96**(1), 82–96 (1987)
52. Anderson, T.E.: Spinal cord contusion injury: experimental dissociation of hemorrhagic necrosis and subacute loss of axonal conduction. J. Neurosurg. **62**(1), 115–119 (1985)
53. Kloos, A.D., et al.: Stepwise motor and all-or-none sensory recovery is associated with nonlinear sparing after incremental spinal cord injury in rats. Exp. Neurol. **191**(2), 251–265 (2005)
54. Choo, A.M., et al.: Modeling spinal cord contusion, dislocation, and distraction: characterization of vertebral clamps, injury severities, and node of Ranvier deformations. J. Neurosci. Methods **181**(1), 6–17 (2009)
55. Jones, C.F., et al.: The effect of cerebrospinal fluid on the biomechanics of spinal cord: an ex vivo bovine model using bovine and physical surrogate spinal cord. Spine (Phila. Pa 1976) **33**(17), E580–E588 (2008)
56. Sparrey, C.J., et al.: The distribution of tissue damage in the spinal cord is influenced by the contusion velocity. Spine (Phila. Pa 1976) **33**(22), E812–E819 (2008)

57. Rivlin, A.S., Tator, C.H.: Effect of duration of acute spinal cord compression in a new acute cord injury model in the rat. Surg. Neurol. **10**(1), 38–43 (1978)
58. Joshi, M., Fehlings, M.G.: Development and characterization of a novel, graded model of clip compressive spinal cord injury in the mouse: Part 1. Clip design, behavioral outcomes, and histopathology. J. Neurotrauma **19**(2), 175–190 (2002)
59. Gruner, J.A., et al.: Histological and functional evaluation of experimental spinal cord injury: evidence of a stepwise response to graded compression. Brain Res. **729**(1), 90–101 (1996)
60. Kim, J.H., et al.: Impact speed does not determine severity of spinal cord injury in mice with fixed impact displacement. J. Neurotrauma (electronic publication ahead of print) (2009)
61. Cargill 2nd, R.S., Thibault, L.E.: Acute alterations in [Ca2+]i in NG108–15 cells subjected to high strain rate deformation and chemical hypoxia: an in vitro model for neural trauma. J. Neurotrauma **13**(7), 395–407 (1996)
62. Geddes, D.M., et al.: Mechanical stretch to neurons results in a strain rate and magnitude-dependent increase in plasma membrane permeability. J. Neurotrauma **20**(10), 1039–1049 (2003)
63. LaPlaca, M.C., et al.: High rate shear strain of three-dimensional neural cell cultures: a new in vitro traumatic brain injury model. J. Biomech. **38**(5), 1093–1105 (2005)
64. Shi, R., Whitebone, J.: Conduction deficits and membrane disruption of spinal cord axons as a function of magnitude and rate of strain. J. Neurophysiol. **95**(6), 3384–3390 (2006)
65. Nightingale, R.W., et al.: Dynamic response of the head and cervical spine to axial impact loading. J. Biomech. **29**(3), 307–318 (1996)
66. Panjabi, M.M., et al.: Dynamic canal encroachment during thoracolumbar burst fractures. J. Spinal Disord. **8**(1), 39–48 (1995)
67. Wilcox, R.K., et al.: Measurement of canal occlusion during the thoracolumbar burst fracture process. J. Biomech. **35**(3), 381–384 (2002)
68. Blight, A.: Mechanical factors in experimental spinal cord injury. J. Am. Paraplegia Soc. **11**(2), 26–34 (1988)
69. Fiford, R.J., et al.: A vertebral dislocation model of spinal cord injury in rats. J. Neurotrauma **21**(4), 451–458 (2004)
70. Clarke, E.C., et al.: Anterior fracture-dislocation is more severe than lateral: a biomechanical and neuropathological comparison in rat thoracolumbar spine. J. Neurotrauma **25**(4), 371–383 (2008)
71. Hodgson, V.R., Thomas, L.M.: Mechanisms of cervical spine injury during impact to the protected head. 24th STAPP car crash conference, Warrendale, PA (1980)
72. Nightingale, R.W., et al.: The influence of end condition on human cervical spine injury mechanisms. Society of Automotive Engineers Transactions, Paper 912915, 391–399 (1991)
73. Halldin, P.: Investigation of conditions that affect neck compression-flexion injuries using numerical techniques. 44th STAPP car crash conference, SAE (2000)
74. Nelson, T.S., Cripton, P.A.: Inducing head motion with a novel helmet during head-first impact can mitigate neck injury metrics: an experimental proof-of-concept investigation using mechanical surrogates. Proceedings of the International Research Council on the Biomechanics of Impacts, Bern, Switzerland, 17–19 September 2008

Modelling of Spinal Cord Biomechanics: In Vitro and Computational Approaches

Cecilia Persson, Jon L. Summers and Richard M. Hall

Abstract In vivo testing of spinal cord biomechanics is surrounded with difficulties. This is due to methodological issues such as reproducibility and visualization problems as well as ethical concerns and the possible lack of correlation between animal models and the human tissue response. Therefore, a number of in vitro and computational studies have been performed in order to increase the understanding of the spinal cord's mechanical behaviour during trauma. Due to its well defined nature, the burst fracture and its effect on the spinal cord have been frequently studied, but investigations on other injuries such as distraction and dislocation are also cited in the literature. In vitro models have utilised a variety of materials including polymers and biological tissue from different animals. Similarly a variety of models have been used in computational studies to represent the cord which range in terms of both geometric complexity and material models. This chapter aims to provide a review of the methods and materials used in vitro and computational models of the spinal cord injury.

C. Persson (✉)
Applied Materials Science, Department of Engineering Sciences, Uppsala University, 751 21 Uppsala, Sweden
e-mail: Cecilia.Persson@angstrom.uu.se

J. L. Summers
Institute of Engineering Thermofluids, Surfaces and Interfaces, School of Mechanical Engineering, University of Leeds, Leeds LS2 9JT, UK
e-mail: J.L.Summers@leeds.ac.uk

R. M. Hall
Institute of Medical and Biological Engineering, School of Mechanical Engineering, University of Leeds, Leeds LS2 9JT, UK
e-mail: R.M.Hall@leeds.ac.uk

Stud Mechanobiol Tissue Eng Biomater (2011) 3: 181–201
DOI: 10.1007/8415_2010_38

Published Online: 30 November 2010

1 Introduction

Traumatic spinal cord injuries are typically caused by falls, traffic accidents, sports or violence and may be the result of different types of mechanical trauma affecting the cord, such as dislocations or burst fractures of vertebral bodies [79]. The initial mechanical insult is followed by a physiological cascade of events, frequently resulting in a neurological deficit [7, 59]. Recent in vivo studies have found a strong correlation between the type of mechanical insult and the resulting neurological damage [19–21], suggesting that the biomechanics of the injury are important in determining the evolution of the cord injury. However, the majority of in vitro and computational studies of spinal cord injury cited in the literature have concentrated on modelling the burst fracture; hence the emphasis is on this injury mechanism within this chapter. As noted, different animal cords as well as entirely artificial models have been used to represent the human spinal cord. There are several advantages related to this, but some issues arise due to potential differences in the mechanical response compared to the in vivo human situation. Unfortunately, the mechanical behaviour of the cord is not yet completely specified and there is a lack of knowledge of the behaviour at traumatic loading rates and large strains, and for different loading situations. This lack of knowledge leads to further issues related to the computational modelling of spinal cord injury, where there is no consensus on the types of models to be used or indeed how such models should be validated for a given set of aims. Whereas other chapters focus on the material properties of the spinal cord and its in vivo behaviour, this chapter is dedicated to the modelling of the cord biomechanics in vitro and computationally.

2 In Vitro Methods

2.1 The Burst Fracture

Since its introduction by Allen in 1911, the drop-weight (or weight-drop) technique has been widely used in attempts to characterise both spinal fractures and cord injuries. As the name suggests, this technique consists of dropping a standardised weight from a known height onto the subject of interest. The impact severity can then be expressed in terms of impact energy, i.e. the kinetic energy achieved by the falling weight at the moment of impact. This technique has been frequently used to impact directly onto the spinal cord in vivo testing [1, 24, 50, 57], but it has also been used on segments of the vertebral column to create burst fractures in vitro [17, 29, 66, 82, 85, 86], as illustrated in Fig. 1. A burst fracture of a vertebra is characterised by the impingement of a bone fragment on the spinal cord, which may cause neurological deficit [22]. This chapter will focus on the latter set of experiments in which the aim of the studies has been to determine the degree of canal compromise caused by the impact [17, 66, 82, 84, 85].

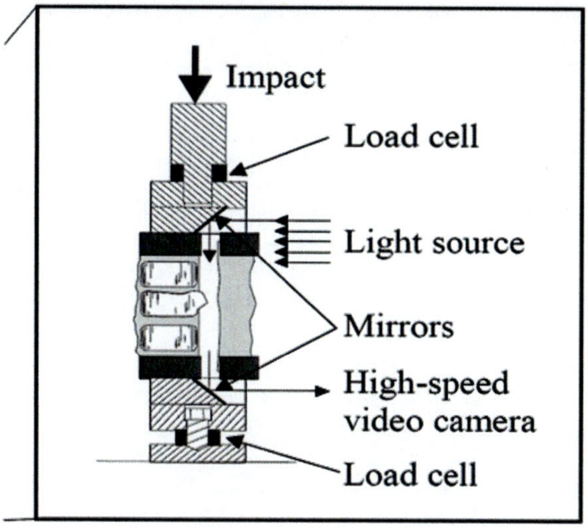

Fig. 1 The drop-weight technique used to simulate and study a burst fracture. Adapted from Wilcox et al. [84, 85]

As shown in Table 1, varying impact energies of between 20 and 140 J have been used to create these burst fractures. Panjabi et al. [66] found that depending on the amount of energy, different types of fractures could be produced: end-plate, wedge and burst fractures, where burst fractures were the most common type only at higher energies of impact (from 84 J and above). However, Panjabi et al. used incremental tests, starting with a lower weight which they gradually increased until the desired degree of fracture was obtained, which is not truly representative of a single impact traumatic event. Wilcox et al. [84, 85] used impact energies of between 20 and 140 J on calf thoracolumbar spine and the final fracture obtained was a burst fracture in every case. It was found that different energies produced different types of burst fractures: at lower impact energies (up to 60 J) mostly Denis type C fractures (fracture of the inferior end-plate [22]) were found compared to more type A fractures (fracture of both end-plates) at higher energies. Tran et al. [82] also used calf thoracolumbar spines, but found the same amount of canal occlusion as Wilcox et al. with lower impact energies. These differences could be due to differences in the rate of loading as well as differences between specimens due to age, breed and natural variability. Tran et al. used a materials testing system to generate low loading rate fractures and a drop-weight test rig to produce higher rate insults of a similar energy (33 and 35 J, respectively). It was found that all high loading rate tests produced burst fractures while the low loading rate tests produced compressive fractures with very little canal compromise.

All of these in vitro studies found a greater transient canal compromise than the final one (Table 1, Fig. 1). However, Panjabi et al. [66] used, as previously mentioned, incremental impacts on the same specimen, which does not reflect an actual injury situation. Chang et al. [16], Tran et al. [82], Ching et al. [18] and Carter et al. [15] successfully produced burst fractures in each single test. They all used the same method to assess the canal compromise: a plastic tube filled with water was placed in the spinal canal, and the change in fluid pressure during impact was

Table 1 In vitro studies assessing the canal compromise during and after a simulated burst fracture (drop-weight method)

Study	Type of vertebrae	Cord material and impact energy	Method of assessment	Results
Chang et al. [16]	Human cervical	Flexible plastic tube (Tygon) containing water/74-89 J	Record pressure changes in the tube, relate to the occlusion area and compare with pre- and post-fracture CT scans.	The transient compromise was higher than the post-fracture one (a recovery of 139% in canal area and 269% of the midsagittal diameter post-fracture). No correlation was found between transient and post-fracture canal geometry.
Panjabi et al. [66]	Human thora-columbar	No cord (empty canal)/ incremental 30–155 J	3 strain gauge transducers within the canal and pre- and post- fracture radiographs.	The transient compromise (> 33% of canal diameter) was significantly higher than the static one (18%) and no correlation was found between the two.
Tran et al. [82]	Calf thora-columbar	As Chang et al. [16]/33–35 J	As Chang et al. [16]	At similar impact energy, higher loading rates (creating burst fractures) gave a transient canal compromise of 47 ± 14% while lower loading rates (creating compression fractures) gave a transient occlusion of (7 ± 6%).
Ching et al. [18]	Human cervical	As Chang et al. [16]/approx 66 J	As Chang et al. [16]	Post-injury canal occlusion was found to increase when a burst-fractured cervical spine was placed into either extension or compression compared with a neutral position.

(continued)

Table 1 (continued)

Study	Type of vertebrae	Cord material and impact energy	Method of assessment	Results
Carter et al. [15]	Human cervical	As Chang et al. [16]/approx 66 J	As Chang et al. [16]	Higher loading rates (creating burst fractures) produced greater transient and static canal occlusions than lower loading rates (creating wedge-compression fractures). The transient canal occlusion was significantly higher than the post-fracture one for the high loading rates only.
Wilcox et al. [85]	Calf thora-columbar	No cord (empty canal)/60–140 J	High-speed video camera recording the transient and final canal occlusion, CT-scans before and after impact.	A high correlation was found between post-impact CT scans and the recorded final position (concordance coefficient of 0.94). The maximum transient canal occlusion was found to be greater than the final one in all cases.
Wilcox et al. [84]	Calf thora-columbar	No cord or synthetic cord (9% gelatine solution)/ 20–140 J	High-speed video camera (no cord), pressure transducer within the synthetic cord, finite element model.	An increase in impact energy gave an increase in canal compromise. The FE model showed that the spinal cord and the dura mater reduced the maximum displacement and increased the level of recoil.

correlated to the canal occlusion. Both Pintar et al. [71] and Raynak et al. [73] developed similar set-ups consisting of a plastic tube filled with saline solution, equipped with sensors to measure the canal occlusion. The set-up by Raynak et al. [73] was later used by Nuckley et al. [61] to assess the effect of maturity on the spinal canal occlusion during injury. They used a custom-built high-rate servohydraulic test frame rather than a drop-weight test, on baboon specimens. It was found that whereas the peak occlusion did not depend on spinal development, the occlusion just before failure decreased with an increase in age, providing a possible explanation for the relatively high incidence of spinal cord injuries without radiographic abnormality (SCIWORA) in paediatric injuries.

Only Wilcox et al. [84, 85] used a visual method of measurement (high speed video of the canal). However, since the cord had to be removed no effect of or on the cord or other spinal tissues could be assessed experimentally. A computational model indicated that the presence of spinal soft tissues in the canal would generally reduce the maximum occlusion of the canal [84]. Further, such a model had to consist of a pure axial load without forward flexion to prevent the apparent occlusion of the canal as the light was occluded due to the bending of the canal.

In conclusion, despite visualisation issues, these dynamic studies support the idea that if the neurological deficit is a consequence of the amount of occlusion of the canal, the dynamic processes at the moment of injury are likely to be more important than the final position of the bone fragment. Also, the type of fracture will depend on the amount of injury energy.

To gain a better appreciation of the actual fragment-cord interaction and overcome some of the issues described previously, Hall et al. [35] developed an in vitro model where a simulated bone fragment was transversely propelled, using a pneumatically controlled impactor, onto the detached bovine spinal cord. The event was recorded with a high-speed video camera to determine the maximum deformation of the cord as well as the duration of impact. The pressure inside the cord was also measured using catheter tip pressure transducers. The authors found a resemblance of the occlusion-time profile determined by Wilcox et al. [84] and an increase in pressure and deformation of the cord with an increasing velocity of the bone fragment. An important factor was found to be the posterior longitudinal ligament, and its state at impact. With a slack ligament, the maximum occlusion, time to maximum occlusion and peak pressure within the cord were all higher (increased by approximately 20%, 11%, and 78%, respectively) than that measured with a tense ligament (pre-strained to 14%) [35]. Without the ligament, all these parameters were even higher (increased by approximately 38%, 16% and 150%, respectively, compared to a tense ligament). However, only the pressure increases were statistically significant at the 0.05 significance level. The model also included the dura mater, for which no significant effect was found. However, the cerebrospinal fluid (CSF) was not included in the model, an element which may have a significant role in protecting the spinal cord. This was later confirmed in the study by Jones et al. [45], in which cerebrospinal fluid was incorporated into the model. Here a reduced maximum deformation of the cord and a longer duration of the impact were observed for those experiments which included the CSF. Jones et al. [45] also performed tests on surrogate cords, which do not experience the rapid degradation in mechanical properties post-retrieval, which is a common and a serious issue in ex vivo specimens [63]. The investigation concluded that these surrogate materials demonstrated similar trends in deformation behaviour compared to bovine cord over different impact conditions (with and without dura, with and without CSF). However, despite these surrogate cords having a similar longitudinal tensile modulus to that of in vivo cords at low strains and strain rates

[17, 38], substantial differences in the absolute deformation values were observed under transverse impact.

In all these ex vivo studies [35, 44, 45, 62] only one size of simulated bone fragment was used, whereas in reality it might vary with fracture energy and between as well as within subjects [23, 32, 77, 84]. Persson et al. [68] used a similar set-up by Hall et al. [35] with bone fragments of different impact areas and concluded that a smaller impact area would lead to a higher degree of spinal cord deformation. Furthermore, the fragment trajectory into the bovine model including the spinal cord, CSF and dura mater, was found to differ substantially from a surrogate model. In addition, the presence of the cerebrospinal fluid and the high fluid content of the spinal cord allow the system to dissipate energy on impact, which may be important to incorporate into computational models, where linear elastic material models have been previously used to represent the soft tissues [33, 62, 83].

2.2 *Other Causes for Spinal Cord Injury*

Other causes of spinal cord injury, such as distraction and fracture dislocations, have not been as extensively studied as the burst fracture. This section summarises a few studies conducted in order to examine spinal cord injury due to clinical treatment [5, 43], facet dislocation [41] and possible aggravation of an injury due to constraints such as automotive belts [12, 87].

Prompted by the observation of spinal cord injuries due to the use of e.g. distraction in the treatment of scoliosis [52], Jarzem et al. [43] and Bassi et al. [5] examined the relationship between cord distraction and cord interstitial pressure in vitro, using dogs, sheep [43] and pigs [5]. Both studies found a high degree of correlation between the applied strain and the interstitial pressure within the cord [5, 43], suggesting an increase in cord distraction may cause a serious reduction in blood flow within the cord. It was also found that the cervical segment was less prone to this mechanism compared to the lumbar segment, which in turn was less susceptible than the thoracic segment [5].

Ivancic et al. [41] studied the mechanism of cervical SCI due to bilateral facet dislocation using a custom made sled apparatus. Compressive muscle forces were simulated though cables attached to springs. They found, similarly to the burst fracture studies, that the dynamic canal occlusion (35–88% depending on original canal size) was significantly greater than the post-injury occlusion.

Several spinal cord injuries occur when the spine is constrained to some extent, such as during traffic accidents, where the safety belt prevents anterior movement of the upper body. However, these types of injuries have rarely been studied. Bilston and Thibault [12] used a plastic skeleton model together with a spinal cord model of silicone and Dacron fibres and simulated dynamic hyperextension and hyperflexion injuries. Strains as high as 15 and 35% were found for extension and flexion, respectively, suggesting temporary or permanent neurological damage could indeed result from this non contact loading [4, 47]. Zhu et al. [87] used the instrumented cord developed by Raynak et al. [73] together with an anteroposterior sliding system in cadaveric spines and found that constrained spines gave rise to dislocations and a substantially higher degree of canal occlusion compared to unconstrained spines. Finally, Nightingale et al. [58] impacted cadaveric heads with intact spines to T3-T4 onto padded and rigid surfaces and found that in the absence of padding, the head mass can itself confer sufficient constraint to cause spine injury.

3 In Vitro Materials

Some of the previously mentioned in vitro studies [35, 82, 84] used bovine spines rather than human ones. Further, in vivo spinal cord injury studies have mainly used smaller animals such as ferrets [2, 3] and rodents [20, 28, 42, 47, 50, 60]. The use of animals instead of humans was naturally unavoidable in the in vivo studies and regarding the in vitro studies there are several advantages of using animal specimens instead of human ones, despite the difference in anatomy. These include the limited availability of human specimens and the use of animal models can be a means of reducing costs, health risks and variation between specimens [84, 85].

Bovine spinal cord has been found to have similar geometry [37, 62] and material properties [11, 63] as the human spinal cord. For example, Oakland et al. [63] investigated the mechanical properties of the bovine spinal cord under uniaxial loading and found a tangent modulus of 1.19 ± 0.13 MPa, similar to that of the human spinal cord, which Bilston and Thibault [11] found to be 1.23 ± 0.51 MPa (at a similar strain rate). As regards the spinal dura mater, Runza et al. [78] performed uniaxial tensile tests on both bovine and human lumbar dura, which showed a similar tangent modulus and tension to failure. However, the bovine dura mater showed a higher failure strain than the human dura, suggesting there are some morphological differences between species. Although rats have been found to show a similar neurological damage pattern to humans [20, 46, 47], some biological reactions have been found to differ [34] and the size of the CSF layer, in both absolute and relative terms, is significantly smaller than that found in humans [37, 54].

In a further attempt to reduce specimen variation and difficulties in handling the material, surrogate materials have been used instead of biological tissues [15, 16, 18, 44, 45, 48, 61, 68, 73, 82, 87]. Surrogate materials that have been used mainly comprise a plastic tube (Tygon) filled with water or saline solution [15, 16, 18, 61, 73, 82, 87], gelatine [62] or a silicon rubber [44, 45, 48, 68]. Even though the material properties may differ considerably (they are usually isotropic rather than anisotropic as is the case of cord tissue), a surrogate material can be useful for preparative or comparative studies and for the early stages of verification of a computational model. Such surrogate models also allow easier instrumentation of the spinal cord/dura/CSF construct allowing for a more detailed analysis of the restriction of the canal during trauma within cadaveric models [15, 16, 18, 87].

When procuring human or animal samples the cerebrospinal fluid is lost in the process. An alternative to this fluid is saline solution: water containing 0.9% NaCl [25]. Both are Newtonian fluids and even though the CSF viscosity depends on the percentage of blood cells and proteins contained, it has been reported that high protein and cell concentrations do not significantly alter the CSF viscosity [13] and that the ratio of CSF viscosity to saline viscosity (relative viscosity) is 1.02 at both 20 and 37°C [25]. Also, the density is similar: at 37°C, a density of 1.0003–1.0007 g/ml has been reported for CSF [75] and 0.9995–0.9996 g/ml for saline solution [76], a maximum difference of 0.1%. In conclusion, saline solution appears to be a suitable substitute for CSF in vitro models.

4 Computational Methods

4.1 The Finite Element Model in Spinal Research

Computational analyses such as finite element (FE) analyses are used in many different fields of engineering. These types of analyses can be particularly useful in studying such a complex system as the spine, since it can facilitate the visualisation of the structural behaviour as well as permit the implementation of tests that are not feasible in vitro or in vivo due to e.g. experimental modelling difficulties or ethical issues. In their most useful form these models, when validated, should provide researchers with information to be utilized in experimental design such that a more complete cycle of validation and prediction is accomplished. The finite element method consists of dividing the structure of interest into smaller parts (elements) which are joined together at points (nodes) [74]. Equations of motion which approximate the behaviour of each part are then developed, assembled and solved simultaneously, which provide key issues in terms of accuracy and/or precision of such models. In spinal biomechanics, computational models have been used to study the behaviour of the spine when subjected to trauma or surgery as well as to simulate the intact and diseased spines at physiological loading rates [53].

However, there are several challenges involved with the computational modelling of spinal cord injury biomechanics:

- Validation and verification of the model may be difficult due to the quality of data available from both clinical and in vitro studies. (For instance, during in vitro experiments on cord-column interaction the deformation of the spinal cord is concealed by the surrounding dura mater—this may be overcome by the use of high speed X-ray cameras in tandem with special radio-opaque agents).
- A vertebral fracture leading to spinal cord injury is a dynamic process and it involves large material deformations over short time spans, all of which adds up to requiring a large number of solution iterations and time steps and thus demanding large computational resources.

- The correlation between the mechanical deformation of the cord and the biological damage to the system is difficult to assess: the cord is often treated as a homogeneous solid whereas in reality it is a complex system made up of components such as nerve cells, fibres and cerebrospinal fluid.
- Combining the CSF mechanics within the model brings about the need for computational simulations of SCI that can capture the fluid–structure interaction (FSI). The dynamics of the structure of the spinal cord is best described in the material (Lagrangian) frame of reference, whereas the CSF dynamics requires a laboratory (Eulerian) frame of reference. For this reason many FSI computational models use an Arbitrary Lagrangian–Eulerian (ALE) description which confines the CSF dynamical equations to the moving boundaries of the bounding structure of the spinal cord.

Despite the dynamic character of the event and the potential importance of the cerebrospinal fluid, most studies have used static solutions and excluded the CSF. However, with the continual increase in computational resources and improved numerical simulation technology, the interest in multiphysics problems such as FSI is gaining momentum [14]. Traditionally, fluid–structure interaction problems have been solved by eliminating one of the fields (solid or fluid), e.g. assuming a specific motion of the confining rigid structures to simulate fluid flow, and prescribing e.g. a pressure to a fluid to influence the structural deformations [6]. However, a fully coupled simulation (where the fluid and the solid system interact) may sometimes be necessary, e.g. when studying the spinal cord/CSF/dura construct during trauma. The finite element method is particularly useful for the simulation of deforming solids, since the computational mesh has the ability to deform. This method is now also well established for fluid problems, although in this case the mesh deforms according to the movement of the fluid boundaries rather than the fluid flow. The finite element method is therefore well suited for constructing an FSI model in order to study spinal cord biomechanics in situations where deformations are small, but the approach will have difficulties coping with large deformation and dislocation in capturing the dynamics of severe trauma.

It is perhaps worth noting that there are several different approaches to FSI modelling, but they can be categorized into one of two main approaches, namely monolithic (where the fluid, structure and interacting boundary are treated as one system of equations and solved simultaneously) or partitioned (where the solid and fluid mechanics equations of motion are solved separately with algebraic constraints on the interacting boundary) [36].

4.2 Computational Models of the Spinal Cord–Column Interaction

Computational models of the burst fracture have assisted in the understanding of the role of different components during the fracture. In terms of the mechanism of vertebral body bursting, the findings have been contradictory [72, 83]. Qiu et al.

[72] found that at maximum vertebral compression, the endplates bulged towards the vertebral body. Also, the highest effective stress was concentrated to the central part of the endplates. From this it was concluded that their results confirmed the hypothesis that under compression, nucleus material is forced into the vertebral body through a fractured endplate, increasing the pressure applied to it, and forcing the fat and marrow contents out of the cancellous bone. This process of increasing the intravertebral pressure creates hoop stresses in the vertebral wall, which subsequently bursts. However, this failed to explain the preferential posterior wall fracture. Contrary to the hypothesis of Qiu et al., Wilcox et al. [83] found in a combined experimental and computational study a reversed order mechanism: since the discs seemed to bulge more in the transverse direction than into the endplates, it was hypothesised that the vertebral body fractures first and the following expulsion of bone and fluid into the canal would reduce the support to the endplates, which would crack. They also suggested that the interaction of the articular processes produces highly localised strains in the posterior part of the vertebral body, providing an explanation as to why the fracture normally occurs in this region.

Computational models have also assisted in the understanding of other cord injury mechanisms: Li and Dai [49] studied hyperextension injuries and found high localised stresses in the grey matter and in particular in the anterior horn, in good agreement with clinical results where a predominance in hand weakness is often found in these types of injuries. Imajo et al. [40] concluded, also using an FE model, that the extent of the bony pincers mechanism which may cause SCIWORA (Spinal Cord Injury Without Radiologic Abnormality) depends on the degree of facet joint inclination. However, the spinal cord tissue was not incorporated in the model.

The incorporation of the spinal cord as well as other surrounding soft tissues may be important to the accuracy of the results. In fact, in the previously discussed FE model developed by Wilcox et al. [83] the importance of the posterior longitudinal ligament was elucidated, in that it limits the deformation of the cord and causes the retropulsion of the fragment back towards the vertebral body. Oakland [62] assessed the effect of the posterior longitudinal ligament in more detail, using a model of the transverse impact of a bone fragment and the detached spinal cord, validated with the experimental work previously described [35]. The model is illustrated in Fig. 2. It was found that the addition of the posterior longitudinal ligament decreased the peak pressure within the cord and that it appeared to distribute the load in the cord, away from the central impact position. However, the temporal features of the model were found to be different to the in vitro model, suggesting accurate modelling of material properties is important.

In fact, Bilston [9] found, using a two-dimensional model of the spinal cord behaviour during flexion, extension and compression, that a change in material properties from linear elastic to a more accurate hyperelastic model gave different results in terms of stresses and strains within the cord. In contrast, the viscoelastic time dependence did not significantly alter the results. This may be due to the chosen mathematical model: experimental studies have yet to show consistent

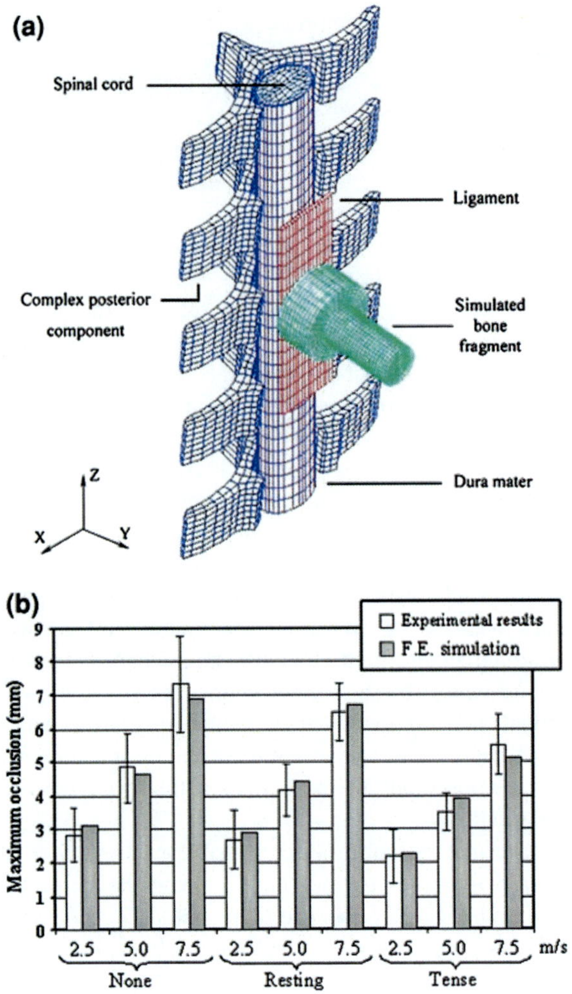

Fig. 2 The Finite Element model created by Oakland [62], studying the effect of the longitudinal ligament. **a** Model geometry and **b** Results showing the main occlusion of the canal, depending on fragment velocity and ligament status (none, resting, tense)

results regarding the effect of strain rate on the material properties of spinal tissues. Studies on the human spinal cord (in the longitudinal direction) [11] and the rat spinal cord [27, 81] have found a trend in increasing stiffness with strain rate, but considerable experimental variability and no statistical significance were reported in two of the three studies [11, 27]. Yet other studies on the human spinal cord [55] and the bovine dura mater [67] found no difference in stiffness due to a hundred-fold increase in strain rate. This indicates that the natural variability in these tissues may mask the effect of a more than hundred-fold increase in strain rate. Further studies are needed in order to provide reliable time-dependent models of the spinal tissues.

Greaves et al. [33] found—when simulating static compression of the cord—that for two different, homogeneous material models of the cord (one linear elastic and one hyperelastic Ogden model [64]) that there was a concentration of stresses peripherally in the cord, whereas for in vivo animal models the damage has been found to be initially concentrated to the grey matter [20, 47, 50]. Greaves suggested this may be due to "a mechanical or biological susceptibility of the grey matter", which may indicate the importance of modelling grey and white matter separately. Ichihara et al. [39] simulated dynamic mechanical compression to the anterior side of the cord at 1 mm/s and found, having modelled the white matter and the grey matter separately with material properties taken from a parallel experimental study, that the highest stresses were located in the grey matter, the grey matter being slightly stiffer than the white matter. Sparrey et al. [81] studied the spinal cord under compression, varying the tangent modulus of white and grey matter as well as the material model (a non-linear model was fitted to the tangent moduli). It was found that the outcome in terms of stresses and pressure within the cord was highly dependent on the material properties, and it was concluded that a change in tissue properties due to age or disease may have a substantial effect on spinal cord biomechanics and this should be taken into account in future research.

Although computational models are frequently used to find stress and strain distributions within the spinal components, very few models have correlated these to biological results. However, those who have, have found that regions of high effective stress [30, 54, 65] and axial [26, 30] or principal strain [54] correspond to the areas of greatest amount of neurological damage in the cord, determined by horseradish peroxidase exclusion [30, 65], extravasation patterns [54], and Hematoxylin and Eosin (H&E) staining [26] in rats [26, 54] or guinea pigs [30, 65]. Ouyang et al. [65] concluded that axon damage would occur at a von Mises stress of 2 kPa. However, axonal damage has been found to increase with strain rate [3] and Fiford's computational model was static and Ouyang et al. [65] performed their tests at non-traumatic strain rates (0.05 mm/s). Furthermore, as previously mentioned, a disadvantage of all of the above models is the absence of the cerebrospinal fluid. The cerebrospinal fluid may have an important effect on the deformation of the cord, as indicated in the experimental studies by Jones et al. [45] and Persson et al. [68]. Although the cerebrospinal fluid has been previously modelled computationally to study e.g. the physiological flow [51] or the origins of a certain disease [8], to the authors' knowledge it has very rarely been included for traumatic conditions. Maikos et al. [54] did model the CSF in their rat model, but not as an actual fluid: it was assigned solid properties with a very low stiffness (see Table 2). Furthermore, the CSF layer in rats is smaller (in relative terms) to that of humans [37, 54]. The only studies including the CSF as an actual fluid in a spinal trauma model were published by Persson et al. [69, 70]. In these studies, it was found, contrary to prior assumptions, that the cord undergoes substantial deformation before complete subdural collapse of the surrounding CSF layer, and that a thicker CSF layer both reduces and alters the displacement pattern of the cord, as shown in Fig. 3. This suggests that the thinner CSF layer in the commonly used rodent models may be a limitation when comparing the results of these studies to a human in vivo situation.

Table 2 Material models used in computational models of spinal cord injuries

Study	Component	Material model	Stress–strain relationship	Poisson's ratio
Bilston [9]	Spinal cord	Linear elastic	E = 1.2 MPa	0.49
		Hyperelastic (Ogden)	G = 100 kPa, $\alpha = 25$	0.49
		Viscoelastic	$G_\infty = 100$ kPa, $\alpha = 25$ $G_1 = 154$ kPa $G_2 = 179$ kPa $G_3 = 559$ kPa $\tau_1 = 4.38$ s $\tau_2 = 0.554$ s $\tau_3 = 234.4$ s	0.49
Ichihara et al. [39]	Spinal cord: White matter	Viscoelastic model (Standard linear solid model with nonlinear component)	$E_e = 0.28$ MPa $E_c = 0.15$ MPa $C_1 = 6.1x10^5$ $C_2 = 1.9x10^4$	0.4
	Grey matter		$E_e = 0.66$ MPa $E_c = 0.36$ MPa $C_1 = 2.2x10^6$ $C_2 = -1.6x10^4$	0.4
Oakland [62]	Spinal cord	Linear elastic	E = 1.25 MPa	0.49
	Dura Mater	Linear elastic	E = 1.44 MPa	0.49
Wilcox et al. [83]	Spinal cord	Linear elastic	E = 1.3 MPa	0.35
	Dura mater	Anisotropic elastic	$E_{rr} = 142$ MPa $E_{\theta\theta} = 142$ MPa $E_{zz} = 0.7$ MPa	N/A
Fiford [26]	Spinal cord	Hyperelastic (Ogden)	G = 176 kPa, $\alpha = 47$	N/A
Galle et al. [30] and Ouyang et al. [65]	Spinal cord White matter	Hyperelastic (Mooney-Rivlin)	$C_{10} = 592$ Pa $C_{01} = 249$ Pa	N/A
Greaves et al. [33]	Spinal cord	Linear elastic	E = 0.26 MPa	0.49
	Dura mater	Linear elastic	E = 5 MPa	0.45
Maikos et al. [54]	Spinal cord	Hyperelastic (Ogden) combined with viscoelastic (two-term Prony series decay)	$G_\infty = 32$ kPa, $\alpha = 4.7$ $G_1 = 99.4$ kPa, $G_2 = 56.8$ kPa, $\tau_1 = 8$ ms, $\tau_2 = 150$ ms	0.45
	Dura mater	As above but four-term Prony series decay	$G_\infty = 1205$ kPa, $\alpha = 16.2$ $G_1 = 1069$ kPa $G_2 = 416$ kPa $G_3 = 335$ kPa $G_4 = 335$ kPa $\tau_1 = 9$ ms $\tau_2 = 81$ ms $\tau_3 = 0.564$ ms $\tau_4 = 4.69$ ms	0.45
	CSF	Hyper-elastic (Mooney-Rivlin)	G = 134 Pa $C_{01} = C_{10} = 33.5$ Pa	N/A

(continued)

Table 2 (continued)

Study	Component	Material model	Stress–strain relationship	Poisson's ratio
Persson et al. [68–70]	Spinal cord	Hyperelastic (Ogden)	$G = 9$ kPa, $\alpha = 9$	0.4
	Dura mater	Linear elastic	$E = 80$ MPa	0.49
	CSF	Newtonian	$\eta = 0.001$Pas	N/A
Sparrey et al. [80]	Spinal cord:			
	White matter	Hyperelastic, Linear elastic*	E = 0.065, 0.09, 0.115, 0.14 and 0.165 MPa	N/A
	Grey matter	Hyperelastic, Linear elastic*	As for the white matter	
	Pia mater	Linear elastic	E = 0.6, 1.2, 1.8, 2.4 and 3 MPa	
Li and Dai [49]	Spinal cord:			
	White matter	Linear elastic	E = 0.277 MPa	0.4
	Grey matter	Linear elastic	E = 0.656 MPa	0.4

*Linear elastic models with different tangent moduli were used, as well as hyperelastic models with parameters based on the fit of a third-order hyperelastic Ogden model to the linear elastic tangent modulus

Another fluid which may have a significant effect on the mechanical response is the blood present in the tissue. In fact, Bilston [10] found in her computational model that the relative amount of blood and the perfusion pressure would have a significant effect on brain tissue deformation. However, a porcine in vivo study found no effect of the perfusion pressure on the mechanical properties of the brain [31].

4.3 Computational Material Models

The material models used for the spinal cord and paraspinal tissues have varied considerably. Table 2 provides a summary of the models used. Although the natural spinal cord presents nonlinear, anisotropic, viscoelastic mechanical properties (see chapter on spinal cord mechanical properties), very few studies have incorporated all factors into the models, most likely due to software limitations, computational costs and the increased complexity in terms of mathematical modelling. Furthermore, the mechanical properties of the soft spinal tissues are not yet completely mapped, as discussed in the chapter on spinal cord mechanical properties. There is a lack of data at traumatic strain rates, under certain loading conditions and in certain loading directions. In fact, in the study by Maikos et al. where a drop-weight test on rats was simulated computationally [54], the mechanical properties of the cord in the model had to be adjusted to fit the experimental trajectory, even though the hyperelastic properties were taken from

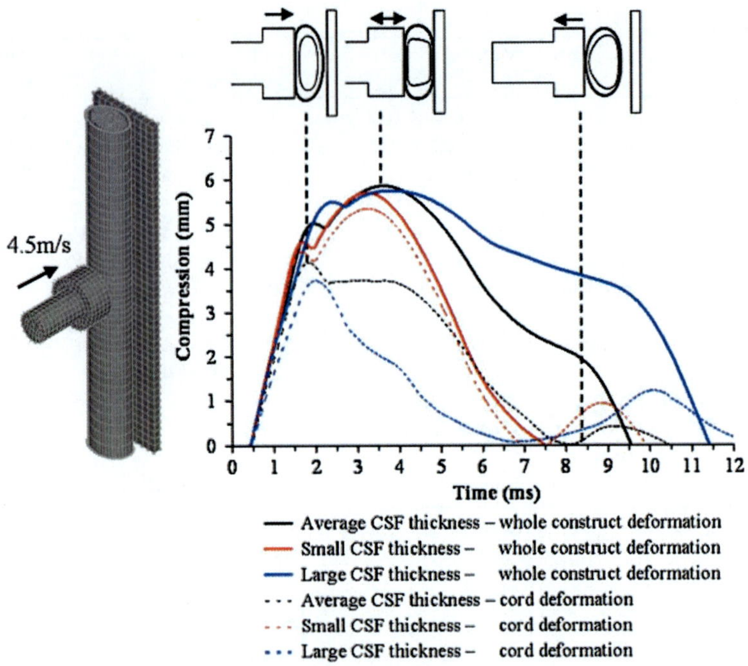

Fig. 3 The Fluid–Structure Interaction model by Persson et al. [69, 70] showed that the cord starts deforming immediately after the impact of the bone fragment with the dura/CSF/cord construct, and that a thicker CSF layer (blue lines) reduces the amount and duration of cord deformation (dashed lines). The cross-sections represent the base model with the average sized bovine CSF layer

an in vitro study on the same species and the model included the viscoelastic response (the viscoelastic model was taken from a study on brain tissue [56]).

Finally, in order to be able to correlate the mechanical process to the biological one, a different approach may be necessary where cells and their reactions are incorporated into the models.

5 Future Work

Although there is a substantial amount of research on simulating spinal cord injury in terms of the mechanical process, the work is far from complete. The neurological damage has been found to depend on the mechanics of the injury [20, 21] and there are several types of injuries that remain to be studied in detail; the burst fracture has been given significantly more attention than other types of injury mechanisms.

Greater clarity is needed on the constitutive relationships that govern the behaviour of the spinal cord under both physiological and injury scenarios (e.g. the influence of strain rate has not been completely clarified albeit repeatedly studied).

Significant improvements in our understanding of the role of the CSF and fluid components of the cord itself are required if the models are going to mimic the complex, interactive behaviour of the cord-CSF-dura complex. Recent studies indicate that the effect of the tissue fluid may be important to the cord deformation pattern.

Biomechanical verification of the computational models requires substantial investment if they are to be accepted as part of the necessary research provision. Here the sensitivity of the models to the boundary conditions as well as the variability in properties of the tissues is required. Initially, the use of surrogate models similar to those developed by Jones et al. [45] have a key role to play in providing the necessary data with which we can determine the suitability or otherwise of the computational models. A key element in this approach is to record the internal deformation of the cord within the dura-CSF during both traumatic and physiological activities.

If these computational models are to be of clinical utility—including animal models—then a greater understanding is required of the link between the mechanical insult experienced by the tissues and the subsequent degradation in performance; this may be correlated to a tissue tolerance. In vivo models combined with multi-scale computational models, where cell reactions can be incorporated are of particular interest for the advancement of our understanding of the injury process.

Finally, future work should also be directed onto new computational formulations that overcome the deficiencies encountered in more traditional FEM, such as large deformations and fluid structure interaction.

References

1. Allen, A.R.: Surgery of experimental lesion of spinal cord equivalent to crush injury of fracture dislocation of spinal column. J. Am. Med. Assoc. **57**, 878–880 (1911)
2. Anderson, T.E.: A controlled pneumatic technique for experimental spinal cord contusion. J. Neurosci. Methods **6**, 327–333 (1982)
3. Anderson, T.E.: Spinal cord contusion injury: experimental dissociation of hemorrhagic necrosis and subacute loss of axonal conduction. J. Neurosurg. **62**, 115–119 (1985)
4. Bain, A.C., Meaney, D.F.: Tissue-level thresholds for axonal damage in an experimental model of central nervous system white matter injury. J. Biomech. Eng. **122**, 615–622 (2000)
5. Bassi, M., Jarzem, P.F., Steibel, M., Barriga, P., Ouellet, J., Reindl, R.: An in vitro study on the effects of freezing, spine segment, repeat measurement, and individual cord properties on cord interstitial pressure. Spine (Phila Pa 1976) **34**, 351–355 (2009)
6. Bathe, K.J., Zhang, H.: Finite element developments for general fluid flows with structural interactions. Int. J. Numer. Method Eng. **60**, 213–232 (2004)
7. Bensch, F.V., Koivikko, M.P., Kiuru, M.J., Koskinen, S.K.: The incidence and distribution of burst fractures. Emerg. Radiol. **12**, 124–129 (2006)

8. Bertram, C.D., Brodbelt, A.R., Stoodley, M.A.: The origins of syringomyelia: numerical models of fluid/structure interactions in the spinal cord. J. Biomech. Eng. **127**, 1099–1109 (2005)
9. Bilston, L.E.: Finite element analysis of some cervical spinal cord injury modes, IRCOBI Conference, Göteborg (1998)
10. Bilston, L.E.: The effect of perfusion on soft tissue mechanical properties: a computational model. Comput. Methods Biomech. Biomed. Eng. **5**, 283–290 (2002)
11. Bilston, L.E., Thibault, L.E.: The mechanical properties of the human cervical spinal cord in vitro. Ann. Biomed. Eng. **24**, 67–74 (1996)
12. Bilston, L.E., Thibault, L.E.: Biomechanics of cervical spinal cord injury in flexion and extension: a physical model to estimate spinal cord deformations. Int. J. Crashworthiness **2**, 207–218 (1997)
13. Bloomfield, I.G., Johnston, I.H., Bilston, L.E.: Effects of proteins, blood cells and glucose on the viscosity of cerebrospinal fluid. Pediatr. Neurosurg. **28**, 246–251 (1998)
14. Bungartz, H.-J., Schäfer, M.: Fluid-Structure Interaction: modelling, simulation, optimization. Berlin, Springer (2006)
15. Carter, J.W., Mirza, S.K., Tencer, A.F., Ching, R.P.: Canal geometry changes associated with axial compressive cervical spine fracture. Spine **25**, 46–54 (2000)
16. Chang, G.L., Hung, T.K., Bleyaert, A., Jannetta, P.J.: Stress-strain measurement of the spinal cord of puppies and their neurological evaluation. J. Trauma **21**, 807–810 (1981)
17. Chang, D.G., Tencer, A.F., Ching, R.P., Treece, B., Senft, D., Anderson, P.A.: Geometric changes in the cervical spinal canal during impact. Spine **19**, 973–980 (1994)
18. Ching, R.P., Watson, N.A., Carter, J.W., Tencer, A.F.: The effect of post-injury spinal position on canal occlusion in a cervical spine burst fracture model. Spine **22**, 1710–1715 (1997)
19. Choo, A.M., Liu, J., Dvorak, M., Tetzlaff, W., Oxland, T.R.: Secondary pathology following contusion, dislocation, and distraction spinal cord injuries. Exp. Neurol. **212**, 490-506 (2008)
20. Choo, A.M., Liu, J., Lam, C.K., Dvorak, M., Tetzlaff, W., Oxland, T.R.: Contusion, dislocation, and distraction: primary hemorrhage and membrane permeability in distinct mechanisms of spinal cord injury. J. Neurosurg. Spine **6**, 255–266 (2007)
21. Clarke, E.C., Choo, A.M., Liu, J., Lam, C.K., Bilston, L.E., Tetzlaff, W., Oxland, T.R.: Anterior fracture-dislocation is more severe than lateral: a biomechanical and neuropathological comparison in rat thoracolumbar spine. J. Neurotrauma **25**, 371–383 (2008)
22. Denis, F.: The three column spine and its significance in the classification of acute thoracolumbar spinal injuries. Spine **8**, 817–831 (1983)
23. Duan, Y., Wang, X.F., Evans, A., Seeman, E.: Structural and biomechanical basis of racial and sex differences in vertebral fragility in Chinese and Caucasians. Bone **36**, 987–998 (2005)
24. Ducker, T.B., Kindt, G.W., Kempf, L.G.: Pathological findings in acute experimental spinal cord trauma. J. Neurosurg. **35**, 700–708 (1971)
25. Ellis, R.W., 3rd, Strauss, L.C., Wiley, J.M., Killmond, T.M., Ellis, R.W., Jr.: A simple method of estimating cerebrospinal fluid pressure during lumbar puncture. Pediatrics **89**, 895–897 (1992)
26. Fiford, R.J.: Biomechanics of spinal cord injury in a novel rat model. School of Aerospace, Mechanical and Mechatronic Engineering. University of Sydney, Sydney, PhD (2005)
27. Fiford, R.J., Bilston, L.E.: The mechanical properties of rat spinal cord in vitro. J. Biomech. **38**, 1509–1515 (2005)
28. Fiford, R.J., Bilston, L.E., Waite, P., Lu, J.: A vertebral dislocation model of spinal cord injury in rats. J. Neurotrauma **21**, 451–458 (2004)
29. Fredrickson, B.E., Edwards, W.T., Rauschning, W., Bayley, J.C., Yuan, H.A.: Vertebral burst fractures: an experimental, morphologic, and radiographic study. Spine **17**, 1012–1021 (1992)

30. Galle, B., Ouyang, H., Shi, R., Nauman, E.: Correlations between tissue-level stresses and strains and cellular damage within the guinea pig spinal cord white matter. J. Biomech. **40**, 3029–3033 (2007)
31. Gefen, A., Margulies, S.S.: Are in vivo and in situ brain tissues mechanically similar? J. Biomech. **37**, 1339–1352 (2004)
32. Grave, B., Brown, T., Townsend, G.: Comparison of cervicovertebral dimensions in Australian Aborigines and Caucasians. Eur. J. Orthod. **21**, 127–135 (1999)
33. Greaves, C.Y., Gadala, M.S., Oxland, T.R.: A three-dimensional finite element model of the cervical spine with spinal cord: an investigation of three injury mechanisms. Ann. Biomed. Eng. **36**, 396–405 (2008)
34. Hagg, T., Oudega, M.: Degenerative and spontaneous regenerative processes after spinal cord injury. J. Neurotrauma **23**, 264–280 (2006)
35. Hall, R.M., Oakland, R.J., Wilcox, R.K., Barton, D.C.: Spinal cord-fragment interactions following burst fracture: an in vitro model. J. Neurosurg. Spine **5**, 243–250 (2006)
36. Heil, M., Hazel, A., Boyle, J.: Solvers for large-displacement fluid–structure interaction problems: segregated versus monolithic approaches. Comput. Mech. **43**, 91–101 (2008)
37. Holsheimer, J., den Boer, J.A., Struijk, J.J., Rozeboom, A.R.: MR assessment of the normal position of the spinal cord in the spinal canal. AJNR Am. J. Neuroradiol. **15**, 951–959 (1994)
38. Hung, T.K., Chang, G.L.: Biomechanical and neurological response of the spinal cord of a puppy to uniaxial tension. J. Biomech. Eng. **103**, 43–47 (1981)
39. Ichihara, K., Taguchi, T., Sakuramoto, I., Kawano, S., Kawai, S.: Mechanism of the spinal cord injury and the cervical spondylotic myelopathy: new approach based on the mechanical features of the spinal cord white and gray matter. J. Neurosurg. **99**, 278–285 (2003)
40. Imajo, Y., Hiiragi, I., Kato, Y., Taguchi, T.: Use of the finite element method to study the mechanism of spinal cord injury without radiological abnormality in the cervical spine. Spine (Phila Pa 1976) **34**, E83–E87 (2009)
41. Ivancic, P.C., Pearson, A.M., Tominaga, Y., Simpson, A.K., Yue, J.J., Panjabi, M.M.: Mechanism of cervical spinal cord injury during bilateral facet dislocation. Spine (Phila Pa 1976) **32**, 2467–2473 (2007)
42. Jakeman, L.B., Guan, Z., Wei, P., Ponnappan, R., Dzwonczyk, R., Popovich, P.G., Stokes, B.T.: Traumatic spinal cord injury produced by controlled contusion in mouse. J. Neurotrauma **17**, 299–319 (2000)
43. Jarzem, P.F., Kostuik, J.P., Filiaggi, M., Doyle, D.J., Ethier, R., Tator, C.H.: Spinal cord distraction: an in vitro study of length, tension, and tissue pressure. J. Spinal Disord. **4**, 177–182 (1991)
44. Jones, C.F.: The effect of cerebrospinal fluid on the biomechanics of spinal cord. School of Mechanical Engineering. University of Leeds, Leeds, MSc (2005)
45. Jones, C.F., Kroeker, S.G., Cripton, P.A., Hall, R.M.: The effect of cerebrospinal fluid on the biomechanics of spinal cord: an ex vivo bovine model using bovine and physical surrogate spinal cord. Spine **33**, E580–E588 (2008)
46. Kakulas, B.A.: Pathology of spinal injuries. Cent. Nerv. Syst. Trauma **1**, 117–129 (1984)
47. Kloos, A.D., Fisher, L.C., Detloff, M.R., Hassenzahl, D.L., Basso, D.M.: Stepwise motor and all-or-none sensory recovery is associated with nonlinear sparing after incremental spinal cord injury in rats. Exp. Neurol. **191**, 251–265 (2005)
48. Kroeker, S.G., Morley, P.L., Jones, C.F., Bilston, L.E., Cripton, P.A.: The development of an improved physical surrogate model of the human spinal cord–tension and transverse compression. J. Biomech. **42**, 878–883 (2009)
49. Li, X.F., Dai, L.Y.: Three-dimensional finite element model of the cervical spinal cord: preliminary results of injury mechanism analysis. Spine(Phila Pa 1976) **34**, 1140–1147 (2009)
50. Liu, X.Z., Xu, X.M., Hu, R., Du, C., Zhang, S.X., McDonald, J.W., Dong, H.X., Wu, Y.J., Fan, G.S., Jacquin, M.F., Hsu, C.Y., Choi, D.W.: Neuronal and glial apoptosis after traumatic spinal cord injury. J. Neurosci. **17**, 5395–5406 (1997)

51. Loth, F., Yardimci, M.A., Alperin, N.: Hydrodynamic modeling of cerebrospinal fluid motion within the spinal cavity. J. Biomech. Eng. **123**, 71–79 (2001)
52. MacEwen, G.D., Bunnell, W.P., Sriram, K.: Acute neurological complications in the treatment of scoliosis. A report of the Scoliosis Research Society. J. Bone Joint Surg. Am. **57**, 404–408 (1975)
53. Mackerle, J.: Finite element modeling and simulations in orthopedics: a bibliography 1998–2005. Comput. Methods Biomech. Biomed. Eng. **9**, 149–199 (2006)
54. Maikos, J.T., Qian, Z., Metaxas, D., Shreiber, D.I.: Finite element analysis of spinal cord injury in the rat. J. Neurotrauma **25**, 795–816 (2008)
55. Mazuchowski, E.L., Thibault, L.E.: Biomechanical properties of the human spinal cord and pia mater. Summer Bioengineering Conference, Sonesta Beach Resort in Key Biscayne, Florida (2003)
56. Mendis, K.K., Stalnaker, R.L., Advani, S.H.: A constitutive relationship for large deformation finite element modeling of brain tissue. J. Biomech. Eng. **117**, 279–285 (1995)
57. Molt, J.T., Nelson, L.R., Poulos, D.A., Bourke, R.S.: Analysis and measurement of some sources of variability in experimental spinal cord trauma. J. Neurosurg. **50**, 784–791 (1979)
58. Nightingale, R.W., McElhaney, J.H., Richardson, W.J., Myers, B.S.: Dynamic responses of the head and cervical spine to axial impact loading. J. Biomech. **29**, 307–318 (1996)
59. Norenberg, M.D., Smith, J., Marcillo, A.: The pathology of human spinal cord injury: defining the problems. J. Neurotrauma **21**, 429–440 (2004)
60. Noyes, D.H.: Electromechanical impactor for producing experimental spinal cord injury in animals. Med. Biol. Eng. Comput. **25**, 335–340 (1987)
61. Nuckley, D.J., Van Nausdle, J.A., Eck, M.P., Ching, R.P.: Neural space and biomechanical integrity of the developing cervical spine in compression. Spine (Phila Pa 1976) **32**, E181–E187 (2007)
62. Oakland, R.J.: A biomechanical study of the spinal cord in the burst fracture process. School of Mechanical Engineering. University of Leeds, Leeds, PhD (2003)
63. Oakland, R.J., Hall, R.M., Wilcox, R.K., Barton, D.C.: The biomechanical response of spinal cord tissue to uniaxial loading. Proc. Inst. Mech. Eng. [H] **220**, 489–492 (2006)
64. Ogden, R.W.: Large deformation isotropic elasticity—on the correlation of theory and experiment for incompressible rubberlike solids. Proc. R. Soc. Lond. Ser. A Math. Phys. Sci. **326**, 565–584 (1972)
65. Ouyang, H., Galle, B., Li, J., Nauman, E., Shi, R.: Biomechanics of spinal cord injury: a multimodal investigation using ex vivo guinea pig spinal cord white matter. J. Neurotrauma **25**, 19–29 (2008)
66. Panjabi, M.M., Kifune, M., Wen, L., Arand, M., Oxland, T.R., Lin, R.M., Yoon, W.S., Vasavada, A.: Dynamic canal encroachment during thoracolumbar burst fractures. J. Spinal Disord. **8**, 39-48 (1995)
67. Persson, C., Evans, S., Marsh, R., Summers, J.L., Hall, R.M.: Poisson's ratio and strain rate dependency of the constitutive behavior of spinal dura mater. Ann. Biomed. Eng. 38, 975–983 (2010)
68. Persson, C., McLure, S.W., Summers, J., Hall, R.M.: The effect of bone fragment size and cerebrospinal fluid on spinal cord deformation during trauma: an ex vivo study. J. Neurosurg. Spine **10**, 315–323 (2009)
69. Persson, C., Summers, J., Hall, R.M.: The effect of cerebrospinal fluid thickness on traumatic spinal cord deformation. 7th Australasian Biomechanics Conference, Parklands, Gold Coast (2009)
70. Persson, C., Summers, J., Hall, R.M.: The role of the cerebrospinal fluid in traumatic spinal cord injuries. 17th Annual Symposium on Computational Methods in Orthopaedic Biomechanics, Las Vegas (2009)
71. Pintar, F.A., Schlick, M.B., Yoganandan, N., Maiman, D.J.: Instrumented artificial spinal cord for human cervical pressure measurement. Biomed. Mater. Eng. **6**, 219–229 (1996)
72. Qiu, T.X., Tan, K.W., Lee, V.S., Teo, E.C.: Investigation of thoracolumbar T12-L1 burst fracture mechanism using finite element method. Med. Eng. Phys. **28**, 656–664 (2006)

73. Raynak, G.C., Nuckley, D.J., Tencer, A.F., Ching, R.P.: Transducers for dynamic measurement of spine neural-space occlusions. J. Biomech. Eng. **120**, 787–791 (1998)
74. Reddy, J.N.: An Introduction to the Finite Element Method. New York, McGraw-Hill (1993)
75. Richardson, M.G., Wissler, R.N.: Density of lumbar cerebrospinal fluid in pregnant and nonpregnant humans. Anesthesiology **85**, 326–330 (1996)
76. Richardson, M.G., Wissler, R.N.: Densities of dextrose-free intrathecal local anesthetics, opioids, and combinations measured at 37 degrees C. Anesth. Analg. **84**, 95–99 (1997)
77. Ross, P.D., Davis, J.W., Epstein, R.S., Wasnich, R.D.: Ability of vertebral dimensions from a single radiograph to identify fractures. Calcif. Tissue Int. **51**, 95–99 (1992)
78. Runza, M., Pietrabissa, R., Mantero, S., Albani, A., Quaglini, V., Contro, R.: Lumbar dura mater biomechanics: experimental characterization and scanning electron microscopy observations. Anesth. Analg. **88**, 1317–1321 (1999)
79. Sekhon, L.H., Fehlings, M.G.: Epidemiology, demographics, and pathophysiology of acute spinal cord injury. Spine **26**, S2–S12 (2001)
80. Sparrey, C.J.M., Choo, A.M.P., Liu, J.M.D., Tetzlaff, W.M.D.P., Oxland, T.R.P.: The distribution of tissue damage in the spinal cord is influenced by the contusion velocity. Spine **33**, E812–E819 (2008)
81. Sparrey, C.J., Manley, G.T., Keaveny, T.M.: Effects of white, grey, and pia mater properties on tissue level stresses and strains in the compressed spinal cord. J. Neurotrauma **26**, 585–595 (2009)
82. Tran, N.T., Watson, N.A., Tencer, A.F., Ching, R.P., Anderson, P.A.: Mechanism of the burst fracture in the thoracolumbar spine. The effect of loading rate. Spine **20**, 1984–1988 (1995)
83. Wilcox, R.K., Allen, D.J., Hall, R.M., Limb, D., Barton, D.C., Dickson, R.A.: A dynamic investigation of the burst fracture process using a combined experimental and finite element approach. Eur. Spine J. **13**, 481–488 (2004)
84. Wilcox, R.K., Boerger, T.O., Allen, D.J., Barton, D.C., Limb, D., Dickson, R.A., Hall, R.M.: A dynamic study of thoracolumbar burst fractures. J. Bone Joint Surg. Am. **85-A**, 2184–2189 (2003)
85. Wilcox, R.K., Boerger, T.O., Hall, R.M., Barton, D.C., Limb, D., Dickson, R.A.: Measurement of canal occlusion during the thoracolumbar burst fracture process. J. Biomech. **35**, 381–384 (2002)
86. Willen, J., Lindahl, S., Irstam, L., Aldman, B., Nordwall, A.: The thoracolumbar crush fracture. An experimental study on instant axial dynamic loading: the resulting fracture type and its stability. Spine **9**, 624–631 (1984)
87. Zhu, Q., Lane, C., Ching, R.P., Gordon, J.D., Fisher, C.G., Dvorak, M.F., Cripton, P.A., Oxland, T.R.: Translational constraint influences dynamic spinal canal occlusion of the thoracic spine: an in vitro experimental study. J. Biomech. **41**, 171–179 (2008)

Nerve and Nerve Root Biomechanics

Kristen J. Nicholson and Beth A. Winkelstein

Abstract Together, the relationship between the mechanical response of neural tissues and the related mechanisms of injury provide a foundation for defining relevant thresholds for injury. The nerves and nerve roots are biologic structures with specific and important functions, and whose response to mechanical loading can have immediate, long-lasting and widespread consequences. In particular, when nerves or nerve roots are mechanically loaded beyond their mechanical tolerance for injury, motor and/or sensory deficits can result. The severity and persistence of the symptoms are modulated by the profile of the mechanical insult. In this chapter, the relevant anatomy, and structure are reviewed in the context of biomechanical data describing the mechanical behavior of nerves and nerve roots. While both nerves and nerve roots have components of their anatomic organization that protect the axons of their neurons, there are distinct differences in their structure and composition. These variations contribute to differences in their mechanical response to loading owing to their strength and stiffness. However, both tissues are time-dependent and exhibit viscoelastic behavior. These time-dependent characteristics in mechanical responses imply that, in addition to the magnitude of force or deformation, the rate of loading and duration of applied mechanical insult can modulate physiologic outcomes associated with mechanical loading to these tissues. This chapter reviews all of these concepts in the context of biomechanics and physiologic outcomes. In addition, a review of current work studying nerve and nerve root loading in animal models is provided in order to relate these outcomes to clinical symptoms.

K. J. Nicholson and B. A. Winkelstein (✉)
Department of Bioengineering, University of Pennsylvania, Philadelphia PA, USA
e-mail: winkelst@seas.upenn.edu

K. J. Nicholson and B. A. Winkelstein
Department of Neurosurgery, University of Pennsylvania, Philadelphia PA, USA

Stud Mechanobiol Tissue Eng Biomater (2011) 3: 203–229
DOI: 10.1007/8415_2010_17

Published Online: 12 June 2010

1 Introduction

The nerve roots and peripheral nerves are the primary neural components of the peripheral nervous system (PNS). These neural structures function together to connect the peripheral nervous system and the central nervous system (CNS), by providing both sensory information and signaling actions such as motor function. Both peripheral nerves and nerve roots are made up of a collection of neurons and connective tissue, but have anatomical and structural organizations that provide different degrees of protection to the neurons from mechanical and chemical injuries. Because of these similarities and differences, in this chapter we will review the anatomical and structural features of nerves and nerve roots that are most-relevant for understanding their roles in mechanical injury. External loading to these neural tissues results in tissue deformation and internal forces within the tissue that can, directly and indirectly, influence physiologic responses. The specific cellular and molecular consequences can depend on the loading conditions and can be different for similar loading to nerves and nerve roots owing to differences in their relative mechanical and structural properties. Developing a clear definition of the mechanical behavior of these two types of neural tissues in response to relevant biomechanical loading conditions is necessary to understand the relationship between trauma to these tissues and the resulting consequences.

Neuronal injury in either peripheral nerves and/or nerve roots can induce sensory or motor deficits, alone or in combination. Neurons connecting the periphery and CNS carry either sensory information from the periphery to the central nervous system or motor information from the CNS out to the periphery [84]. The neurons that relay information from the periphery to the central nervous system are termed *afferent neurons* and their injury can produce sensory deficits. Conversely, the neurons that relay information from the spinal cord to the periphery are *efferent* and injury to efferent neurons results in motor deficits. Axons project from the neuron's cell body and conduct the electrical impulses to a target site. While nerves contain both afferent and efferent axons, nerve roots only contain one or the other type specifically; detailed anatomy for each neural structure will be discussed in this chapter. From a physiologic standpoint, however, this implies that an injury to a single nerve has the potential to induce both sensory and motor deficits, while an injury to a single nerve *root* will only have either sensory or motor consequences.

This chapter covers the biomechanics of both nerve and nerve root and addresses physiological context of such loading. Accordingly, it begins with a brief overview of their anatomy that is relevant for discussing the mechanical properties and modes of injury for each. In addition, a brief summary of relevant principles of soft tissue mechanics highlights specific concepts pertinent for understanding how nerve and nerve root tissues behave under compressive and tensile loading. With that foundation in biomechanics, we then present a more detailed discussion of specific mechanical properties of nerve and nerve root tissue, separately. The anatomy and biomechanics of these tissues provide the context for the discussions of injury mechanisms for these neural structures that follows and specific outcomes are discussed with respect to the mechanics that modulate their effects. Finally, a brief discussion is presented that provides broader comparisons of biomechanics of tissues from the central nervous system.

2 Relevant Anatomy

The vertebral column consists of 33 vertebrae, with five separate anatomic regions: cervical, thoracic, lumbar, sacral, and coccygeal. There are seven vertebrae in the cervical region, 12 in the thoracic, five in the lumbar, five fused vertebrae in the sacral region, and four in the coccygeal region, also fused. The spinal canal is formed by the vertebral foramen of each bony vertebra which encloses and surrounds the spinal cord (Fig. 1). The pedicles lie along the posterolateral border of each vertebral foramen and the pedicles of adjacent vertebrae form intervertebral foramina which enclose the afferent and efferent nerve roots as they extend from the spinal cord to the periphery (Fig. 1). At each spinal level, a pair of afferent and

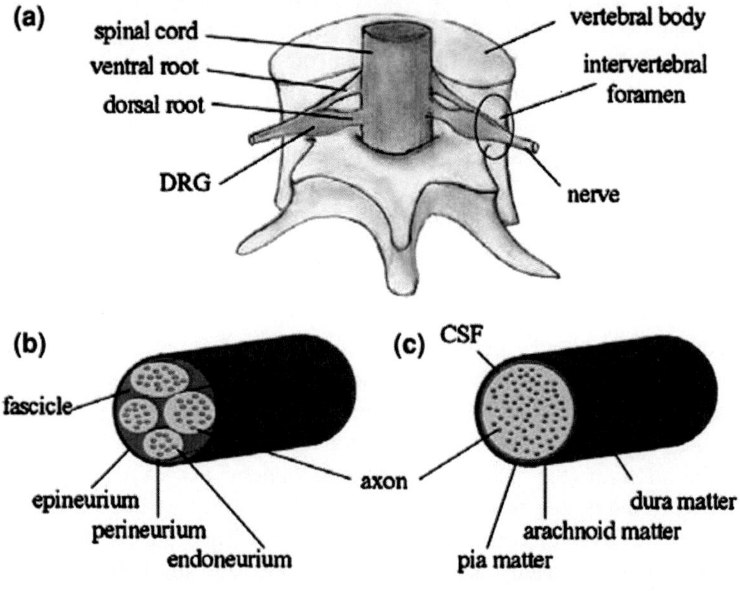

Fig. 1 Relevant anatomy. **a** Dorsal view of the ventral and dorsal roots exiting the spinal cord and forming a nerve as they pass through the intervertebral foramen formed by the adjacent vertebrae. Also shown are oblique views of generic **b** nerve and **c** nerve root structures illustrating the organization of the axons within the fascicles. In the nerve, **b** endoneurium surrounds each axon and the fascicles are surrounded by perineurium. The outermost layer of connective tissue of the nerve is the epineurium. **c** The nerve root is enclosed by three layers of meninges (pia matter, arachnoid matter, dura matter) and cerebrospinal fluid (CSF) lies between the arachnoid and pia matter

efferent nerve roots fuse together to form a nerve as they pass through the intervertebral foramen. Once the nerve enters the periphery, it branches out as the individual axons within the nerve continue towards their specific site of innervation.

The number of pairs of nerve roots does not match the number of vertebrae exactly. There are 31 pairs of nerve roots: eight cervical, 12 thoracic, five lumbar, five sacral and one coccygeal. By convention, in the cervical region, each nerve root is named for the spinal level of the vertebra inferior to (i.e. below) it, although in the remaining regions of the spine the nerve roots are named for their superior vertebra. Despite there being only seven cervical vertebrae, the nerve root between the lowest cervical vertebra (C7) and the next first thoracic vertebra (T1) is called the C8 nerve root.

At each spinal level, there is a ventral nerve root and a dorsal nerve root that exit the spinal cord anteriorly and posteriorly, respectively (Fig. 1). The efferent neurons make up the ventral nerve root and the afferents are contained in the dorsal nerve. As the ventral and dorsal nerve roots pass through the intervertebral foramen, they join together to form a spinal nerve. Spinal nerves can branch out and, on occasion, recombine as they course through the body. Nerves and nerve roots are mechanically and structurally distinct from each other. Although the axons of nerves are organized into bundles, called *fascicles*, the axons of nerve roots are enclosed together (Fig. 1). There are also distinct differences in the composition of the connective tissue that surrounds and protects nerve and nerve tissue. The next subsections will review the anatomy of nerve and nerve root, separately, with particular focus on highlighting those aspects that are relevant to understand their mechanical behavior and mechanisms for injury.

2.1 Nerve

For the most part, nerves in the body are relatively unprotected by any bony structures in comparison to the brain and spinal cord that have the mechanical protection by the skull and spinal column, respectively. As such, nerves are an organized collection of axons with connective tissue that provides compressive and tensile strength (Fig. 1). The axons in peripheral nerves are organized into fascicles, or bundles that collectively make up a nerve. At each organizational level, there is a layer of connective tissue (Fig. 1). The endoneurium envelopes each axon, the perineurium encloses each fascicle, and the epineurium forms a tough, fibrous sheath around the entire structure of the nerve.

Other cells, such as Schwann cells support the chemical and homeostatic environment of the neurons to facilitate axonal survivability and electrical conduction along the axons. While those support cells provide an environment that is conducive to and maintains the normal function of axons, they provide minimal mechanical contributions for the structural integrity of axons. As such, they will not be discussed in this chapter. Within each nerve, the fascicular bundles and the axons course through the nerve in an undulating manner; this undulation allows for some elongation along the nerve without imposing significant loading along the lengths of the fascicles and axons as they may become straightened when the overall nerve is pulled in tension [75, 80]. The structural organization and mechanical properties of the connective tissue of the nerve are able to preserve the integrity of the relatively fragile axons in the nerves under normal physiologic conditions such as joint motions and muscle elongation.

Together, the epineurium, perineurium, and endoneurium provide mechanical protection when tension and/or compression are applied to the overall nerve structure. Type I and Type III collagen, as well as elastic fibers, comprise the connective tissue of the epineurium [31, 80]. The dense network of collagen and elastic fibers that make up epineurium tissue can absorb shock by dissipating the compressive forces and protecting nerves from such compression to the nerve structure as a whole [75]. While it has been hypothesized that the epineurium does provide some protection to the nerve in tension, it does not appear to be the primary contributor to the tensile mechanical strength of the nerve [23, 66]. In all but the most-peripheral of nerves, the nerve contains more than one fascicle [71]. For nerves with more than one fascicle, the epineurium is divided into the epifascicular epineurium, surrounding the entire nerve, and the interfascicular epineurium that separates the individual nerve fascicles. The perineurium surrounding each fascicle is the primary contributor to the nerve's tensile strength and elasticity [66, 75]. This layer contains Type I and Type II collagen fibers, as well as elastic fibers oriented in circumferential, oblique, and longitudinal orientations, which together can potentially provide multidirectional tensile strength to the fascicles [31, 80]. The endoneurium that surrounds each axon is composed of collagen (Types I, II, IV) and provides a nominal degree tensile strength [75, 79, 80]. The three layers of connective tissue that surround nerve tissue act in concert across scales and directions to provide the compressive and tensile strength collectively necessary to protect the axons they surround during normal motions and physiologic loading scenarios.

2.2 *Nerve Root*

Nerve roots, like nerves, are also a collection of axons. At the distal, more-peripheral, end of the dorsal nerve root there is a visible enlargement called the dorsal root ganglion (DRG) that contains the cell bodies of the afferent neurons (Fig. 1). The cell bodies of the efferent neurons in the ventral nerve root lie within the spinal cord. The axons of the nerve root are, as a group, enveloped within three layers, called *meninges*. These are organized, from deep to superficial, the pia mater, the arachnoid mater, and the dura mater. Cerebrospinal fluid (CSF) flows within the subarachnoid space, between the arachnoid mater and pia mater. The CSF protects the neural tissue it encloses, including the brain and spinal cord, by damping any blunt forces to these structures [55]. However, it has also been hypothesized that the pressure gradients that can be established within the CSF surrounding the spinal cord and nerve roots during rapid motions of the spine can be sufficiently high to cause injury to the axons within the nerve root [5, 77]. Thus, while both the meninges and the CSF do provide mechanical protection to the nerve roots, CSF and associated fluid-induced loading should not be ignored when considering mechanisms of injury to these axons.

The dura mater of the nerve roots is continuous with the epineurium of the nerve, as well as with the dura mater of the spinal cord. Moreover, the dura is believed to provide much of the tensile support for the nerve roots [4, 43]. Tensile testing of the dura mater of rat cranial and spinal dura mater have shown it to be 2–1,000 times stiffer than the neural tissue it surrounds, suggesting a similar role for the dura of the nerve roots [43]. However, in that same study, differences were also noted in the mechanical properties between these two separate anatomical regions of dura due to variations in their structural organization and composition [43]. Together, those findings preclude any direct conclusion to be made regarding the protective function and the mechanical properties of the dura of the nerve roots. Regardless, the dura mater is likely the primary contributor to the tensile strength of nerve roots.

In addition to differences in the structural organization between nerve and nerve root, there are substantial differences in tissue composition and the orientation of the axons within the nerve root that contribute to differences in the mechanical behavior of these tissues. Unlike the axons in a nerve, nerve root axons are largely parallel to each other, with few undulations and as a result, contribute minimally to the tensile stretch to the nerve root ([75]; Fig. 1). Yet, the fibers of the dura mater likely do undulate along the length of the nerve root, as has been demonstrated in spinal dura mater tissue [43]. This wave-like pattern of the fibers, along with the elastin content of the dura, may be the primary contributors to the demonstrated ability of nerve roots to stretch for a substantial distance before developing significant resistive forces [68].

The collagen content of the connective tissue in both nerve and nerve root tissue provides much of the mechanical strength of these tissues. Yet, the collagen content of the nerve root is 3–7 times less than that of the peripheral nerve and this difference has been attributed to the characteristically stiffer properties of nerves compared to nerve roots [72]. In a mechanical study that directly compared the tensile strength and stiffness of the sciatic nerve and lumbar nerve roots from mice, the nerve root was found to be only 10% as strong and 17% as stiff as the nerve [4]. However, anatomically, nerve roots are more protected by the bony elements which surround them, compared to their peripheral nerve counterparts. Together, the organization and the composition of the tissues within the nerve contribute to the differences in the mechanical properties between nerve and nerve root.

Nerves and nerve roots are conduits of connective tissue that provide a protective pathway for axons to reach the periphery. This protection is accomplished via layers of connective tissue whose compressive and tensile strengths mitigate the specific forces applied to these structures. Yet, distinct differences in anatomy between nerves and nerve roots, including the organization and composition of this connective tissue leads to distinct differences in their mechanical behavior. While it is recognized that the inner perineurium of the nerve and the outer dura mater of the nerve roots are the primary load bearers of tension for their respective structures [43, 66], the significance of such observations is quantified using mechanical tests to measure the tissue's response to specific loading conditions. The following section discusses the mechanical characteristics of these tissues, starting with a general discussion on soft-tissue mechanics and then going into a more detailed review of the mechanical behavior of nerves and nerve roots.

3 Relevant Tissue Mechanics Background

As with other soft tissues in the body, the response of neural tissue to mechanical loading depends on the size, shape, organization, and composition of that tissue. Additionally, the mechanical properties depend on the specific conditions related to loading, such as the rate and duration of applied load or deformation. Previous loading will also alter the tissue's response to a given insult. Beel et al. [3] demonstrated that, within 2 weeks following crush of a sciatic nerve in the mouse, the nerve becomes stronger, stiffer, and exhibits behavior that is less elastic when it is pulled in tension. When describing the response of tissue to mechanical loading, the behavior and structure of each material comprising the overall tissue, as well as their local and organization and structure all contribute to the overall biomechanical response. *Structural properties* describe the behavior of the tissue as a whole, and depend on the geometry and material properties of the entire structure, in this case, of the tissue organ (i.e. nerve or nerve root or spinal cord). By extension, specific regions of the brain, such as the corpus callosum and the outer gray matter cortex, have been shown to have different mechanical properties [59], which will together contribute to the brain's overall response if both areas are involved in the loading scenario. On the other hand, *material properties* describe the intrinsic properties of a material and are independent of the shape of the material. As such, material properties are often more useful for comparing the mechanical characteristics between biologic materials.

The simplest description of a material's response to mechanical loading relates the amount of stress in the material to the strain. Stress and strain are scaled measures of the force and deformation within a material that are normalized with respect to the geometry (Fig. 2). As such, they inherently account for shape effects as they contribute to mechanical responses. *Stress* (σ) is defined as the amount of force over a given cross-sectional area. Accordingly, it can vary with direction and at different points in a structure. *Strain* (ε) represents the amount of deformation a material undergoes with respect to its original shape. For a linear elastic material, the amount of stress is linearly related to the amount of strain it experiences, and the slope of the line relating the two is the *modulus of elasticity* (E) of the material. Similarly, the relationship between the applied force and the resulting deformation of a structure is described by the *stiffness* of that structure. Many tissues, including

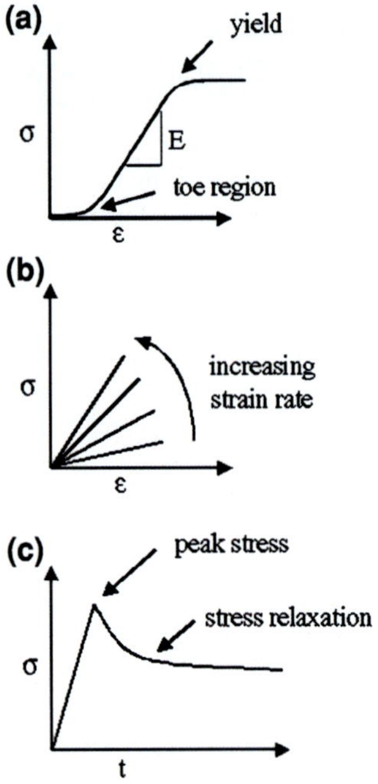

Fig. 2 Characteristic plots of the mechanical response of biologic tissue under load. **a** For a linear elastic material, stress (σ) is related to strain (ε) by a constant modulus of elasticity (E), until the tissue begins to yield. **b** The stress–strain behavior exhibits increased stiffness with strain rate for viscoelastic materials. **c** When a viscoelastic material is held under constant strain, the stress in the tissue relaxes over time

neural tissue exhibit complicated biomechanical responses that are not easily described by simple linear relationships. *Elastic* materials deform instantaneously with an applied force and, likewise, instantaneously return to their original shape after the applied force is removed. For this type of material, the amount of stress is only a function of strain. *Viscous* fluids demonstrate time-dependent deformation to an applied load and the stress that develops in response to that deformation depends on the rate of strain. In general, neural tissue responds to loading partly as if it were an elastic material, but also by undergoing viscous fluid flow. This type of mechanical behavior is *viscoelastic* and describes many biologic tissues [2, 20, 50, 60].

Nerves and nerve roots exhibit viscoelastic responses when loaded and this is due to the mechanical behavior of the materials they contain as well as the interaction between those materials. Viscoelastic materials exhibit several time-dependent responses to load; most basically, the stress–strain relationship varies with the rate at of loading and with the length of time that a tissue is held in deformation (Fig. 2). For these tissues, the modulus of elasticity varies with the rate at which the tissue is strained. Additionally, when the tissue is held under constant deformation, the amount of applied stress experienced in the tissue actually decreases with time (Fig. 2). Conversely, the tissue will also continue to deform over time when it is held under a constant stress. These two behaviors of the material are referred to as *stress–relaxation* and *creep*, respectively. The specific loading conditions applied to nerves and nerve roots are, therefore, important to bear in mind when describing the mechanical response of these tissues.

Given the viscoelastic nature of nerves and nerve roots, it is important to consider testing parameters such as rate of loading and magnitude of strain when comparing the properties that are measured experimentally for these types of biologic tissues. Recent advances in measurement techniques and computational approaches have made it possible to mathematically describe the time-dependent properties of these tissues, by incorporating both the elastic and the fluid components of the tissue's response. In a theory known as *quasi-linear viscoelasticity* (QLV), first developed by Fung [20] and more recently adapted for other neural tissues, the viscoelastic response can be described by a set of parameters that quantify both the stress–strain and time-dependent characteristics of the tissue [9, 19, 22, 47]. In the QLV model, force is estimated as a function of displacement, $F(d)$, and is described by a polynomial representing the elastic component of the tissue response (Eq. 1). The viscous, time-dependent component, $G(t)$, can be represented as a series of exponentials, where G_0 is the relaxation value as time, t, approaches infinity. This describes the long-term behavior of the tissue as it reaches a state of equilibrium. G_i is the relaxation constant for each relaxation term and β_i describes the time-dependent effects of relaxation (Eq. 2), with G_i describing the magnitude of the relaxation and β_i describing the rate at which the relaxation occurs.

$$F(d) = Ad^4 + Bd^3 + Cd^2 + Dd + E \tag{1}$$

$$G(t) = G_0 + \sum_{i=1}^{n} G_i e^{-\beta_i t}. \tag{2}$$

Equations 1 and 2 can be combined into a single time-dependent integral to capture the overall time-dependent relaxation component within the force–displacement relationship (Eq. 3).

$$F(t) = \int_0^t G(t-\tau)\frac{\partial F^{\varepsilon}(\delta)}{\partial \delta}\frac{\partial \delta}{\partial \tau}d\tau. \tag{3}$$

The relaxation function, $G(t)$, and elastic function, $F^{\varepsilon}(\delta)$, can be expressed as functions of the elastic parameters (A, B) and the relaxation parameters (C, τ_1, τ_2) (Eqs. 4, 5), where E_1 is the exponential integral.

$$G(t) = \frac{1 + C[E_1(t/\tau_2) - E_1(t/\tau_1)]}{1 + C\ln(\tau_2/\tau_1)} \tag{4}$$

$$F^{\varepsilon}(\delta) = A(e^{B\delta} - 1). \tag{5}$$

Together, these parameters capture the time-dependent properties of viscoelastic materials. Unlike the modulus of elasticity, which varies with the rate of loading for these materials, the parameters for the QLV model describe the material's response for a broad range of loading conditions. In addition to considering the time-dependent properties of neural tissue, it is also important to keep in mind that the material properties exhibit sensitivity to direction, or the tissue's orientation with respect to the direction of applied load. Materials whose mechanical behaviors vary with the direction are *anisotropic*; materials with no dependence on orientation are *isotropic*.

The longitudinal orientation of the axons and surrounding connective tissue within nerves and nerve roots (Fig. 1) suggest that these biologic structures behave differently when loaded along their long-axis than when loaded in any other direction. Also, some studies report differences in the mechanical properties of the brain when loaded in compression rather than tension [9]. Together, the anatomy of the nerve and nerve roots along with the reported behavior of brain tissue, suggest that the mechanical properties of nerves or nerve roots are not isotropic and do not necessarily behave similarly in compression or tension. The available data on nerves and nerve roots, however, is limited to tensile testing along their length and to compression in the radial direction or across their cross-sections. These two loading conditions are also the most clinically relevant as they also represent the most likely injury modes sustained by these two neural structures; they will be discussed in more detail in the section describing 'Injury Mechanisms' later in this chapter.

Nonetheless, it is important to provide relevant context for experimental testing approaches before reviewing specific biomechanical responses for nerves and nerve roots. Accordingly, most mechanical testing of these neural tissues has been performed for isolated compression and tension. In tension, the nerve or nerve root is elongated by anchoring the two most-extreme ends of a sample of tissue to enable stretching and the load and deformation are measured. In compression, a rigid applicator typically applies force through the width at a specific location along the length of the sample. In both of these types of tests, there is a loading phase, in which a load is applied to the tissue that results in deformation of that tissue. If the deformation applied to the tissue is held constant, there will then also be a stress–relaxation phase, as discussed above. In the following discussion on the mechanical properties of nerves and nerve roots, the loading phase will be discussed for tension and compression, separately. This will then lead to a discussion of the stress–relaxation properties observed for both tension and compression.

3.1 Tensile Stress–Strain Properties

The mechanical behavior of nerves and nerve roots in tension is typically characterized by reporting the load–deformation or stress–strain behavior as described above, but for specific rates of loading. The typical stress–strain curve that describes the elastic response of these tissues has three distinct regions; a toe region, a linear region, and a yield or failure region ([4, 23, 38, 68]; Fig. 2). In the toe region, the tissue can undergo substantial deformation with the generation of only minimal force. For example, the isolated tibial nerve of rabbit can sustain tensile strains as large as 15% before any appreciable stress develops in that tissue [38]. As neural tissue is further deformed under tension beyond the initial toe region, more force is required to continue deformation and the slope of the load–deformation curves becomes approximately linear. Finally, as deformation continues to be applied and the nerve or nerve root is stretched beyond this linear elastic region, the tissue will begin to yield, marked by a reduction in slope of the stress–strain slope. As with other materials, yield indicates the initiation of sub-structural failures and is followed by a complete rupture of the tissue that occurs when deformation continues beyond the tissue's ability to support the resulting forces [44]. Both the stiffness and yield points of nerve and nerve root are important in characterizing thresholds for injury to these tissues. As such, much attention has been devoted to understanding these two properties of nerves and nerve roots and each will be discussed here.

The tensile stiffness of both nerve and nerve root is characterized by measuring the slope of the linear region of the stress–strain plot during elongation applied at a constant rate. As mentioned above, nerve roots are considerably less stiff than nerve [4, 75]. In two companion studies, Beel et al. [3, 4] reported the tensile modulus of elasticity for comparable loading rates of the lumbar nerve roots and the sciatic nerve of mice to be 1.2 and 7 MPa, respectively, corresponding to a nearly sixfold difference between the two structures of the same species. However, significant differences have also been reported to exist among different nerves of the body. Under similar tensile loading conditions, nerve tissue taken from the brachial plexus of the rabbit has a modulus of elasticity of 28.5 MPa, whereas the tibial nerve in that same species was reported to be considerably less stiff, with a modulus of 0.58 MPa [38, 78]. Given these differences in the mechanical behavior, thresholds for producing injury likely vary between specific nerves throughout the body, as well as between nerves and nerve roots.

Due to the viscoelastic nature of nerves and nerve roots, the shape of the stress–strain plot depends on the rate at which load is applied, with the slope of the linear region becoming more steep as the rate increases ([27, 68]; Fig. 2). The nerve root in tension, for example, has a modulus of elasticity of 1.3 MPa when stretched at a rate of 0.01 mm/s, which doubles to 2.9 MPa when stretched at a rate of 15 mm/s [68]. In general, at higher rates of loading, tissues deform less than for lower rates at a given magnitude of applied stress, exhibiting more solid-like behavior. It is possible, therefore, that for a given tissue, the threshold for injury at high rates of loading may be defined by a certain set of mechanical conditions, which are different for lower rates of loading. Alternatively, the *rate* of loading may actually serve as the criterion for injury, as has been demonstrated for other soft tissues of the body [48].

In addition to the stress–strain relationship, the point at which tissue failure or rupture occurs is also important in serving as specific thresholds for injury. Studies of rabbit tibial nerve report failure to occur at 73% strain when the nerve is stretched at 0.5 mm/s and at 39% strain when stretched at slower rate of 0.17 mm/s; the ultimate tensile strength of the nerve stretched at 0.17 mm/s was 11.7 MPa [23, 38]. While these two studies appear to indicate that the degree of strain that nerve tissue can tolerate before failure is dependent on the rate of loading, Singh et al. [68]

reports no significant difference in the strain at failure for nerve roots loaded at 0.01 or 10 mm/s, despite finding differences in the stiffness of the roots as described above. While the nerve roots in both groups failed at strains of 24–37%, those loaded at the higher rate exhibited a greater ultimate tensile strength of 625 kPa, whereas the nerve roots loaded at 0.01 mm/s failed at 258 kPa. For lumbar nerve roots of mice, Beel et al. [4] reported that failure occurred at 302 kPa at a loading rate of 1 mm/s. This, taken together with the results reported by Singh et al. [68], indicates that nerve roots can support greater tensile forces before failure at higher rates of loading. Human nerve roots also may be stronger at higher rates of loading; when stretched at 0.17 mm/s human spinal nerve roots have been reported to have an ultimate tensile strength of 1.71 MPa [37], which increases to 3.9–29.4 MPa for a loading rate of 1.27 mm/s [76]. Although gross failure of the tissue certainly represents an injury with severe physiologic consequences, it should be kept in mind that normal function of the nerves and nerve roots can be disrupted at loads that do not produce complete rupture of the tissue or even have any evidence of mechanical injury. Therefore, the subfailure stress–strain relationship may be as important in understanding the onset of "injury" to nerves and nerve roots. This is discussed further in later sections of this chapter.

3.2 Compressive Stress–Strain Properties

The generalized mechanical behavior of nerves and nerve roots in compression is similar to that in tension, with an initial toe region, followed by a linear elastic region ([26, 27]; Fig. 2). However, to date, there has been very limited experimental work defining such properties, owing largely to experimental challenges that exist with making such measurements to these small and very soft materials. Regardless, a study in the rat has quantified the mechanical properties of nerve roots in compression, but no such corresponding data exist for the nerves [27]. In that study, the lumbar nerve roots of rats were compressed in vivo using varying rates to develop a quasi-linear viscoelastic model (as presented in Eqs. 1–5). The corresponding parameters for that model are: $A = 2.432$ mN, $B = 6.319$, $C = 0.555$, $\tau_1 = 0.190$ s, $\tau_2 = 261.7$ s (Eqs. 4, 5) [27]. While the mechanical properties of nerves and nerve roots in compression is limited, many studies have reported that the physiologic consequences of compression do depend on the magnitude, duration, and rate of loading and these will be discussed in detail.

3.3 Stress–Relaxation

Both nerves and nerve roots exhibit a reduction in stress when held under constant deformation, and this occurs for both tension and compression. As the tissue is deformed, the stress increases until the deformation reaches its peak and the stress also peaks; under constant deformation, though the stress begins to decrease (Fig. 2). For nerves and nerve roots, there is an initially rapid decrease in the applied load, followed by a leveling off or slowing down of this decrease.

The specific stress–relaxation characteristics depend on the direction of the applied load and the degree of deformation [20]. In a study of cervical nerve roots compressed in vivo, those that were compressed to a peak load of 22 mN, relaxed to 50% of their peak load, whereas those compressed to a higher peak load of 108 mN, relaxed less and reached a final load that was 91% of the peak load [26]. Similarly, in a study of rat tibial nerves in tension, nerves held at 6% strain demonstrated a significantly greater degree of stress relaxation than those strained by larger amounts (i.e. 9 and 12%) [38]. At 6% strain, the nerve relaxed to 52% of the peak applied stress, while those held at 9 or 12% strain only relaxed to 67% of the peak stress. Driscoll et al. [14], however, reported no significant difference in the amount of stress–relaxation between rat sciatic nerves held in tension to 8.8 or 16.1% in vivo, with both conditions resulting in stress relaxation of approximately 60% of the peak stress [14]. Interestingly, Kwan et al. [38] also found no difference in stress–relaxation between nerves strained to 9 and 12%, which suggests that the effect of tissue strain on the magnitude of the stress relaxation may only be relevant for smaller strains. In studies of the nerve in tension, the stress relaxes rapidly within the first 10–20 min [14, 32, 38]. However, Hubbard et al. [26] reported that the applied compression load for a nerve root reaches a steady-state within the first 5 min. The different rates of relaxation may be attributed to differences in the properties of nerves and nerve roots and/or differences in how these tissues behave in tension compare to in compression. By extension, these studies also suggest that a relationship may exist between the physiologic consequences of injury to nerve or nerve root tissue and the duration for which these tissues are loaded. This suggestion is supported by recent work demonstrating that both behavioral and inflammatory changes following compression to the nerve root are, indeed, modulated by the duration of the held compression [64].

This brief presentation characterizes the mechanical properties of nerve and nerve root tissue and highlights the fact that many factors must be considered when discussing the constitutive relationships of these tissues. Certainly, nerves behave differently than nerve roots; however, each nerve of the body may also have unique mechanical characteristics. Additionally, both the rate and duration of loading affect the mechanical response. These phenomena imply that nerve and nerve root injuries need to be evaluated within the context of these conditions (location, rate, duration) when determining the types of forces and/or deformation that induce injury to the tissue.

4 Injury Mechanisms

While both nerve and nerve root structures are susceptible to injury from tensile and compressive loading, the anatomy of the human body places each in particular risk for certain types of loading based on their proximity to adjacent structures that can provide protection and/or injurious loading. The specific location of nerves and nerve roots make each structure susceptible to different sources and types of loading. Nerves are not typically enclosed within any bony structures and so are directly susceptible to loading by external forces and deformation, as well as those caused by adjacent structures of the body. Nerve roots, on the other hand, are located within the vertebral column, which does provide protection from externally applied loads. However, because nerve roots are situated between two vertebrae, motion of one vertebra relative to the other has the potential to change the shape of the intervertebral foramen that can potentially compress the nerve roots. The unique sources of potential loading for nerves and nerve roots place each at different risks for experiencing injurious loading.

Because tissue loading imposes both forces and deformations to tissues, it is important to consider both mechanical parameters when considering neural tissue tolerance to injury. As discussed in the previous section, at high rates of loading tissues are stiffer and also experience less deformation for given loading magnitudes. For such a scenario, it may be possible that the magnitude of the applied load may be the determining factor in what the physiologic consequences of such loading are. However, at slower rates the tissue is able to deform quite a bit without establishing substantially high loads, and in those cases, it is possible that the injury threshold(s) may be more appropriately or realistically determined by the degree of deformation the tissue experiences and not the magnitude of the corresponding load. Finally, because nerves and nerve roots do not exist in isolation but are part of a larger functioning biological system that requires a continuous flow of nutrients, loads that disrupt this flow may also result in injury. Therefore, when considering the relationship between biomechanics and physiology for nerves and nerve roots, the rate, extent (i.e. magnitude of force and/or deformation), direction, duration, and resulting structural compositions of loading must each be considered.

While a complete rupture or separation of nerve or nerve roots is certainly an injury, less severe damage to the tissue also result in physiologic consequences [11, 15, 21, 69]. For that reason, injuries to nerves and nerve roots must also be understood in the context of the physiological responses that are induced, as well as the resulting tissue changes. Specific clinical symptoms of injuries to nerves and nerve roots include weakness, paralysis, pain and sensitivity [1, 30, 88]. However, the severity of these symptoms is not always easily quantifiable, making it difficult for clinicians to diagnose and treat the condition. Pain and sensitivity are, in particular, difficult to evaluate objectively because the sensation and verbal description of pain can vary from one individual to another. Although pain is subjective and is not typically quantified, increased sensitivity to stimuli can be evaluated in humans and in animal models. Allodynia is one measure of behavioral sensitivity that provides a quantitative assessment of increased sensitivity to a non-painful stimulus and can be measured by counting the frequency of responses that is evoked for a given stimulus. For example, increased sensitivity in the forepaw of rats has been demonstrated following a transient compression of the C7 nerve root

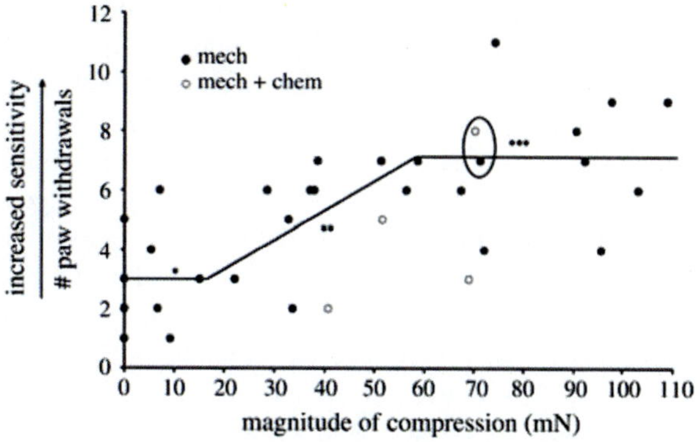

Fig. 3 Behavioral sensitivity (i.e. pain) as evaluated by the number of paw withdrawals is modulated by the magnitude of the compressive load and the presence of a chemical insult. Each data point represents a rat in which the nerve root underwent either a mechanical compression alone (*mech*) of magnitude indicated or a combined mechanical compression with exposure to inflammatory components (*mech + chem*). Superimposed on this plot is a model fit indicating that compression loads below 16 mN did not induce sensitivity different from sham (*), a linear increase in responses with load magnitude (**), and the maximum allodynia response (***). Variability in the pain responses with and without chemical insult is also apparent (*circle*)

(Fig. 3). Within one day of the applied compression, allodynia is elevated and the magnitude of the number of paw withdrawals is also modulated by the magnitude of the applied compression [26]. Interestingly, for lower amplitude compressions, there is no change in the behavioral response (Fig. 3). Yet, beyond a load-threshold, allodynia increases linearly until a maximum allodynia response is reached for this injury (Fig. 3). Other indicators of physiologic dysfunction following neural trauma include edema, altered electrical conduction along the axons, and inflammation [8, 42, 52, 86], which can also provide insight into outcomes related to nerve and nerve root injury when the overall structure of the tissue remains intact. In the following sections, mechanisms of injury to nerves and nerve roots are discussed in the context of the consequences that are observed both clinically and in experimental models using in vivo systems.

4.1 Mechanisms of Nerve Injury

Nerve injuries can be the result of a host of different causes and changes in the normal environment of the tissues that surround them, such as direct mechanical insults and/or changes in the chemical milieu of the local environment. Here we focus primarily on mechanical insults. Compressive and tensile loads on the nerve can occur by an abnormal condition in an adjacent structure as in the case of a lesion compressing the nerve or by forces transmitted through the body during a traumatic event, such as a motor vehicle accident.

A common example of a nerve injury related to abnormal biologic changes is carpal tunnel syndrome (CTS) that results from compression of the median nerve in the wrist [45]. Symptoms of CTS include numbness and tingling that manifest in the thumb, index, and middle finger, which are the areas innervated by the axons of the median nerve [45, 70]. CTS is often the result of repetitive compression of the nerve due to the flexion–extension motion of the wrist that changes the geometry of the carpal tunnel through which the median nerve runs; nerve compression can also occur in other regions of the body like the peroneal nerve of the leg or the radial nerve of the arm when nerve become compressed by adjacent tumors and cysts, and by increases in the pressure of surrounding fluids, as occurs in some pregnancies and obesity [70]. In each of these examples, the mechanical forces on the nerve develop slowly over time. Yet, injury to the nerve can also be the consequence of an acute, traumatic event in which the nerve experiences a sudden onset of rapidly applied loads. Trauma-related nerve injuries are often associated with bone fractures and dislocations and are often the result of a motor vehicle accident or a fall [16, 36]. Transection of the nerve due to penetrating trauma also accounts for a significant number of trauma-related nerve injuries [16, 36].

Trauma-related nerve injuries and injuries with a gradual onset can result in similar symptoms despite their coming about by distinctly different mechanical loading conditions. To begin to define relationships between loading conditions, tissue damage, and clinical symptoms, an understanding of the type and severity of symptoms that manifest is important. When applying the term *injury* to nerves we must consider then any evidence of direct tissue damage, such as tears to the epineurium, or any outcome that includes any one of a host ensuing physiological dysfunctions, such as weakness or sensitivity.

Altered physiologic responses are likely the result of specific damage within the nerve structure and/or changes in the cellular function and phenotypic responses. Likewise, the degree of dysfunction that presents clinically is likely modulated by the degree of damage within the tissue or that which is produced more globally in the entire biologic system as a result of the local tissue biomechanics. In order to distinguish specific nerve injuries from a clinical perspective, these injuries are often categorized according to a scheme first proposed by Seddon [67]. In that paradigm, nerve injuries are categorized as either neurapraxia, axonotmesis, or neurotmesis [62]. *Neurapraxia* is the most-mild injury and is marked by a temporary loss of function that returns within hours-to-weeks. Although nerve conduction may be temporarily disrupted or altered, the nerve itself remains intact and axons are not irreparably modified; this disorder often results from a subthreshold compression and/or ischemia [62]. *Axonotmesis* is marked by a physical disruption in the axons of the nerve, with the connective tissue remaining in intact, allowing for possible axonal regeneration along the intact endoneurial tubes. While Wallerian degeneration of the axons does occur in this case, the overall nerve function may return within weeks or years. Axonotmesis is believed to result from crushed or stretched nerves. The most severe nerve injuries are classified as *neurotmesis*; both the axons and the connective tissue are disrupted and function will likely not return without any surgical intervention, and perhaps even never. These injuries often occur as the result of contusions, severe stretch, or laceration of the nerve and the clinical outcomes demonstrate the range of severity of symptoms that nerve injuries can produce. Although the relationships between injury and clinical outcome have been hypothesized in this way, the specific mechanical and *biomechanical* mechanisms and thresholds that induce these injuries were not described. Such mechanisms and tolerances require in vivo studies to define specific links between mechanical loading and cellular and molecular mechanisms that contribute to the physiologic symptoms.

Many in vivo animal models have been developed and used to study nerve injury and to define mechanisms by which injury mechanics modulate physiology. Broadly, those models have demonstrated that compression applied to nerves induces decreased blood flow within the nerve, impairs axonal transport, and changes the nerve's structure and function [39, 61], and that the severity of each of these changes is modulated by the magnitude of the applied force [8, 15, 41, 42, 58, 65]. For example, blood flow in the rabbit tibial nerve is reduced by pressures of 2.7 kPa, but is completely blocked at pressures above 10.7 kPa [65]. Similarly, pressure applied to the sciatic nerve of rats produces axonal degeneration at pressures of 4.0 kPa that is markedly more severe following compression at 10.7 kPa of pressure [58]. Indicators of neural structure and toxicity, such as demyelination, degeneration, and myelin thickening, have been shown to increase in severity when compression to the peroneal nerve of rats increases in magnitude from 50 to 300 mmHg [15]. That same study found that the duration of applied compression also modulates these changes, with longer durations resulting in more severe changes [15]. Nerve compression also alters electrical conduction along axons; sciatic nerve compression in the rat induced decreased conduction velocity along the nerve that was more pronounced for greater magnitudes of deformation [42].

In addition to these compression-induced changes, nerve tension also reduces blood flow and nerve conduction velocity [11, 14, 40, 83]. At strains of 6–8% blood flow within the tibial and sciatic rat nerves begins to decrease and is completely blocked at strains of 14–16% [11, 40]. Conversely, when rabbit sciatic nerve is stretched to either 9 or 16%, both groups demonstrate a significant 70–78% decrease in blood flow that is not significantly different from each other [14]. The contradictory results may be attributed to advances in technology that allowed the more recent study by Driscoll et al. [14] to measure the blood flow in an intact nerve while the previous studies tested a nerve that was cut at one end. Driscoll et al. [14] also measured the peak nerve conduction velocity in the rabbit sciatic nerve at strains of 9 or 16%. Unlike the observations about blood flow, peak nerve conduction velocity did decrease with increasing strain. At 16% strain, the peak nerve conduction velocity decreased by 44% but no significant change was reported at 9% strain, despite the significant drop in blood flow. This study demonstrates that strain differentially modulated blood flow and nerve function, but no attempt was made in this study or those previously presented to correlate these tissue-specific changes to behavioral outcomes. As such, it remains unclear which physiologic changes preferentially drive the severity of the pain following nerve injury.

Animal studies enable definition of mechanistic relationships between mechanical tissue loading and physiologic outcomes and provide a test platform to investigate how aspects of the loading parameters modulate the severity of these outcomes. Yet, currently studies of nerve injury do not yet bridge the gap between local, tissue-level changes in the nerve to the general sensation of pain. In addition to providing insight into the changes that occur locally at the site of injury, animal models can also be used to define the relationship between local tissue mechanics and behavioral changes that are indicative of pain, such as sensitivity to thermal, chemical or mechanical stimulations [13]. Future studies can build on the present models of nerve injury to identify the relationship between the mechanics and physiology at the tissue-level to the sensation of pain.

4.2 Mechanisms of Nerve Root Injury

Like nerves, nerve roots are susceptible to injuries from a variety of trauma-related causes, as well as additional local changes in the surrounding environment that result from progression of other diseases and/or disorders. For example, nerve roots can be injured from a slow-onset foraminal narrowing due to aging or from a disc herniation that can impart both mechanical and inflammatory perturbations to the nerve root. In addition, traumatic events of the spine can compress the nerve root by altering the geometry of the intervertebral foramen or can put the nerve root under traction when stretching of the spinal cord generates tension along the nerve roots [74]. Brachial plexopathies, commonly referred to as *burners* or *stingers*, is a collection of trauma-related nerve root injuries that are hypothesized to be caused by traction or compression to the nerve roots in the cervical region of the spine [10, 73]. Traction to the brachial plexus, which is a network of nerves formed by the lower cervical and upper thoracic nerve roots, is often associated with lateral motion of the neck away from the side of injury. Compression-related burners occur when there is compression of the nerve root by the intervertebral foramen as a result of cervical spine extension that is combined with rotation, lateral flexion, and/or compression of the spine [73]. Alternatively, the nerve root can be injured by pathologic changes such as foraminal stenosis or a herniated intervertebral disc that can not only compress the nerve root but can also tether it and place it under tension for certain spinal motions. Traumatic nerve root injuries occur over rapid timeframes with high-magnitude loads applied to the tissue; in contrast, those associated with local pathologies often have a slower-onset of loading that is more chronic in nature.

Both trauma-related and pathology-induced nerve root injuries can produce symptoms of varying severity and persistence. Clinically, nerve root injuries can manifest as radicular pain and numbness, and/or deficits in motor and reflex functions [1, 82]. The severity of the symptoms can be classified as acute, subacute, or chronic [1]. Acute radiculopathy typically results from a traumatic event and presents with the most severe pain, often described as "sharp" or "burning" and that pain can be persistent. The symptoms of subacute radiculopathy, on the other hand, are typically short-lived, but may be reoccurring. Subacute radiculopathy is most often associated with a pre-existing condition of the spine such as spondylosis. Both acute and subacute radiculopathy can develop into chronic radiculopathy when the symptoms do not respond to treatment. The specific location of the presenting symptoms after a nerve root injury depends on which nerve root is injured. Injuries to the C7 nerve root, for example, result in sensitivity and numbness in the arm, forearm, and middle finger, while injuries to the C4 nerve root affect the neurons that innervate the back of the neck and shoulder regions [1]. However, the pathologic sequelae of such injury can be localized to the region of injury (i.e. in the root tissue and/or cell bodies of the DRG) or be more-widespread with modifications throughout the spinal cord, despite its being remote from the injury itself (Fig. 4).

As with nerves, nerve roots are subject to many loading conditions that modulate their physiology. It has been demonstrated in animal models that compression to the nerve root can produce a host of biologic responses, including behavioral hypersensitivity, decreased axonal conduction velocity and blood flow, as well as edema and swelling [21, 25, 29, 52, 53]. Moreover, many of these physiologic responses depend on the magnitude, duration and rate of applied tissue compression [33–35, 52, 53, 57]. In a study of compression to the cauda equina in pigs, blood flow in the capillaries of the nerve roots was stopped for pressures above 40 mmHg [53]. In a similar model, compression of 100 mmHg resulted in a decrease in axonal conduction amplitude after 30 min of applied compression, while 200 mmHg produced an immediate decrease in axonal conduction amplitude with complete blockage of axonal conduction if applied for 100 min [21].

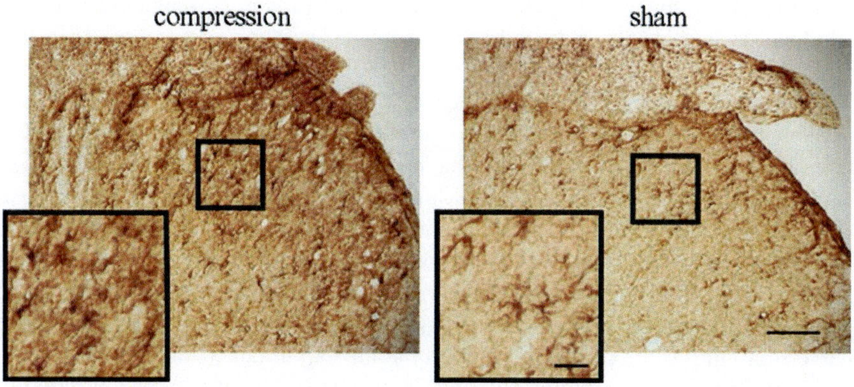

Fig. 4 Lumbar spinal cord dorsal horns stained for astrocyte activation using GFAP following application of a transient compression to the C7 cervical nerve root or a sham control procedure (*scale bar* = 100 μm). Even a transient mechanical injury in the cervical region is sufficient to induce changes in GFAP in the lumbar region of the spinal cord, with increased activation in compression compared to sham. *Insets* show the highlighted regions at higher magnification, with *scale bar* = 20 μm

Similarly, the rate of compression has also been shown to affect neuronal function. Olmarker et al. [54] demonstrated a more pronounced decrease in conduction velocities if the porcine nerve root is compressed at rapid rates when compared to lower compression rates. These findings indicate a multidimensional response of the nerve root physiology to compressive loading that that is sensitive to magnitude, duration, and rate.

There is a paucity of data for nerve root tension. However, one study to-date has investigated the relationship between tensile nerve root strain and conduction velocity [69]. Singh et al. [69] demonstrated that conduction velocity decreases with respect to both the amount of strain and the strain rate, suggesting physiologic changes following tension and compression are modulated by similar mechanical factors. At tensile strains below 10%, nerve conduction velocity decreased by 40% when the lumbar nerve root of rat was elongated at 0.01 mm/s, yet the velocity decreased by 80% when the nerve root was elongated at 15 mm/s. In contrast, at strains greater than 20% elongation, nerve conduction velocity was completely blocked regardless of the strain rate. Given the physiologic responses of the nerve root to compression and tension, the threshold for injury is likely a function of more than one mechanical parameter (i.e. a combination of deformation magnitude and rate of deformation). Before teasing out the specific mechanical parameters that modulate the local changes within the nerve root, the mechanics that drive the global sensation of pain should also be considered.

Studies of nerve root injury in the rat demonstrate functional and sensory behavioral changes are modulated by the degree of tissue deformation, the magnitude of load, and the duration of the applied compression [26, 64, 87]. Hubbard et al. [26] demonstrated that a compression load of only 26 mN to the rat C7 nerve root for 15 min will elicit sensitivity in the forepaw within 24 h but 38 mN of compression is required to induce sensitivity that is sustained for 1 week. Using the same model, Rothman et al. [64] demonstrated that for similar magnitudes of load allodynia is significantly elevated after 15 min of compression compared to that produced for a compression applied for only 30 s. In that study, the loads applied were above the threshold identified for inducing sustained allodynia [26], but 30 s of compression was not sufficient to produce any elevation in allodynia. The threshold for producing changes in behavioral sensitivity at this shorter duration of compression was hypothesized to be higher than that for a longer duration of compression. The amount of strain applied to the nerve root in compression also modulates allodynia. Both 19 and 42% of strain to the L5 nerve root of rats elicit increased sensitivity in the hind paw, with more sensitivity for the higher strain group [87]. Identifying the mechanical basis by which transient, chronic, and severe nerve root injuries are produced is an important step in understanding the mechanisms by why clinical symptoms vary with respect to the severity and duration.

Certainly, the loading parameters of a mechanical insult modulate both the local physiologic changes at the root as well as pain-related behavioral responses. Likewise, chemical insults can also injure the nerve root and, when applied in combination to a mechanical injury, can modulate the mechanical thresholds for injury [51, 56, 63]. Disc herniations, a common source of nerve root injuries, impose both chemical and mechanical insults to the nerve root due to the inflammatory material of the disc tissue [85]. One study that specifically evaluated the combined effects of chemical and mechanical insults, reports that behavioral sensitivity is elevated to similar levels following either a mechanical or a chemical injury to the nerve root, and is further elevated by the combination of these two injury inputs [63]. Another example of this combined mechanical and chemical effect is observed when allodynia is plotted as a function of compression magnitude (Fig. 3). While the combined injury group has a small sample size which limits the ability to draw specific conclusions, the results do suggest chemical injuries may modify thresholds for mechanical injury. Some rats that sustained the combined injury demonstrated an increased in sensitivity over those that received the same mechanical injury alone. As future studies define the mechanical threshold for nerve root injury, the role of chemical insults in raising, or lowering, these thresholds will give clinicians and drug-developers key insights to proceed in preventing and treating these injuries.

5 Comparisons of Biomechanics of Nerve and Nerve Root CNS Tissues

Understanding the mechanical behavior and injury mechanisms of the tissues of the central nervous system (CNS) can help to provide additional insight into the biomechanics of peripheral nervous tissues. As with the PNS, the CNS contains a collection of neurons and other cells that respond similarly to external stimuli. Yet, differences in the structural organization of these nervous systems preclude direct comparisons of injury mechanisms between these two systems.

The brain and spinal cord are protected by the vertebral column and the skull, respectively. These two bony structures provide protection from global external forces that may affect those important tissues of the CNS. Like nerve roots, both the spinal cord and brain are enveloped by the meninges. Yet, there are several differences within the neural tissue based on the organization of the neurons. Myelinated and unmyelinated neurons are organized into distinct regions, referred to as white matter and gray matter in the CNS. Under certain loading conditions, each region has a unique mechanical response. For example, Ichihara et al. [28] reported that at compressive strains higher than 10% spinal gray matter is more rigid and weaker than the spinal white matter under compression. Due to these regional differences, there may be more inhomogeneity of the mechanical behavior within the brain and spinal cord, compared to that of the nerve and nerve roots. The brain also has distinct structural regions, such as the cerebellum and corpus callosum, in which tissue is oriented in different directions, creating further regions of inhomogeneity. The details of such anatomy and structure of the brain are beyond the scope of this chapter. However, given these structural differences, the stresses and strains within the brain and spinal cord are distributed differently than in nerves and nerve roots. For example, in finite element simulations of head impacts, tissue strains were found to be highest in the corpus callosum and the brain stem than any other area of the brain [89]. While no study has explicitly measured the stress distribution within nerves, Sunderland [75] suggests based primarily on anatomic inferences that compressive loads are more evenly distributed among the fascicles and epineurium, implying that regional stress concentrations would likely not exist in the nerve. Clearly, the structural organization of the neural tissue within brain and spinal cord suggests the specific mechanical properties of CNS tissue to be different than PNS tissue.

As with other neural tissues, brain and spinal cord demonstrate rate- and time-dependent mechanical properties. For example, both CNS tissues become stiffer in tension with increasing rates of loading [19, 46, 81]. Brain has also been shown to stiffen under compression with increasing strain rates [17, 47]. Specifically, when the porcine brain is compressed to 20% strain, the resulting stress increases from 220 to 1000 kPa as the rate of compression increases from 0.005 to 500 mm/min [47]. Stress relaxation has also been observed for brain and spinal cord under both compression as well as tension [6, 7, 22]. Because the mechanical response of the tissues from the PNS and CNS are similar, mathematical models developed for one tissue, may similarly be appropriate to apply to another tissue. Indeed, the QLV model presented earlier in this chapter also shows a good fit for brain tissue [22]. While that model is able to describe the overall behavior of both brain and nerve root tissues, the specific parameters of the model are different for each tissue. Gefen and Margulies [22] reported a short-term time constant of approximately 2 s for brain tissue in compression, whereas Hubbard et al. [27] reported the short-term time constant for nerve root tissue in compression to be 0.19 s. Nonetheless, just as injury mechanism research in CNS and PNS tissues complement each other, advances in modeling the mechanics of each of these tissues will also lead to advances in the other.

6 Conclusions

It was the intent of this chapter to integrate the mechanics and physiology of nerve and nerve root injuries. Accordingly we summarized the relationship between the local anatomy, the mechanical behavior, and the mechanisms of injury following a mechanical insult to the nerve and nerve root, and indicated similarities and differences where applicable. The focus of this chapter was to present how the two structures provide mechanical protection to their axons and to discuss the consequences of exceeding their protective abilities. When mechanically loaded, there is a load–deformation relationship of the tissue that varies with the rate at which the tissue is loaded and for how long the load is applied. In addition to this time-dependent and load–deformation response, there are physiologic responses that vary with the magnitude of the load and deformation, the rate of loading, and the duration of the load. Animal studies that investigate the physiologic changes in the context of the loading profile begin to provide the link between the severity and duration symptoms of nerve and nerve root injury and the responsible mechanical insults.

Keeping in mind that each axon is part of a neuron that synapses in the spinal cord, an injury that causes axonal dysfunction has broader consequences for activities throughout the central nervous system. Indeed, chronic pain following injuries to the peripheral nervous system is due, at least in part, to an over-reaction of glial activation in the spinal cord [18]. Under normal conditions, glial cells function to maintain a chemical environment for neuronal survivability [13]. In response to an injury, these cells become activated and release factors to promote neuronal repair and to recruit immune cells. Glial activation in the spinal cord following nerve root compression and mechanical loading to peripheral nerves has been reported by many groups [12, 24, 49], and also that this response is modulated by the loading parameters [25, 64, 86]. Moreover, further complicating the relationship between nerve biomechanics and injury is the fact that this CNS response is often not only restricted to the local region at the site or spinal level of injury. For example, widespread modifications of glial activation in response to a peripheral injury may be evident in regions of the spinal cord far from the level of injury (Fig. 4). Given the interdependence of tissues in the CNS and PNS to maintain a healthy environment for neuronal survivability and to promote neuronal repair, studies of the injury mechanisms in each system must also consider the global response of both systems.

Additional studies that investigate the physiologic changes that occur in the central nervous system following an injury to the peripheral nervous system will complement the collection of reports presented in this chapter to provide a foundation for understanding the mechanistic link between the mechanical insult to the nerve and nerve root and the symptoms that follow. Understanding the biomechanics of these neural tissues will ultimately help in the development of preventative measures, improved diagnostic tools, and more effective treatments following injury. Accordingly, future work in nerve and nerve root injury mechanics must focus on defining the mechanical parameters that produce injury, defining what "injury" is for these tissues, and identify which local and widespread physiologic mechanisms drive the production of functional, sensory- and motor-deficits. Certainly, none of these efforts can be performed in isolation, but must integrate across fields to provide meaningful relationships between biomechanics, tissue-level physiology, cellular dysfunction, and clinical outcomes for nerve and nerve root injuries.

Acknowledgments The authors would like to acknowledge the following for providing funds to support this work: the National Institutes of Health (NINDS, NIDCR), the National Science Foundation, and the Catharine Sharpe Foundation. The authors also thank Benjamin Guarino for assistance with figure preparation.

References

1. Abbed, K.M., Coumans, J.: Cervical radiculopathy: pathology, presentation, and clinical evaluation. Neurosurgery **60**(1), 28–34 (2007)
2. Bedford, A.M., Liechti, K.M.: Mechanics of Materials. Prentice-Hall, Upper Saddle River, New Jersey (2000)

3. Beel, J.A., Groswald, D.E., Luttges, M.W.: Alterations in the mechanical properties of peripheral nerve following crush injury. J. Biomech. **17**(3), 185–193 (1984)
4. Beel, J.A., Stodieck, L.S., Luttges, M.W.: Structural properties of spinal nerve roots: biomechanics. Exp. Neurol. **91**(1), 30–40 (1986)
5. Bostrom, O., Svensson, M.Y., Aldman, B., et al.: A new neck injury criterion candidate-based on injury findings in the cervical spinal ganglia after experimental neck extension trauma. Proc. IRCOBI Conf. 123–136 (1996)
6. Carlson, G.D., Warden, K.E., Barbeau, J.M., et al.: Viscoelastic relaxation and regional blood flow response to spinal cord compression and decompression. Spine **22**(12), 1285–1291 (1997)
7. Chang, G., Hung, T., Feng, W.W.: An in vivo measurement and analysis of viscoelastic properties of the spinal cord of cats. J. Biomech. Eng. **110**, 115–122 (1988)
8. Chen, L., Seaber, A.V., Urbaniak, J.R.: The influence of magnitude and duration of crush load on functional recovery of the peripheral nerve. J. Reconstr. Microsurg. **9**(4), 299–307 (1993)
9. Cheng, S., Clarke, E.C., Bilston, L.E.: Rheological properties of the tissues of the central nervous system: a review. Med. Eng. Phys. **30**(10), 1318–1337 (2007)
10. Clancy, W.G., Brand, R.L., Bergfield, J.A.: Upper trunk brachial plexus injuries in contact sports. Am. J. Sports Med. **5**(5), 209–216 (1977)
11. Clark, W.L., Trumble, T.E., Swiontkowski, M.F., et al.: Nerve tension and blood flow in a rat model of immediate and delayed repairs. J. Hand Surg. **17A**(4), 677–687 (1992)
12. Colburn, R., Rickman, A., DeLeo, J.: The effect of site and type of nerve injury on spinal glial activation and neuropathic pain behavior. Exp. Neurol. **157**, 289–304 (1999)
13. DeLeo, J.A., Winkelstein, B.A.: Physiology of chronic spinal pain syndromes: from animal models to biomechanics. Spine **27**(22), 2526–2537 (2002)
14. Driscoll, P.J., Glasby, M.A., Lawson, G.M.: An in vivo study of peripheral nerves in continuity: biomechanical and physiological responses to elongation. J. Orhop. Res. **20**(2), 370–375 (2002)
15. Dyck, P.J., Lais, A.C., Giannini, C., et al. Structural alterations of nerve during cuff compression. Proc. Natl. Acad. Sci. U.S.A. **87**(24), 9828–9832 (1990)
16. Eser, F., Aktekin, L.A., Bodur, H., et al.: Etiological factors of traumatic peripheral nerve injuries. Neurol. India **47**(4), 434–437 (2009)
17. Estes, M.S., McElhane, J.H.: Response of brain tissue to compressive loading. Proc. 4th ASME Biomech. Conf. 70-BHF-13 (1970)
18. Fields, R.D.: New culprits in chronic pain. Sci. Am. **301**, 50–57 (2009)
19. Fiford, R.J., Bilston, L.E.: The mechanical properties of rat spinal cord in vitro. J. Biomech. **38**(7), 1509–1515 (2005)
20. Fung, Y.C.: Elasticity of soft tissues in simple elongation. Am. J. Phys. **213**(6), 1532–1544 (1967)
21. Garfin, S.R., Cohen, M.S., Massie, J.B., et al.: Nerve-roots of the cauda-equina: the effect of hypotension and acute graded compression on function. J. Bone Joint Surg. **72A**(8), 1185–1192 (1990)
22. Gefen, A., Margulies, S.S.: Are in vivo and in situ brain tissues mechanically similar? J. Biomech. **37**(9), 1339–1352 (2004)
23. Haftek, J.: Stretch injury of peripheral nerve: acute effects of stretching on rabbit nerve. J. Bone Joint Surg. **52B**(2), 354–365 (1970)
24. Hashizume, H., DeLeo, J.A., Colburn, R.W., et al.: Spinal glial activation and cytokine expression after lumbar root injury in the rat. Spine **25**(10), 1206–1217 (2000)
25. Hubbard, R.D., Winkelstein, B.A.: Transient cervical nerve root compression in the rat induces bilateral forepaw allodynia and spinal glial activation: mechanical factors in painful neck injuries. Spine **30**(17), 1924–1932 (2005)
26. Hubbard, R.D., Chen, Z., Winkelstein, B.A.: Transient cervical nerve root compression modulates pain: load thresholds for allodynia and sustained changes in spinal neuropeptide expression. J. Biomech. **41**(3), 677–685 (2008a)

27. Hubbard, R.D., Quinn, K.P., Martinez, J.J., et al.: The role of graded nerve root compression on axonal damage, neuropeptide changes, and pain-related behaviors. Stapp Car Crash J. **52**, 33–58 (2008b)
28. Ichihara, K., Taguchi, T., Shimada, Y., et al. Gray matter of the bovine cervical spinal cord is mechanically more rigid and fragile than the white matter. J. Neurotrauma **18**(3), 361–367 (2001)
29. Igarashi, T., Yabuki, S., Kikuchi, S., et al.: Effect of acute nerve root compression on endoneurial fluid pressure and blood flow in rat dorsal root ganglia. J. Orthop. Res. **23**(2), 420–424 (2005)
30. Jensen, T.S., Baron, R.: Translation of symptoms and signs into mechanisms in neuropathic pain. Pain **102**, 1–8 (2003)
31. Junqueira, L.C.U., Montes, G.S., Krisztan, R.M.: Collagen of the vertebrate peripheral nervous-system. Cell Tissue Res. **202**(3), 453–460 (1979)
32. Kendall, J.P., Stokes, I.A.F., Ohara, J.P., et al.: Tension and creep phenomena in peripheral-nerve. Acta Orthop. Scand. **50**(6), 721–725 (1979)
33. Kobayashi, S., Yoshizawa, H., Hachiya, Y.: Vasogenic edema induced by compression injury to the spinal nerve root: distribution of intravenously injected protein tracers and gadolinium-enhanced magnetic-resonance-imaging. Spine **18**(11), 1410–1424 (1993)
34. Kobayashi, S., Kokubo, Y., Uchida, K.: Effect of lumbar nerve root compression on primary sensory neurons and their central branches. Spine **30**(3), 276–282 (2005a)
35. Kobayashi, S., Meir, A., Baba, H., et al.: Imaging of intraneural edema by using gadolinium-enhanced MR imaging: experimental compression injury. Am. J. Neuroradiol. **26**, 973–980 (2005b)
36. Kouyoumdjian, J.A.: Peripheral nerve injuries: a retrospective survey of 456 cases. Muscle Nerve **34**, 785–788 (2006)
37. Kwan, M.K., Rydevik, B., Myers, R.R., et al.: Biomechanical and histological assessment of human lumbosacral spinal nerve roots. Trans. Orthop. Res. Soc. **14**, 348 (1989)
38. Kwan, M.K., Wall, E.J., Massi, J., et al.: Strain, stress, and stretch of peripheral nerve rabbit experiments in vitro and in vivo. Acta Orthop. **63**(3), 267–272 (1992)
39. Lundborg, G.: Structure and function of the intraneural microvessels as related trauma, edema formation, and nerve function. J. Bone Joint Surg. **57**(7), 938–948 (1975)
40. Lundborg, G., Rydevik, B.: Effects of stretching the tibial nerve of the rabbit: a preliminary study of the intraneural circulation and the barrier function of the perineurium. J. Bone Joint Surg. **55**(2), 390–401 (1973)
41. Lundborg, G., Myers, R., Powell, H.: Nerve compression injury and increased endoneurial fluid pressure: a "miniature compartment syndrome." J. Neurol. Neurosurg. Psychiatry **46**, 1119–1124 (1983)
42. Mackinnon, S.E., Dellon, A.L., Hudson, A.R., Hunter, D.A.: Chronic nerve compression—an experimental model in the rat. Ann. Plast. Surg. **13**(2), 112–120 (1984)
43. Maikos, J.T., Elias, R.A.I., Shreiber, D.I.: Mechanical properties of dura mater from the rat brain and spinal cord. J. Neurotrauma **25**, 38–51 (2008)
44. Martin, R.B., Burr, D.B., Sharkey, N.A.: Skeletal Tissue Mechanics. Springer, New York (1998)
45. Martyn, C.N., Hughes, R.A.C.: Epidemiology of peripheral neuropathy. Curr. Opin. Neurol. **8**(5), 335–338 (1997)
46. Miller, K.: Mechanical properties of brain tissue in tension. J. Biomech. **35**(4), 483–490 (2002)
47. Miller, K., Chinzei, K.: Constitutive modeling of brain tissue: experiment and theory. J. Biomech. **30**(11), 1115–1121 (1997)
48. Miller, M.A.: The biomechanical response of the lower abdomen to belt restraint loading. J. Trauma **29**(11), 1571–1584 (1989)
49. Milligan, E.D., Watkins, L.R.: Pathological and protective roles of glia in chronic pain. Neuroscience **10**, 23–36 (2009)
50. Navajas, D., Maksym, G.N., Bates, J.H.: Dynamic viscoelastic nonlinearity of lung parenchymal tissue. J. Appl. Physiol. **79**(1), 348–356 (1995)

51. Olmarker, K., Myers, R.R.: Pathogenesis of sciatic pain: role of herniated nucleus pulposus and deformation of spinal nerve root and dorsal root ganglion. Pain **78**, 99–105 (1998)
52. Olmarker, K., Rydevik, B., Holm, S.: Edema formation in spinal nerve roots induced by experimental graded compression: an experimental study on the pig cauda equina with special reference to differences in effects between rapid and slow onset of compression. Spine **14**(6), 569–573 (1989a)
53. Olmarker, K., Rydevik, B., Holm, S., et al.: Effects of experimental graded compression on blood flow in spinal nerve roots: a vital microscopic study on the porcine cauda equine. J. Orthop. Res. **7**(6), 817–823 (1989b)
54. Olmarker, K., Holm, S., Rydevik, B.: Importance of compression onset rate for the degree of impairment of impulse propagation in experimental compression injury of the porcine cauda equina. Spine **15**(5), 416–419 (1990)
55. Omaya, A.K.: Mechanical properties of tissues of the nervous system. J. Biomech. **1**(2), 127–138 (1968)
56. Onda, A., Murata, Y., Rydevik, B., et al.: Nerve growth factor content in dorsal root ganglion as related to changes in pain behavior in a rat model of experimental lumbar disc herniation. Spine **30**, 188–193 (2005)
57. Pedowitz, R.A., Garfin, S., Massie, J., et al.: Effect of magnitude and duration of compression on spinal nerve root conduction. Spine **17**(2), 194–199 (1992)
58. Powell, H.C., Myers, R.R.: Pathology of experimental nerve crush. Lab. Invest. **55**(1), 91–100 (1986)
59. Prange, M.T., Margulies, S.S.: Regional, directional, and age-dependent properties of the brain undergoing large deformation. J. Biomech. Eng. **124**(2), 244–252 (2002)
60. Provenzano, P., Lakes, R.S., Keenan, T., Vanderby, R. Jr.: Non-linear ligament viscoelasticity. Ann. Biomed. Eng. **29**, 908–914 (2001)
61. Rempel, D., Dahlin, L., Lundborg, G.: Pathophysiology of nerve compression syndromes: response of peripheral nerves to loading. J. Bone Joint Surg. **81A**(11), 1600–1610 (1999)
62. Robinson, L.R.: Traumatic injury to peripheral nerves. Muscle Nerve **23**, 863–873 (2000)
63. Rothman, S.M., Winkelstein, B.A.: Chemical and mechanical nerve root insults induce differential behavioral sensitivity and glial activation that are enhanced in combination. Brain Res. **1181**, 30–43 (2007)
64. Rothman, S.M., Nicholson. K.J., Winkelstein. B.A.: Time-dependent mechanics and measures of glial and behavioral sensitivity in a rodent model of radiculopathy. J. Neurotrauma **27**(5), 803–814 (2010)
65. Rydevik, B., Lundborg, G., Bagge, U.: Effects of graded compression on intraneural blood flow. An in vivo study on rabbit tibial nerve. J. Hand Surg. Am. **6**(1), 3–12 (1981)
66. Rydevik, B., Kwan, M.K., Myers, R.R., et al.: An in vitro mechanical and histological study of acute stretching on rabbit tibial nerve. J. Orthop. Res. **8**(5), 694–701 (1990)
67. Seddon, H.J.: Three types of nerve injury. Brain **66**(4), 237–288 (1943)
68. Singh, A., Lu, Y., Chen, C., et al.: Mechanical properties of spinal nerve roots subjected to tension at different strain rates. J. Biomech. **39**(9), 1669–1676 (2006)
69. Singh, A., Kallakuri, S., Chen, C., et al.: Structural and functional changes in nerve roots due to tension at various strains and strain rates: an in vivo study. J. Neurotrauma **26**, 627–640 (2009)
70. Sternbach, G.: The carpal tunnel syndrome. J. Emerg. Med. **17**(3), 519–523 (1999)
71. Stewart, J.D.: Peripheral nerve fascicles: anatomy and clinical relevance. Muscle Nerve **28**, 525–541 (2003)
72. Stodieck, L.S., Beel, J.A., Luttges, M.W.: Structural properties of spinal nerve roots: protein composition. Exp. Neurol. **91**, 41–51 (1986)
73. Stuber, K.: Cervical collar and braces in athletic brachial plexus injury and excessive cervical motion prevention: a review of the literature. J. Can. Chiropr. Assoc. **49**(3), 216–222 (2005)
74. Sunderland, S.: Mechanisms of cervical nerve root avulsion in injuries of the neck and shoulder. J. Neurosurg. **41**, 705–714 (1974)
75. Sunderland, S.: The anatomy and physiology of nerve injury. Muscle Nerve **13**(9), 771–784 (1990)

76. Sunderland, S., Bradley, K.C.: Stress-strain phenomena in human peripheral nerve trunks. Brain **84**, 102–119 (1961)
77. Svensson, M.Y., Aldman, B., Hansson, H.A., et al.: Pressure effects in the spinal canal during whiplash extension motion: a possible cause of injury to the cervical spinal ganglia. Proc. IRCOBI Conf. 189–200 (1993)
78. Takai, S., Dohno, H., Watanabe, Y., et al.: In situ strain and stress of nerve conduction blocking in the brachial plexus. J. Orthop. Res. **20**(6), 1311–1314 (2002)
79. Thomas, P.K.: The connective tissue of peripheral nerve: an electron microscope study. J. Anat. **97**(1), 34–44 (1963)
80. Topp, K.S., Boyd, B.S.: Structure and biomechanics of peripheral nerves: nerve responses to physical stresses and implications for physical therapist practice. Phys. Ther. **86**(1), 92–109 (2006)
81. Velardi, F., Fraternali, F., Angelillo, M.: Anisotropic constitutive equations and experimental tensile behavior of brain tissue. Biomech. Model Mechanbiol. **5**, 53–61 (2006)
82. Wainner, R.S., Gill, H.: Diagnosis and nonoperative management of cervical radiculopathy. J. Orthop. Sports Phys. Ther. **30**(12), 728–744 (2000)
83. Wall, E.J., Massie, J.B., Kwan, M.K., et al.: Experimental stretch neuropathy. J. Bone Joint Surg. **74B**, 126–129 (1992)
84. Wall, P., Melzack, R.: Textbook of Pain. Churchill Livingstone, London (1994)
85. Weiler, C., Nerlich, A.G., Bachmeier, B.E., et al.: Expression and distribution of tumor necrosis factor alpha in human lumbar intervertebral discs: a study in surgical specimen and autopsy controls. Spine **30**, 44–54 (2004)
86. Winkelstein, B.A., DeLeo, J.A.: Nerve root injury severity differentially modulates spinal glial activation in a rat lumbar radiculopathy model: considerations for persistent pain. Brain Res. **956**(2), 294–301(2002)
87. Winkelstein, B.A., Weinstein, J.N., DeLeo, J.A.: The role of mechanical deformation in lumbar radiculopathy: an in vivo model. Spine **27**(1), 27–33 (2002)
88. Woolf, C.J., Mannion, R.J.: Neuropathic pain: aetiology, symptoms, mechanisms, and management. Pain **353**, 1959–1964 (1999)
89. Zhang, L., Yang, K.H., King, A.I.: Comparison of brain responses between frontal and lateral impacts by finite element modeling. J. Neurotrauma **18**(1), 21–30 (2001)

Mechanotransduction in the Nervous System

Lynne E. Bilston and Cheryl L. Stucky

Abstract Mechanotransduction is the process by which cells convert mechanical load into cellular responses in the body. It is essential for interacting with the physical world around us and crucial for many internal physiological functions. Within the nervous system, mechanotransduction is how mechanical force or tissue deformation is converted to neural signals, typically action potentials. Mechanosensation is the subset of mechanotransduction processes that lead to the perception of a tactile stimulus to the body.

Mechanotransduction is a ubiquitous process within the body, and it underlies many key physiological functions, such as the sense of touch, proprioception, hearing, vestibular function, and gustation. There is also recent evidence that mechanical signalling is involved in cell differentiation and development in many organs, axon guidance during development of the nervous system [1] and metastasis of cancer cells [2]. During intense mechanical pressure that damages or threatens to damage tissue, mechanotransduction in nociceptors is involved in acute and persistent pain sensation [3]. Mechanical signals also play a key role in the autonomic nervous system, including regulation of gut motility [4], baroreceptor stimulation and regulation of blood pressure [5], and bladder function [6]. Mechanotransduction is a fundamental mechanism in most, if not all, living cells [7].

A prominent site for mechanotransduction is in the skin, the largest sensory organ in the body. The skin is exquisitely sensitive to different qualities, properties

L. E. Bilston (✉)
Neuroscience Research Australia and Prince of Wales Clinical School,
University of NSW, Sydney, Australia
e-mail: L.Bilston@neura.edu.au

C. L. Stucky
Department of Cell Biology, Neurobiology and Anatomy,
Medical College of Wisconsin, Milwaukee, USA

Stud Mechanobiol Tissue Eng Biomater (2011) 3: 231–245
DOI: 10.1007/8415_2011_83

Published Online: 1 June 2011

and intensities of tactile stimuli. Certain areas of the skin are highly sensitive to tiny changes in mechanical force, such as blowing air on the palm or the lightest touch to the lips. The physiological site for mechanotransduction is still unclear. It likely occurs in nerve fibre endings that innervate the skin, such as free nerve endings of pain receptors. However, mechanotransduction may also take place in, or be modulated by, specialized receptor ending accessory cells or structures such as Merkel cells, Pacinian corpuscles, or Meissner corpuscles, or even keratinocytes adjacent to free nerve endings in the epidermis [8–11].

In this chapter, we will briefly review the breadth of mechanosensory processes in the nervous system, discuss the current understanding of how mechanotransduction works, and eludicate how abnormalities in mechanotransduction can result in disease.

1 Mechanosensation

Human sensory perception of physical environmental stimuli on a daily basis depends extensively on mechanotransduction. Touch transduction is the first way in which a newborn baby bonds with its parent, typing on a keyboard and playing musical instruments depend on touch-initiated sensory feedback to motor systems, feedback from skin stretch and limb position modulate how we accurately pick up a pencil or coffee cup, and mechanotransduction in joint, tendon and muscle nociceptors provide essential sensory feedback to protect limbs and muscles from overextension and damage. Indeed, the absence of mechanotransduction signaling in nociceptors and the resulting lack of unconscious acute pain protection in individuals with chronic insensitivity to pain results in frequent serious injures that debilitate these individuals and severely shorten their lifespan and quality of life. Sensory neurons in the skin convert pressure and deformation of the skin surface into afferent neural signals that are interpreted by the brain as light touch, vibration, brush, deep pressure or frank pain sensations. The ability to localize and discriminate a sensation is affected by the number and density of receptors and their receptive field size in the stimulated region of tissue. There are many different types of receptors that tune the afferent to respond best to specific types of stimuli, and their neural response can adapt to give either tonic (continuous) or transient (on/off) signal input to the CNS (spinal cord, brainstem and cortex).

1.1 Touch

Tactile sensation includes detection of surface texture, pressure, vibration, limb position and object shape. Mechanical pain sensation from pinch, pinprick or damaging pressure will be discussed later in this chapter. These diverse sensations are mediated by many different types of sensory neurons and their associated receptors that are tuned to detect specific qualities of stimuli. Environmental touch stimuli are largely detected by the skin and significant research on mechanotransduction mechanisms has been carried out in cutaneous sensory neurons. However, is important to bear in mind that mechanotransduction is not limited to tactile sensations in the skin, as there are mechanosensitive afferents and force transduction mechanisms throughout the body in most organs and systems, including interoreceptor sites in the gut, bladder, colon, muscles, endothelial cells.

Sensory neurons come in various shapes and sizes, with peripheral axons that are Aβ, Aδ and C fibres. These nerve fibres have their cells bodies in the dorsal root ganglia (neck to toes) or trigeminal ganglia (face and head), and their processes extend to peripheral targets including skin, muscle, viscera, bone, tendon and fascia. The type of nerve fibre is classified on the basis of the myelination of the axon and the speed of action potential transmission along the nerve. Aα fibers are the most heavily myelinated, conduct the fastest (80–120 m/s) and innervate muscle spindles and Golgi tendon organs. Aβ fibres are the next most myelinated and conduct rapidly (35–75 m/s), followed by the thinly myelinated Aδ fibres (5–30 m/s), and finally the unmyelinated C fibres (0.5–2 m/s) [12].

In the skin, light touch is transmitted by the Aβ fibres, which have a low threshold for detection of mechanical stimuli, by some Aδ fibers, and also by some low-threshold C fibres. These C fibres appear to project to regions of the brain involved in emotional processing and result in a pleasant sensation [13]. Aβ fibres are associated with diverse specialized endings, including Merkel cells, Pacinian corpuscles, Meissner corpuscles, Ruffini corpuscles, and hair follicles, providing specific qualities of touch sensation (Fig. 1). Noxious pressure is transmitted by Aδ and C-fibre nociceptors, along with other modalities of pain sensations (see below).

Meissner corpuscles are dynamic mechanoreceptors found commonly in the superficial glabrous (non-hairy) skin, such as the fingertips and lips. Pacinian corpuscles are also found in hairy skin, and are located more deeply than the Meissner corpuscles. Both produce transient responses to changing (on/off) mechanical stimulation (i.e. they are rapidly adapting). Meissner corpuscle afferents help sense edges, bumps, corners and textures of objects. For example, they help adjust one's grip and release when lifting objects (such as a coffee cup) with the hands. They detect newly donned clothing and then adapt such that the clothing is no longer noticed. Pacinian corpuscles are more sensitive to high frequency vibration (peak sensitivity $\sim$300 Hz), whereas the Meissner corpuscles respond more to lower frequency stimulation (peak sensitivity $\sim$50 Hz) [12]. Pacinian corpuscles in foot pads of cats are essential for helping cats detect, at a considerable distance, mice running across the ground or barn floor boards.

The Merkel cells and Ruffini corpuscles are slowly adapting receptors that provide tonic responses to sustained, continuous mechanical stimuli. Merkel cells are located quite superficially in the skin, in the basal layer of the epidermis [14]. Recent evidence indicates that Merkel cells are essential for the proper encoding of the mechanical response by their slowly adapting sensory afferents [10]. Merkel cells are thought to detect mechanical stimuli by deformation of microvilli and potentially release neurotransmitters, including glutamate, that activate the

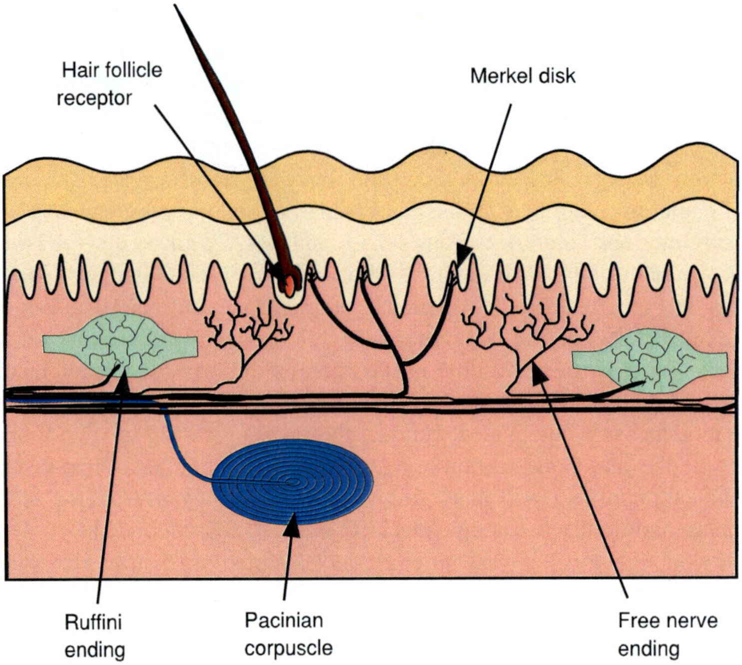

Fig. 1 Receptors that can transduce mechanical signals

associated sensory afferent [14, 15]. Although a direct demonstration of synaptic release of neurotransmitter from the Merkel cell is still lacking. Furthermore, whereas Merkel cells are required for normal response of the afferent, it is still not clear whether the actual site of mechanotransduction is in the Merkel cell, the somatosensory afferent or both [10]. Merkel cells are the best at detecting 2-point discrimination or the shortest distance between two distinct stimuli, and they give the sharpest resolution of texture and the surface of objects (bumpy, grooved, rough, smooth). Ruffini corpuscles are located more deeply in the dermis, and are a complex arrangement of sensory nerve endings that are encapsulated by connective tissue (collagen fibers). Individual Ruffini endings are sensitive to skin stretch in a particular orientation [16]. They are especially tuned to convey information about the shape of a grasped object.

1.2 Proprioception

Proprioception is the normal awareness of one's posture, balance, movement and location. It is the sense of where your body is in space, where your limbs are and how the limbs and body are located relative to each other. Proprioception involves detection of joint angles and head position. Proprioceptive inputs can be classified into those that sense static limb position and those that sense limb or body movement. Mechanoreceptors in the joints, skin, muscles and tendons sense limb and joint position, and the visual and vestibular system detect head position and orientation. Ultimately, higher processing areas in the cortex integrate somatosensory stimuli and visual information to encode where objects and atmospheric stimuli are in space with respect to the head and body.

In the musculotendon unit, muscle spindles within the muscle belly detect muscle stretch (length), whereas the Golgi tendon organs detect force (tension) in the tendon attached to a muscle belly. There are also stretch receptors in the joint capsules that detect joint angle. Skin mechanoreceptors (discussed above) also contribute substantially to both postural and joint position detection by conveying information about how skin stretches around joints, limbs and body parts [17].

The vestibular system detects linear and angular acceleration, and is housed in the vestibular labyrinth adjacent to the cochlea on each side of the head. There are five receptor organs in the vestibular system, allowing us to detect linear and angular acceleration along three orthogonal axes. Linear accelerations are detected in the utricle and the saccule, while angular accelerations are detected in the three semicircular canals. All of these are filled with endolymph fluid, which moves through the vestibular system, deflecting hair cells that transduce these mechanical stimuli into neural signals. In the vestibular labyrinth, these hair cells have a number of small hair-like stereocilia, whose mechanical deflection in one direction hyperpolarizes the cell membrane. This reduces neurotransmitter release into the synapse with the associated neuron and thereby decreases the firing rate of the neuron. Deflection in the opposite direction depolarizes the membrane, increases neurotransmitter release, thereby increasing the neural firing rate. This elegant system enables the firing rate of individual neurons to encode information about the direction of fluid flow through the vestibular system [12]. This mechanism of encoding mechanotransduction is similar to the process by which sound vibration is detected in the cochlea. The cellular mechanisms of this transduction are discussed further below in Sect. 5.

2 Pain

Pain can be evoked by tissue-damaging mechanical stimuli, intense thermal stimuli (extreme hot or cold), and by a plethora of chemical stimuli such as endogenous inflammatory mediators. These sensations are detected by *nociceptors* or injury detectors. In the case of strong mechanical stimuli such as a needle stick or pinch of the skin, intense mechanical pressure on the terminal endings of Aδ fibres gives rise to sensations of sharp, pricking pain [12]. These are transmitted quickly to the spinal cord or brainstem due to the fast conduction speed of the lightly myelinated Aδ fibres. High intensity mechanical stimulation can also activate small unmyelinated C fibres, and give rise to longer-lasting deep burning pain sensations. Both Aδ and C fibre nociceptors are located in throughout hairy and glabrous skin, as well as in deep tissues including muscles, tendons and bone.

Moreover, Aδ and C fiber nociceptors innervate many visceral organs (stomach, gut bladder, uterus) where they encode intense mechanical pain stimuli such as distention and overstretching. In contrast to somatic stimuli in skin or muscle which are easily localized, visceral pain is often difficult to pinpoint (poorly localized), and visceral pain sensations are often referred to somatic regions of the skin.

3 Hearing

The detection of sound waves, which are essentially mechanical vibrations of the air, occurs in the ear. Sound waves vibrate the ear drum, which in turns transmits these vibrations through the bones of the middle ear and onto the oval window. The oval window transmits these deformations to the basilar membrane which then transmits the vibration to the hair cells of the cochlea.

Vibration of the basilar membrane results in deformation of the hair cells that are attached to its surface. These cells detect the magnitude of the vibration in their specific location, and are tuned to a specific frequency, based on the wave propagation and mechanical characteristics of the basilar membrane and the hair cell bundles. Thus, the resonant frequencies of the hair cells are correlated with their physical size and mechanical properties. For example, taller sterocilia are sensitive to lower frequencies whereas shorter stereocilia are sensitive to higher frequencies. An ideal mechanical transduction system would place each of these at the spatial location along the basilar membrane where the maximal vibration at their resonant frequency occurs. Indeed, frequency selectivity is contributed to by the spatial location along the cochlea, as different frequency waves result in maximal vibration at different locations along the basilar membrane [18, 19]. There are two types of hair cells, aligned in one row of inner cells and three rows of outer hair cells. The inner row cells are thought to be the primary sensory cells, while the outer rows are thought to act as amplifiers and to have efferent inputs that can either inhibit or excite these cells to adjust their gain [20].

Small tip links between the tips of the stereocilia are thought to directly activate a mechanically gated ion channel that allows calcium and other cations to enter the hair cell, depolarising the cell membrane and triggering a receptor potential. Despite many elegant molecular and physiological studies in the hair cell field (see [21, 22] for review), the ion channel is still not known (see Sect. 5 below).

4 Autonomic

While conscious detection of mechanical stimuli in the gut are often confined to discomfort due to abnormal gut function (being over-full after eating a large meal or excessive gas from indigestion), the mechanical signalling in the gut plays an important role in normal gut function on a daily basis. Gut distension is sensed by stretch receptors in the muscle cells surrounding the gastrointestinal tract. Peristalsis, the rhythmic contraction of regions of the gut that propels food through the gastrointestinal tract, is triggered by stretching of the intestinal wall, activating neural circuits that cause contraction of that region of the intestine while inhibiting contraction in the region ahead to allow transport of the gut contents. Furthermore, distension of the rectum signals the need for defecation.

A range of signalling systems in the gut are involved in mechanosensation. These include enteric neurons that respond directly to stretch of their axons or soma, and vagal sensory neurons and spinal sensory neurons that sense neuronal deformation. Afferents are also tuned to sense mucosal deformation and stretch of the gut musculature [4].

Mechanoreceptors in the bladder are responsible for the detection of bladder filling [6], and detection of bladder wall stretch excites sympathetic nervous system reflexes that relax the bladder smooth muscle to allow bladder expansion, and simultaneously excite the sphincter muscles to ensure no leakage occurs [12].

The cardiovascular system also extensively depends on mechanically-mediated neural inputs that are incorporated into its complex control system. These include pressure-sensitive neurons in the aortic arch and carotid sinus (baroreceptors). Activation of these neurons, signalling high blood pressure, results in neural signalling that results in vasodilation and a decrease in heart rate [23, 24].

5 Mechanisms of Mechanotransduction

As described briefly above, a wide variety of different forms of mechanical signals are detected and transmitted within the nervous system. Indeed, mechanotransduction is a fundamental mechanism in virtually all living cells and organisms. In recent years there has been an explosion of interest in understanding the detailed mechanisms of mechanotransduction. This has included a concerted effort to identify the molecular mechanisms, channels and structures involved. Although some promising molecules have been identified, much remains to be identified and many pressing questions remain. The field of mechanotransduction mechanisms encompasses some of the most exciting biological puzzles to yet solve (see [25, 26], for review).

Due to the rapidity of mechanotransduction (typically less than 5 ms) [27], it has been widely assumed that mechanotransduction is likely to be mediated by direct gating of a transmembrane channel of some form in the mechanoreceptor cell membrane, rather than occurring as the secondary result of a chemical message [28, 29]. This rapid response was first observed in hair cells in the inner ear, but studies in other mechanoreceptors (such as the touch receptor organ in the nematode worm, *C. elegans*, and the bristle touch organ in drosophila) have found similarly short latencies [30–32].

There is still much debate in the mechanotransduction field about exactly how such channels are gated—whether deformation of the lipid bilayer of the cell membrane directly activates the embedded ion channel, opening the pore of the channel to allow ions to pass across the cell membrane, or whether the mechanical loading acts indirectly through associated proteins and structures, either extracellular or intracellular or a combination of both (Fig. 2).

In some bacterial cells, it is known that identified mechanically gated channels (MscL and MscS) that open in response to osmotic swelling are gated directly by the lipid bilayer without the need for other accessory structures [33]. These channels consist of five or seven repeated proteins in a pentagonal arrangement, with a central pore. Recent studies suggest that tension in the membrane causes the transmembrane proteins that make up the channel to tilt relative to each other and change configuration, thus opening a pore through which ions can flow to relieve the osmotic gradient [34]. However, mammalian cells do not express homologues of MscL or MscS channels, and the identity of mechanotransduction ion channels in mammalian cells is largely unknown.

Many other types of mechanosensitive ion channels are thought to be activated by deflection of extracellular structures or the cytoskeleton which are coupled to the channel proteins. For example, the activation of channels in inner ear hair cells involves deflection of tip link proteins joining the ends of the adjacent stereocilia. Both the extracellular matrix and cytoskeleton have been implicated in mechanotransduction in dorsal root ganglion neurons, at least in cell culture [35].

Despite much recent research, the specific molecules that transduce mechanical signals in neurons are still not known. These are likely to vary between different receptors and body regions (Table 1). Identified candidate mechanosensory channels come from four families: the Transient Receptor Potential (TRP) channel families (TRPA, TRPC, TRPM, TRPN, TRPV), the degenerin/epithelial Na+ channel (DEG/ENaC) family, the two pore potassium channel (KCNK) family, and most recently a novel class of mechanosensitive channels, Piezo1 and Piezo2 [37].

The TRP channel family appears to be involved in several types of mechanotransduction. They were first identified as being necessary for nose touch sensation in the nematode worm, *C. elegans* (osm-9 gene). These worms are often used for such studies due to the ease of genetic manipulations and access to the nerves for neurophysiological studies. In the fruit fly, the NOMPC gene, which codes for the TRPN1 channel, is required for bristle cell mechanosensation, but the role of this channel in other species is less clear. There is some evidence that TRPN1 is involved in proprioception [38] in *C. elegans*, and in vestibular and auditory function in zebrafish [39], however this channel is not present in mammals. The TRPV channels are thought to be involved in hearing and proprioception in fruit flies, but their major role appears to be in transduction of thermal (heat) stimuli in mammals [40–43]. TRPV1 is thought to be involved in stretch-evoked bladder reflexes [44], and mediates mechanical hypersensitivity after tissue injury [45–47]. TRPV2 is expressed in sensory neurons, and is activated by stretch and osmotic stimuli in vitro, but its function in vivo is not yet clear [48]. TRPA1 channels appear to mediate certain mechanically-activated currents in DRG sensory neurons

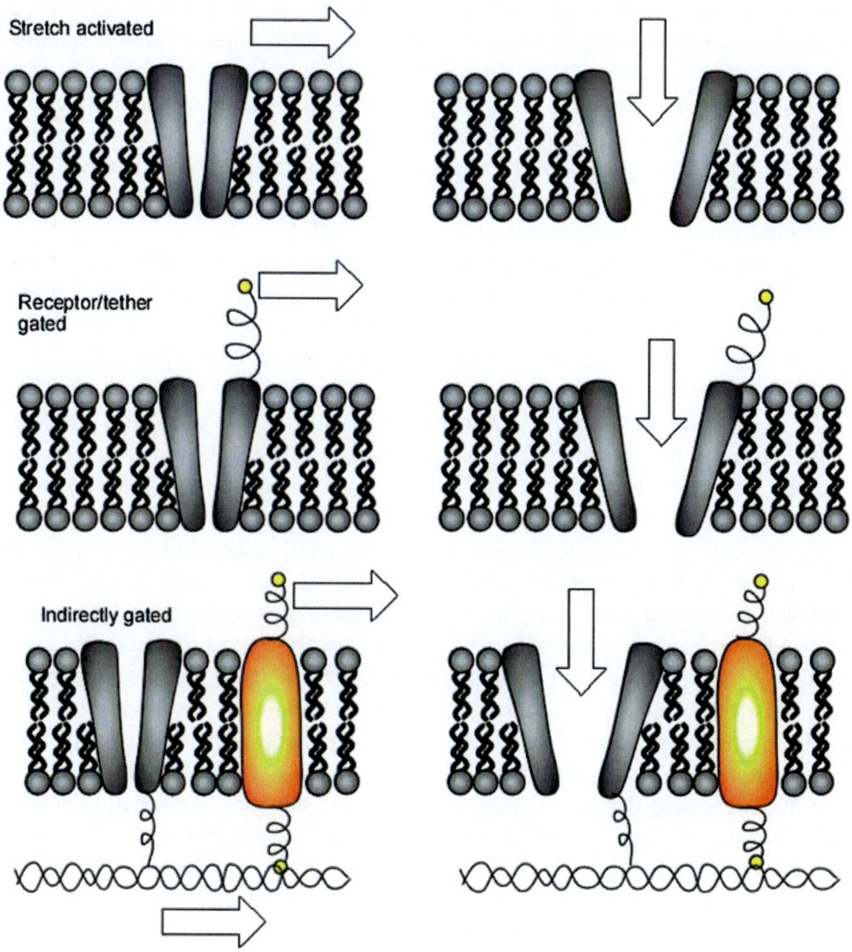

Fig. 2 Proposed mechanisms of gating/activation of mechanosensitive channels—direct stretch activation through the lipid bilayer (*top*), activation via receptors or tethers (extracellular or intracellular) that are coupled to mechanosensitive channel (*middle*), or indirectly activated through the cytoskeleton after activation of a nearby mechanosensitive protein (*bottom*). Adapted from [36]

[49], amplify the normal nociceptor response to intense force [50–52] and mediate the mechanical hypersensitivity after inflammation and nerve injury [53–55]. TRPC1, which is expressed in sensory neurons in mammals, was suggested to be a stretch-sensitive channel, but its function has not been confirmed, and it is now thought to be indirectly activated rather than a direct mechanosensory channel and it collaborates with other TRP channels including TRPV4 [56–58].

The degenerin/epithelial Na+ channel (DEG/ENaC) family include the MEC-4 and MEC-10 proteins, and the acid sensing ion channels ASIC1, ASIC2, ASIC3. MEC4 and MEC-10 have been shown to form the channel pore and MEC-2 and

MEC-6 accessory proteins link the channel to the cytoskeleton, and together act as mechanotransducers [31, 59]. Knocking out these proteins results in abnormalities in touch sensation. The ASIC channels, however, appear to be involved in cutaneous sensation, but are likely not the key mechanotransducers, since their absence has a small effect on mechanical response force detection and touch sensation in mice [60, 61]. Absence of ASIC3, in fact, can cause hypersensitivity to mechanical stimuli in mice, so it may be a regulatory molecule, or it might work in concert with other channels during mechanotransduction [62].

The two pore potassium channel (KCNK) family includes the TREK-1, TREK-2 and TRAAK channels. TREK-1 has been shown to be mechanically activated in mammals [63], but its precise role in mechanotransduction is not yet clear. Knocking out TREK-1 in mice results in a higher sensitivity to light touch [64]. TRAAK is expressed in mammalian neurons, and has been proposed to be involved in neurite elongation and neuronal growth [65], inhibition of mechanically induced pain, as well as in thermal sensation. Recent work suggests that TREK-1 and TRAAK work together with an as yet unidentified sodium channel to control the threshold for mechanical stimulation in nociceptors [66].

Piezo1 and Piezo2 are members of a recently identified novel class of mechanosensitive channels in mammals. Piezo1 (gene Fam38A) and Piezo2 (gene Fam38B) were identified in the neuroblastoma cell line Neuro2A and found to mediate rapidly-adapting mechanically-gated channels [37]. Overexpression of Piezo1 or Piezo2 in multiple heterologous cell lines confers robust mechanically-sensitive cation currents that are blocked by gadolinium and ruthenium red, known inhibitors of mechanosensitive channels. Piezo1 and Piezo2 are expressed in many diverse tissues including bladder, colon and lung. Piezo2 is particularly enriched in DRG sensory neurons, suggesting that Piezo2 may be involved in mechano transduction in these somatosensory neurons [37, 67]. Piezo1 and Piezo 2 are pioneers of a new class of membrane proteins. They appear to be unrelated to any other known ion channels, and are unique in that they contain 24–36 putative transmembrane segments, somewhat similar to the structure of voltage-gated sodium and calcium channels.

6 Diseases/Disorders

Alterations in mechanotransduction can have a variety of outcomes, and disturbances in mechanotransduction are involved in a broad range of diseases and disorders, both in the nervous system and other tissues, such as the heart, skeletal muscles, lungs, and bone.

6.1 Mechanical Hypersensitivity

Mechanical hypersensitivity is an exaggerated response to a mechanical stimulus that is normally non-painful (allodynia) or painful (hyperalgesia). It is common

Table 1 Putative mechanically activated ion channels

Family	Channel	Function
TRP	TRPA1	Mechanosensation
TRP	Painless	Mechanosensation
TRP	TPRC1	Mechanosensation
TRP	TRPN1	Hearing
TRP	NOMPC	Hearing, touch, proprioception
TRP	TRPV4	Touch, proprioception
TRP	TRPV2	Mechanosensation, osmosensation
TRP	NAN	Hearing, proprioception
DEG/ENaC	ASIC1	Touch
DEG/ENaC	ASIC2	Touch
DEG/ENaC	ASIC3	Touch, nociception
DEG/ENaC	MEC4/MEC10	Touch
KCNK	TREK1	Touch, mechanosensation
Piezo	Piezo1	Mechanotransduction
Piezo	Piezo2	Mechanotransduction

Adapted and extended from [36]

after tissue injury (mechanical, thermal, or chemical), inflammation or other disease conditions. Chronic mechanical allodynia is frequently observed in patients that suffer from neuropathic pain following nerve damage. Some of these issues are discussed in the chapter on nerve and nerve root biomechanics. There are several changes that are involved in hyperalgesia, including increased expression of sodium and potassium channels in nociceptor axons, which increase the excitability of the nerves. There have also been reported to be changes in TRP channels that are involved in mechanosensation in hyperalgesia caused by chemical and inflammatory conditions as well as neuropathy [53, 68, 69]. In genetically-modified mouse models where specific TRP channels can be knocked out, ablation of select TRP channels has been shown to substantially reduce or eliminate hyperalgesia in some circumstances. The precise mechanisms are still being elucidated, as some of the TRP channels are involved not only in mechanotransduction but also in the response to inflammation and tissue injury.

6.2 Genetic Disorders

There are a range of genetic and other disorders that include abnormalities in mechanosensation as part of their classic symptomatology. These include Tangier disease, Charcot-Marie-Tooth disease, hereditary sensory and autonomic neuropathy (HSAN), Machado-Joseph disease, Kanzaki disease, Prader-Willi Syndrome, Bardet-Biedel Syndrome and more. Some of these conditions have been associated with abnormalities in specific mechanosensory ion channel proteins [70].

Experiments with genetically modified animals used as models for a range of neurological disorders rely on knockouts of specific proteins, the absence of which is associated with abnormalities in sensory perception. Abnormalities in, or absence of, genes associated with primary cilia that play a role in mechanotransduction, e.g. in hearing, are also associated with altered perception of mechanical stimuli and/or pain (either increases or decreases) [70].

7 Conclusions

Mechanotransduction is a ubiquitous process throughout most organs and cell types within the body. It has a particularly important role in the nervous system where environmental stimuli and tactile sensations are used daily for function and survival. While the molecular mechanisms are still not well understood, the importance of mechanotransduction beyond the traditional area of somatosensory transduction is becoming better appreciated. Abnormalities in the cellular mechanotransduction apparatus and their role in a range of neurological and other disorders are slowly becoming clearer. Research on mechanotransduction mechanisms continues to be a cutting-edge area in the fields of hearing, cardiovascular mechanism and sensory neurobiology. Indeed, there is much knowledge about mechanotransduction to uncover that is far more than skin deep.

References

1. Engler, A.J., Sen, S., Sweeney, H.L., et al.: Matrix elasticity directs stem cell lineage specification. Cell **126**, 677–689 (2006)
2. Jaalouk, D.E., Lammerding, J.: Mechanotransduction gone awry. Nat. Rev. Mol. Cell Biol. **10**, 63–73 (2009)
3. Lewin, G.R., Moshourab, R.: Mechanosensation and pain. J. Neurobiol. **61**, 30–44 (2004)
4. Blackshaw, L.A., Brookes, S.J.H., Grundy, D., et al.: Sensory transmission in the gastrointestinal tract. Neurogastroenterol. Motil. **19**, 1–19 (2007)
5. Drummond, H.A., Price, M.P., Welsh, M.J., et al.: A molecular component of the arterial baroreceptor mechanotransducer. Neuron **21**, 1435–1441 (1998)
6. Daly, D., Rong, W., Chess Williams, R., et al.: Bladder afferent sensitivity in wild type and TRPV1 knockout mice. J. Physiol. **583**, 663–674 (2007)
7. Wozniak, M.A., Chen, C.S.: Mechanotransduction in development: a growing role for contractility. Nat. Rev. Mol. Cell Biol. **10**, 34–43 (2009)
8. Boulais, N., Pennec, J.P., Lebonvallet, N., et al.: Rat Merkel cells are mechanoreceptors and osmoreceptors. PLoS One **4**, e7759 (2009)
9. Koizumi, S., Fujishita, K., Inoue, K., et al.: Ca2+ waves in keratinocytes are transmitted to sensory neurons: the involvement of extracellular ATP and P2Y2 receptor activation. Biochem. J. **380**, 329 (2004)
10. Maricich, S.M., Wellnitz, S.A., Nelson, A.M., et al.: Merkel cells are essential for light-touch responses. Science **324**, 1580 (2009)
11. McCarter, G.C., Reichling, D.B., Levine, J.D.: Mechanical transduction by rat dorsal root ganglion neurons in vitro. Neurosci. Lett. **273**, 179–182 (1999)
12. Kandel, E., Schwartz, J., Jessell, T. (eds.): Principles of Neural Science, 4th edn. McGraw-Hill, New York (2000)

13. Olausson, H., Lamarre, Y., Backlund, H., et al.: Unmyelinated tactile afferents signal touch and project to insular cortex. Nat. Neurosci. **5**, 900–904 (2002)
14. Lucarz, A., Brand, G.: Current considerations about Merkel cells. Eur. J. Cell Biol. **86**, 243–251 (2007)
15. Haeberle, H., Lumpkin, E.: Merkel cells in somatosensation. Chemosensory Perception **1**, 110–118 (2008)
16. Olausson, H., Wessberg, J., Kakuda, N.: Tactile directional sensibility: peripheral neural mechanisms in man. Brain Res. **866**, 178–187 (2000)
17. Proske, U., Gandevia, S.C.: The kinaesthetic senses. J. Physiol. **587**, 4139–4146 (2009)
18. Hudspeth, A.: Hearing. In: Kandel, E.R., Schwartz, J.H., Jessel, T.M. (eds.) Principles of Neural Science, 4th edn, pp. 590–613. McGraw-Hill, New York (2000)
19. Hudspeth, A.: Sensory transduction in the ear. In: Kandel, E.R., Schwartz, J.H., Jessel, T.M. (eds.) Principles of Neural Science, 4th edn, pp. 614–624. McGraw-Hill, New York (2000)
20. Evans, M.G., Kros, C.J.: The cochlea–new insights into the conversion of sound into electrical signals. J. Physiol. **576**, 3–5 (2006)
21. Gillespie, P.G., Müller, U.: Mechanotransduction by hair cells: models, molecules, and mechanisms. Cell **139**, 33–44 (2009)
22. Schwander, M., Kachar, B., Müller, U.: The cell biology of hearing. J. Cell Biol. **190**, 9–20 (2010)
23. Hahn, C., Schwartz, M.A.: Mechanotransduction in vascular physiology and atherogenesis. Nat. Rev. Mol. Cell Biol. **10**, 53–62 (2009)
24. Tavi, P., Laine, M., Weckström, M., et al.: Cardiac mechanotransduction: from sensing to disease and treatment. Trends Pharmacol. Sci. **22**, 254–260 (2001)
25. Delmas, P., Hao, J., Rodat-Despoix, L.: Molecular mechanisms of mechanotransduction in mammalian sensory neurons. Nat. Rev. Neurosci. **12**, 139–153 (2011)
26. Lumpkin, E.A., Marshall, K.L., Nelson, A.M.: The cell biology of touch. J. Cell Biol. **191**, 237–248 (2010)
27. Christensen, A.P., Corey, D.P.: TRP channels in mechanosensation: direct or indirect activation? Nat. Rev. Neurosci. **8**, 510–521 (2007)
28. Chalfie, M.: Neurosensory mechanotransduction. Nat. Rev. Mol. Cell Biol. **10**, 44–52 (2009)
29. Corey, D., Hudspeth, A.: Response latency of vertebrate hair cells. Biophys. J. **26**, 499–506 (1979)
30. Albert, J., Nadrowski, B., Göpfert, M.: Mechanical signatures of transducer gating in the Drosophila ear. Curr. Biol. **17**, 1000–1006 (2007)
31. O'Hagan, R., Chalfie, M., Goodman, M.B.: The MEC-4 DEG/ENaC channel of *Caenorhabditis elegans* touch receptor neurons transduces mechanical signals. Nat. Neurosci. **8**, 43–50 (2005)
32. Walker, R., Willingham, A., Zuker, C.: A Drosophila mechanosensory transduction channel. Science **287**, 2229 (2000)
33. Martinac, B., Adler, J., Kung, C.: Mechanosensitive ion channels of *E. coli* activated by amphipaths. Nature **348**, 261–263 (1990)
34. Perozo, E., Cortes, D.M., Sompornpisut, P., et al.: Open channel structure of MscL and the gating mechanism of mechanosensitive channels. Nature **418**, 942–948 (2002)
35. Lin, Y., Cheng, C., LeDuc, P., et al.: Understanding sensory nerve mechanotransduction through localized elastomeric matrix control. PLoS One **4**, 4293 (2009)
36. Lumpkin, E.A., Caterina, M.J.: Mechanisms of sensory transduction in the skin. Nature **445**, 858–865 (2007)
37. Coste, B., Mathur, J., Schmidt, M., et al.: Piezo1 and Piezo2 are essential components of distinct mechanically activated cation channels. Science **330**, 55 (2010)
38. Li, W., Feng, Z., Sternberg, P., et al.: A *C. elegans* stretch receptor neuron revealed by a mechanosensitive TRP channel homologue. Nature **440**, 684–687 (2006)
39. Sidi, S., Friedrich, R., Nicolson, T.: NompC TRP channel required for vertebrate sensory hair cell mechanotransduction. Science **301**, 96 (2003)

40. Caterina, M.J., Leffler, A., Malmberg, A., et al.: Impaired nociception and pain sensation in mice lacking the capsaicin receptor. Science **288**, 306 (2000)
41. Caterina, M.J., Schumacher, M.A., Tominaga, M., et al.: The capsaicin receptor: a heat-activated ion channel in the pain pathway. Nature **389**, 816–824 (1997)
42. Chung, M.-K., Lee, H., Mizuno, A., et al.: TRPV3 and TRPV4 Mediate warmth-evoked currents in primary mouse keratinocytes. J. Biol. Chem. **279**, 21569–21575 (2004)
43. Moqrich, A., Hwang, S.W., Earley, T.J., et al.: Impaired thermosensation in mice lacking TRPV3, a heat and camphor sensor in the skin. Science **307**, 1468 (2005)
44. Birder, L., Nakamura, Y., Kiss, S., et al.: Altered urinary bladder function in mice lacking the vanilloid receptor TRPV1. Nat. Neurosci. **5**, 856–860 (2002)
45. Ghilardi, J.R., Röhrich, H., Lindsay, T.H., et al.: Selective blockade of the capsaicin receptor TRPV1 attenuates bone cancer pain. J. Neurosci. **25**, 3126 (2005)
46. Jones, R.C.W., Xu, L., Gebhart, G.: The mechanosensitivity of mouse colon afferent fibers and their sensitization by inflammatory mediators require transient receptor potential vanilloid 1 and acid-sensing ion channel 3. J. Neurosci. **25**, 10981 (2005)
47. McGaraughty, S., Chu, K.L., Scanio, M.J.C., et al.: A Selective Nav1. 8 sodium channel blocker, A-803467 [5-(4-chlorophenyl-N-(3, 5-dimethoxyphenyl) furan-2-carboxamide], attenuates spinal neuronal activity in neuropathic rats. J. Pharmacol. Exp. Ther. **324**, 1204 (2008)
48. Shibasaki, K., Murayama, N., Ono, K., et al.: TRPV2 enhances axon outgrowth through its activation by membrane stretch in developing sensory and motor neurons. J. Neurosci. **30**, 4601 (2010)
49. Vilceanu, D., Stucky, C.L.: TRPA1 mediates mechanical currents in the plasma membrane of mouse sensory neurons. PLoS One **5**, 15959–15964 (2010)
50. Brierley, S.M., Hughes, P.A., Page, A.J., et al.: The ion channel TRPA1 is required for normal mechanosensation and is modulated by algesic stimuli. Gastroenterology **137**, 2084–2095 (2009). e2083
51. Kerstein, P.C., Del Camino, D., Moran, M.M., et al.: Pharmacological blockade of TRPA 1 inhibits mechanical firing in nociceptors. Mol. Pain **5**, 19 (2009)
52. Kwan, K.Y., Glazer, J.M., Corey, D.P., et al.: TRPA1 modulates mechanotransduction in cutaneous sensory neurons. J. Neurosci. **29**, 4808–4819 (2009)
53. Eid, S.R., Crown, E.D., Moore, E.L., et al.: HC-030031, a TRPA1 selective antagonist, attenuates inflammatory-and neuropathy-induced mechanical hypersensitivity. Mol. Pain **4**, 48 (2008)
54. McGaraughty, S., Chu, K.L., Perner, R.J., et al.: TRPA1 modulation of spontaneous and mechanically evoked firing of spinal neurons in uninjured, osteoarthritic, and inflamed rats. Mol. Pain **6**, 14 (2010)
55. Petrus, M., Peier, A.M., Bandell, M., et al.: A role of TRPA1 in mechanical hyperalgesia is revealed by pharmacological inhibition. Mol. Pain **3**, 40 (2007)
56. Alessandri-Haber, N., Dina, O.A., Chen, X., et al.: TRPC1 and TRPC6 channels cooperate with TRPV4 to mediate mechanical hyperalgesia and nociceptor sensitization. J. Neurosci. **29**, 6217 (2009)
57. Maroto, R., Raso, A., Wood, T.G., et al.: TRPC1 forms the stretch-activated cation channel in vertebrate cells. Nat. Cell Biol. **7**, 179–185 (2005)
58. Patel, A., Sharif-Naeini, R., Folgering, J., et al.: Canonical TRP channels and mechanotransduction: from physiology to disease states. Pflügers Arch. Eur. J. Physiol. **460**, 571–581 (2010)
59. Suzuki, H., Kerr, R., Bianchi, L., et al.: In vivo imaging of *C. elegans* mechanosensory neurons demonstrates a specific role for the MEC-4 channel in the process of gentle touch sensation. Neuron **39**, 1005–1017 (2003)
60. Price, M.P., Lewin, G.R., McIlwrath, S.L., et al.: The mammalian sodium channel BNC1 is required for normal touch sensation. Nature **407**, 1007–1011 (2000)
61. Price, M.P., McIlwrath, S.L., Xie, J., et al.: The DRASIC cation channel contributes to the detection of cutaneous touch and acid stimuli in mice. Neuron **32**, 1071–1083 (2001)

62. Mogil, J., Breese, N., Witty, M., et al.: Transgenic expression of a dominant-negative ASIC3 subunit leads to increased sensitivity to mechanical and inflammatory stimuli. J. Neurosci. **25**, 9893 (2005)
63. Dedman, A., Sharif-Naeini, R., Folgering, J.H.A., et al.: The mechano-gated K 2P channel TREK-1. Eur. Biophys. J. **38**, 293–303 (2009)
64. Alloui, A., Zimmermann, K., Mamet, J., et al.: TREK-1, a K^+; channel involved in polymodal pain perception. EMBO J. **25**, 2368–2376 (2006)
65. Maingret, F., Fosset, M., Lesage, F., et al.: TRAAK is a mammalian neuronal mechano-gated K^+ channel. J. Biol. Chem. **274**, 1381–1387 (1999)
66. Noel, J., Zimmermann, K., Busserolles, J., et al.: The mechano-activated K^+ channels TRAAK and TREK-1 control both warm and cold perception. EMBO J. **28**, 1308–1318 (2009)
67. Xiao, R., Xu, X.Z.S.: Mechanosensitive channels: in touch with Piezo. Curr. Biol. **20**, R936–R938 (2010)
68. Ji, G., Zhou, S., Carlton, S.: Intact A [delta]-fibers up-regulate transient receptor potential A1 and contribute to cold hypersensitivity in neuropathic rats. Neuroscience **154**, 1054–1066 (2008)
69. Woolf, C., Ma, Q.: Nociceptors–noxious stimulus detectors. Neuron **55**, 353–364 (2007)
70. Tan, P.L., Katsanis, N.: Thermosensory and mechanosensory perception in human genetic disease. Hum. Mol. Genet. **18**, R146–R155 (2009)

In Vitro Models for Biomechanical Studies of Neural Tissues

Barclay Morrison III, D. Kacy Cullen and Michelle LaPlaca

Abstract In vitro models are invaluable tools for studying cell behavior in a highly controlled setting. Cell and tissue culture models of the nervous system can be utilized to elucidate neurobiological phenomena that are difficult to observe, manipulate, or measure in vivo. In the context of biomechanics, culture models that accurately mimic specific brain features can be used to determine tissue properties and tolerances to mechanical loading. There are several criteria that culture models must meet in order to complement in vivo and macroscopic biomechanical studies. In addition to providing an environment that is conducive to cell survival, cell type and source are critical to the interpretation of results. In this review, we present design criteria for ideal cultures, the current state of the art in neural cell and tissue culturing methods, and the advantages and limitations to using culture mimics. We will further present what insights in vitro models can provide to complement in vivo and macroscopic biomechanics in terms of meso- to microscale material properties and tissue-level tolerance criteria. The discussion will focus primarily on central nervous system (CNS) tissue, which is inherently complex in cytoarchitecture and organization. In addition, the CNS is not typically exposed to mechanical loading beyond physiological motion; therefore, it is expected that cell death and functional failure may be particularly prominent at large deformations and high loading rates. These and other

B. Morrison III (✉)
Biomedical Engineering, Columbia University, 351 Engineering Terrace,
MC 8904, 1210 Amsterdam Avenue, New York, NY 10027, USA
e-mail: bm2119@columbia.edu

D. K. Cullen
Neurosurgery, University of Pennsylvania, 105 Hayden Hall, 3320 Smith Walk,
Philadelphia, PA 19104, USA

M. LaPlaca
Biomedical Engineering, Georgia Institute of Technology, 313 Ferst Drive,
Atlanta, GA 30332-0535, USA

Stud Mechanobiol Tissue Eng Biomater (2011) 3: 247–285
DOI: 10.1007/8415_2011_79

Published Online: 30 April 2011

factors must be considered when attempting to extract and culture CNS tissue or its components for studying neurobiological or neuromechanical phenomena.

1 Neural Tissue Structure and Composition

1.1 Heterogeneity of Brain Tissue

Brain is a heterogeneous tissue that is divided into distinct anatomical and functional regions. The gray and white matter comprise the cellular constituents, and the blood supply, cerebrospinal fluid/ventricular system, interstitial fluid, and extracellular matrix (ECM) make up the extracellular components. The intracellular space (ICS) and the extracellular space (ECS) can be considered as one scale of heterogeneity. The ICS is a composite with neurons and their extensions, glia cells, vascular cells, and other support cells. The orientation is quite organized in some areas and apparently random in others. Most neuronal communication in the brain exists on a local level (i.e. shorter interneurons, which comprise most neurons in the brain), with 1,000s of synapses on some neurons, creating an extremely complex network [1]. The brain has a very high cellularity compared to other organs and a very diverse cell population. Cell sizes range from <10 to >100 μm in cell soma diameter with axons ranging from microns to 100s of microns long. Dendrites are highly branched, permitting diversification and maximization of interactions between communicating neurons [1].

Brain tissue can be divided further by anatomical regions. Just rostral to the spinal cord is the brain stem, a deep, well-protected portion of the brain comprised of the medulla oblongata, pons, and midbrain, which houses the controls to many homeostatic functions such as respiratory and cardiovascular regulation. The cerebellum, dorsal to the brainstem, helps control balance, posture, motor coordination, sensory and motor relaying. The diencephalon is the deep region of the brain that contains the thalamus (primary relay for sensory input and motor output) and the hypothalamus (neuroendocrine structure critical for maintaining homeostasis). Traveling toward the brain surface are the white matter tracts, the cerebral nuclei (basal ganglia, hippocampus, amygdala) and the cerebral cortex, comprised of six distinct, organized, interconnected layers. The isolation of any brain region (for primary culture) should consider the connections with other brain areas not represented in the culture and the possible effects (e.g., activation of compensatory responses) harvest and dissection may have. Interneuronal connections and interregional junctions are disrupted during tissue dissociation as are projection axons for most culture techniques, therefore the culturing procedures must consider these microinjuries and provide conditions conducive for repair and healthy culture maturation and reconnection. Some regions, such as the hippocampus are particularly vulnerable to excitotoxicity and the pH of the medium and constituents should allow for normal receptor function and neurotransmitter reuptake. Yet, this region is ideal for studying phenomena related to excitotoxicity, such as ischemia and trauma.

1.2 The Extracellular Space of Brain Tissue

In addition to the cell composition, the ECS is critical for communication, transport, and is essential for normal homeostatic function. It is approximately 20% of the total intracranial volume with half of this volume attributed to the blood and half to the cerebrospinal fluid (CSF) [2]. Some macroscopic parameters used to characterize brain tissue structure are the volume fraction of ECS and effective diffusivity or tortuosity. The ECS is composed of ECM proteins, as well as ions, neurotransmitters, metabolites, and peptides among other molecules. As most cultures are derived from prenatal or perinatal tissue, one must consider that the ECS is larger in the developing brain (36–46% by volume) than the adult, which may play an important role in the diffusion of nutrients and waste. The size of the ECS is especially important, since in most tissue culture systems it is essentially infinite. From a functional point of view, the anisotropy and inhomogeneity of brain tissue dictates the ECS shape, volume, and tortuosity, and hence intraneural (or interneuronal) diffusion parameters. For example, excess neurotransmitters, in addition to cellular reuptake, may diffuse to nearby capillaries or along elongated structures such as white matter tracts [3, 4]. Diffusion of extracellular substances will be discussed below in light of brain tissue dynamics and the artificial culture environment.

2 Neural Culture Models

Cells and tissues of the body may be explanted and kept alive ex vivo for extended periods of time, a process termed cell or tissue culture. Keeping the explanted tissue alive (i.e. metabolically active and functional) and preserving its phenotypic state similar to the in vivo state requires specialized culture conditions and nutrient medium reproducing vital conditions found in vivo. Specialized equipment required includes a tissue culture incubator which maintains physiological temperature (37°C) and gas tension (5% CO_2), laminar flow cabinets to maintain sterility, and in some cases bioreactor perfusion systems as described below. Explanted tissues may be cultured as thin slices of tissue, called organotypic slice cultures, or dissociated into suspensions of single cells to be plated as dissociated cultures. Cultures grown immediately after explantation are referred to as primary cultures.

2.1 Culture Model Utility

Cell and tissue culture models of the nervous system are desirable platforms for studying cell responses to various stimuli, including biochemical and pharmacological investigations, electrophysiological assessment, and mechanical stimulation or injury. These types of manipulations are not mutually exclusive and therefore it is critical to take into account the multimodal reactions that may follow a seemingly simple input. Well-characterized culture models, therefore, must mimic the cell states of brain tissue, including cell phenotype(s), cell–cell connections, cell-extracellular communication, mechanical properties, chemical composition, and electrochemical balance. The advantages of a well-defined culture preparation are that the complexity of an intact organism is drastically reduced and the culture (both extra- and intracellular environments) can be manipulated much more readily than in an animal or human. Despite the inherent simplicity of culture systems, they can be invaluable for use in understanding neural function as well as supra-physiological states. In order to assemble realistic brain tissue surrogates in a dish, one must rigorously prioritize the above characteristics and properties in a non-biased manner. Simply put: do existing cell and tissue culture models mimic brain tissue? What key parameters are to be mimicked?

2.2 Design Criteria for Brain Mimetics

In order to recapitulate brain tissue in culture models, several criteria are critical and are represented in various degrees of fidelity among culture models. The source of the tissue must be considered in terms of species, developmental age, and brain regions. The type of culture is equally important: primary cells versus cell lines, dissociated cells plated in a constraint-free culture vessel or in a reaggregate or constrained (e.g., scaffold) format, cultured explants or slices (see Table 1). In some instances, a single cell type is desired, yet multi-typic cultures are becoming more commonplace given the complex intercellular interactions that contribute to cellular response. In addition, as discussed above, the immediate extracellular environment must be defined in terms of medium, medium supplements (serum vs. defined supplements), substrate, dimension (2D or 3D), gas composition, pH, and temperature (Table 1).

In addition to these basic design considerations, specialty applications, such as mechanical property testing, mechanical and electrical stimulation, and fluidic control may require further manipulation. Geometric control will preserve the volume fraction of cells to extracellular space and relies on representative cell–cell connections and tissue-like cellular orientation. To this end, slice cultures meet this criterion. Topological features, such as soma shape, neuritic branching and glial-neuronal interactions are often lost with cell dissociation, yet can be maintained through culture manipulation (e.g., scaffolding).

While most brain regions can be cultured, we will focus on hippocampal and cortical dissociated and organotypic cultures in the subsequent sections. The particular purpose of a culture study should consider the cell type(s) to be cultured, as well as the source, age of animal, and other experimental variables. For an

Table 1 Neural culture variables and considerations. Adapted from Ref. [5]

Design Consideration	Culture Options
Cell source/cell type Convenience Adapted developmental age Similarities with human nervous system	Rodent fetal and neonatal Cortex, hippocampus, cerebellum, spinal cord, and sensory ganglia are common
Culture type/culture configuration Preservation of anatomy/histiotypic organization Electrophysiological integrity Ability to culture from developing & adult animals Replicate cultures, cell accessibility, cell subset purification	Organ, slice, and explant (fragment) cultures Disaggregated or reaggregated cells and cell lines
Culture Environment Extracellular fluid composition Temperature, pH, gas phase Substrate Dimension	37°C, 100% humidity, 5% CO_2/95% air Culture medium such as DMEM, DMEM/F12, or Neurobasal as neural cells require specific concentrations of ions, amino acids, vitamins, cofactors, hormones, mitogens (for proliferating cells), and other metabolites Additional supplements include fetal bovine or horse serum, or chemically defined supplements

exhaustive methods-based description of culture methods across the regions of the nervous systems, the reader is referred to [5].

2.3 *Dissociated Cell Cultures*

Specific cell populations may be isolated from animals or humans and dissociated, or separated from each other, to acquire a suspension consisting of numerous single cells. The classes of dissociated cell cultures consist of primary cells, secondary cells, and/or cell lines, which are then plated in some configuration (see Table 1) based on experimental objectives. Primary cell cultures consist of cells that are isolated directly from tissues and plated for experimentation. Primary cells therefore may closely represent the endogenous phenotype; however, many primary neural cell types are difficult to keep alive for extended periods of time. Secondary cell cultures consist of cells that were permitted to proliferate for several cycles in vitro prior to experimental use. This process is referred to as "passaging", and is typically done to amplify the cell number or purify a particular cell population; however, absent genetic manipulation passaging is limited by a finite number of cell divisions (Hayflick principle) [6]. Cell lines are primary cells that have been "immortalized", typically by genetic mutations or insertions, and may represent a clone of a particular type of cell, which can, in theory, be kept in culture indefinitely. As cell lines may essentially be cultured indefinitely, there is typically an abundant supply with uniform/consistent behavior and ease of experimental manipulation (e.g., transfection). However, issues exist with "clonal variation", and cell behaviors may not always be representative of a population of cells in the body (i.e. discrepancies between primary cells and cell lines of the same lineage) [7].

Dissociated neural cells may be plated on planar (2D) surfaces, on or within three-dimensional (3D) matrices, or formed into 3D spheres, referred to as reaggregate cultures. These general configurations present a spectrum of critical factors that profoundly influence neural cell growth and function, such as culture architecture (2D vs. 3D; affecting morphology/cytoarchitecture), ECM constituents (affecting cell survival and process extension), and cell density (affecting degree of cell–cell interactions). Additionally, these factors are interrelated; for example the three-dimensionality, ECM properties, and cell density affect the ICS-ECS ratio as well as mass transport requirements (see below).

Planar cultures consist of dissociated neural cells plated directly on a rigid substrate, typically polystyrene or glass, with surface modifications to alter charge or present bioadhesive ligands to facilitate cell attachment [8–10]. Also, 3D neural cell cultures have been developed from dissociated cells as reaggregated sphere cultures [11, 12] or by distribution throughout a matrix [13–17]. For many neurobiological questions, the choice of a planar versus 3D culture configuration offers trade-offs between experimental complexity and fidelity in representing a specific response. In particular, important differences exist between cells cultured in planar versus 3D configurations in terms of the distribution and types of cell–cell/cell–matrix interactions, access to soluble factors, and cell morphology that may drastically affect critical neural cell responses [18]. Cells within a 3D bioactive matrix present different types, quantities, and distribution of cell–cell and cell–matrix interactions [19–22] compared to 2D counterparts. Specifically, cells in 3D contact ECM and experience cell–cell interactions (e.g., receptor-mediated, synaptic junctions) in all spatial directions. Alternatively, cells in 2D typically have a majority of the cell surface directly exposed to media, and may only experience cell–cell interactions in a single plane and are exposed to a much larger ECS as compared to the intact tissue organization. Also, the presence of a 3D matrix alters the cellular microenvironment, affecting concentrations of trophic and signaling factors secreted by cells. Not surprisingly, cells cultured in 2D have shown altered responses to exogenous factors independent of changes in surface area [23]. In addition, cells grown in 3D versus 2D environments have a starkly different morphology and cytoskeletal structure [24]. In the case of neurons, somata and growth cones in 2D are flatter compared to cells in 3D which present a rounder, more bulbous shape [25]. Thus, inherent differences in cell–cell/cell–matrix interactions coupled with corresponding alterations in cell morphology and alterations in the cellular microenvironment may have an impact on intracellular signaling and gene expression.

Given that all neural cells in vivo interact within a 3D environment at relatively high cell densities, quasi-planar neural cell culture models have been developed consisting of dissociated primary cells above a 3D matrix material [26–28]. These systems effectively approximate a 3D morphology/orientation and may support high 2D cell densities. However, cell–cell and cell–matrix interactions are still somewhat constrained, as cells are not distributed throughout the full thickness of the matrix. Conversely, complete three-dimensionality is achieved in reaggregate neural cultures that are developed by rotation-induced reassociation [11, 12, 29]. These systems produce spheres at high 3D cell densities where diffusion-based mass transport is enhanced by convection due to circulating media. In reaggregate culture systems, the volume available for growth is inherently limited based on surface area to volume ratios that permit survival, thus limiting the scope of 3D interactions. Moreover, the ECM components and cellular distribution are difficult to control in reaggregate cultures; however, these models are extremely useful for studying cell–cell interactions, growth, and function at cell densities that closely match those found in vivo. Three-dimensional cell culture models consisting of neural cells distributed throughout a matrix have also been developed [13, 15–17, 30–32]. In these systems, trade-offs exist between culture thickness (i.e., surface area to volume ratio), and hence the scope of 3D spatial interactions, and cell density, necessitating that relatively thick (>500 μm) cultures use cell densities at least an order of magnitude lower than that found in brain cortices [32, 33]. However, using these 3D scaffold-based neural cultures, key parameters governing neuronal survival and neurite outgrowth based on scaffold physical/biological properties and mass transport phenomena have been uncovered. For example, primary dorsal root ganglion (DRG) neurons survive and extend neurites through hydrogel matrices dependent on the (1) physical properties (e.g., agarose pore size [34], stiffness [25]), (2) ligand concentration (e.g., collagen [35], RGD peptides in fibrin [36]), and (3) substrate geometry [37]. Also, in 3D matrices, DRG neurite growth was inhibited by both certain biochemical and mechanical transitions [38], whereas outgrowth was enhanced in engineered matrices by the presence of specific peptide sequences [39]. The survival of primary neurons from the cerebral cortex has been demonstrated within 3D matrices of collagen and/or various hydrogels (e.g., poly[N-(2-hydroxypropyl)-methacrylamide] [16], poly(acrylate) [13], agarose mixed with collagen [15], collagen covalently linked to agarose [32], and Matrigel [31]), In general, primary cortical neuronal survival and the extent of neurite outgrowth are improved by the addition of specific bioactive cell–matrix interactions [13, 15, 16, 32]. However, neuronal survival and neurite outgrowth in 3D matrices are influenced by intrinsic (e.g., neuronal maturation, receptor expression) as well as extrinsic (e.g., matrix mechanical properties, ligand concentration) signals. Complex, non-linear, and often synergistic relationships exist between matrix mechanical properties and the presence and density of a specific ligand. However, these culture systems present a broad scope of 3D cell–cell/cell–matrix interactions, over a length-scale of millimeters. Importantly, these systems may be engineered from the bottom-up, presenting exquisite control over neural cell populations and scaffold parameters to enable optimization based on the individual neural cell populations and cell densities.

Additionally, neural culture models consisting of multiple cell types closer approximate the heterogeneity of in vivo neural tissue. This heterogeneity in cell-type composition may be particularly important to represent such interactions as physical support and metabolic coupling between neurons and glial cells [40–42]. Co-cultures consisting of neurons and glia are typically 2D models, although multi-cellular 3D reaggregate cultures [43, 44] and 3D neuronal-astrocytic co-cultures [14, 45] have been developed. In 2D neuronal-astrocytic co-cultures, the cells typically self-organize into a base layer of astrocytes with neurons on top [46]. Although this distribution spatially constrains neuron-astrocyte and neuron-neuron interactions, planar co-cultures have established a pivotal role for astrocytes in neuronal survival and synapse formation [47, 48]. These neural cell culture models consisting of multiple cell types are capable of maintaining many positive aspects of in vitro modeling while closer approximating additional aspects of neural cytoarchitecture.

2.4 *The Effects of Cell Density on Neuronal Survival in 3D Culture*

A direct relationship between neuronal plating density and neuronal survival has been found [32, 49]. Primary cerebral cortical neurons were plated in thick (500–600 μm) bioactive matrices at cell densities varying over an order of magnitude (1,250–12,500 cells/mm^3), and culture viability was assessed at 2 and 7 days post-plating. At 2 days, neuronal survival was high in lower density cultures <5,000 cells/mm^3, but poor in the higher density cultures. However, by 7 days, there was a parabolic relationship between cell plating density and cell viability, as cultures plated at either lower densities ≤2,500 cells/mm^3 or higher densities ≥6,250 cells/mm^3 exhibited extremely poor viability (<50% for each). However, neuronal cultures plated at 3,750–5,000 cells/mm^3 produced an optimum viability of ~90%. It was postulated that neuronal death in low cell density 3D culture reflected some minimum threshold for neuron-neuron interactions, both physical and chemical, and was likely independent of mass transport limitations. However, mass transport limitations were likely the predominant reason that 3D neuronal cultures exhibited poor viability at high cell densities (≥6,250 cells/mm^3). Taken together, these observations suggest an important balance between diffusional requirements (affecting the higher density cultures) and a potential threshold for cell–cell interactions (affecting the low density cultures). Increasing the cell density in 3D effectively decreases the available space for diffusion, increases tortuosity factors, and increases the overall rates of nutrient consumption and waste production, creating system-specific relationships between culture thickness, cross-sectional area for diffusion, cell type and cell density. Overall, viability in 3D neuronal cultures was highly dependent on cell density, but an optimized cell density range (3,750–5,000 cells/mm^3 for 500–600 μm thick cultures) yielded cultures with extensive neurite arborization, robust neuronal survival and active outgrowth.

2.5 Theoretical Analysis of Diffusional Parameters

Cell–cell and cell–matrix interactions are important for the reconstitution of brain tissue properties in a culture environment. The cell density is often an important parameter considered for dissociated cultures, directly related to the intercellular spacing, and thus crucial to function. The brain has low porosity and diffusivity compared to many other tissues. This is due, in part, to a high cell density and the tortuosity. Tortuosity is a measure of hindrance due to cellular obstructions to molecular diffusion. The volume fraction of the tissue ($\alpha =$ volume of ECS/ volume of tissue) and the tortuosity (λ) have an inverse relationship. The volume fraction is therefore a function of the relative amounts of extracellular space, pore space, and void fractions [50], and is relatively low in brain tissue. For sufficient mass transport, the brain overcomes low diffusivity by maintaining tightly coupled neuronal-glial interactions, with astrocytes in particular having an extension in direct contact with the vasculature. In addition, as with most tissue in the body, a dense vasculature network leaves brain cells in close proximity to a capillary, thus maintaining relatively short distances (typically $<100\ \mu m$) to the convective nutrient source.

However, simple 3D neural cell cultures have no such vasculature, and therefore typically rely on passive diffusion, which alters the cellular micro-environment by driving nutrients in and waste products out. Thus, mass transport phenomena become crucial in affecting the health and viability of neural cells cultured in 3D. Fick's 2nd Law of Diffusion provides a mathematical framework to describe passive diffusion in 3D neural constructs and, in particular, assess the effects of increased cell density. For simplicity, this analysis was performed in one-dimension (e.g., 3D cell culture adhered to a substrate), approximating the cell-containing matrix as a disc (with specified cross-sectional area and thickness in the z direction), and initially assuming a constant diffusion coefficient, D.

$$\frac{\partial C}{\partial t} = D\frac{\partial^2 C}{\partial z^2} \tag{1}$$

This relationship accounts for passive diffusion into the matrix; however, the cells within the cultures are not passive, therefore the rate at which a particular nutrient is consumed must be considered. For this purpose, consumption includes a given nutrient being metabolized or immobilized on or within a cell, which if this process proceeds rapidly, local equilibrium may be assumed to exist between the extracellular (free) and consumed components of the nutrient. In this case, the concentration S of consumed substance is assumed to be first order and will be proportional to the concentration C of free (diffusing) nutrient by a consumption rate constant, k, described by the relationship $S = kC$. This consumption rate constant will be specific to a particular compound, as some nutrients freely cross the cell membrane while others exhibit specific mechanisms of entry. Also, consumption will vary as a function of cell number, cell type and level of metabolic activity. Accordingly, Fick's 2nd Law may be modified to become

$$\frac{\partial C}{\partial t} = D\frac{\partial^2 C}{\partial z^2} - \frac{\partial S}{\partial t} = D_c\frac{\partial^2 C}{\partial z^2}, \text{ where } D_c = \frac{D}{k+1} \tag{2}$$

where a new diffusion coefficient, D_c, is the diffusion coefficient given nutrient consumption, and is a synthesis of the free diffusion coefficient and the consumption rate constant. Assuming a large nutrient source (i.e. bulk medium volume above the matrix), the resulting concentration profile, C (z, t), within the cell culture may be calculated. The mass entering the matrix may then be described by the integration of the concentration profile times the effective area for diffusion (A_{eff}) over the thickness (w) through the matrix:

$$M = \int_0^w C(z,t)A_{eff}dz \tag{3}$$

In acellular constructs, A_{eff} would be equal to the overall cross-sectional area of the matrices; however, in cell-containing matrices, A_{eff} is the total cross-sectional area of the matrix minus the average area (per plane) occupied by cells:

$$A_{eff} = A_{total} - A_{cells} \tag{4}$$

Thus, mass transport in this system is (1) governed by the diffusion/consumption characteristics of a particular compound, (2) proportional to the effective cross-sectional area for diffusion, and (3) effectively delayed by the culture thickness. For this analysis, the effective area for diffusion may be considered inversely proportional to cell density, as the open porosity of the matrix was assumed to be constant and thus not affected by cell density. Changes in diffusional area are demonstrated in confocal reconstructions of 3D neuronal cultures plated at various cell densities [49]. Increases in cell density will increase the tortuosity of the system as an enhanced cell-neurite network necessitates increasingly convoluted pathways for diffusion. Although this analysis addressed the rate of nutrient consumption, this relationship may be modified to describe the rate of metabolic waste production and subsequent diffusion out of the cellular microenvironment. This process will also be adversely affected by increases in cell density as this will increase the rate of waste production, possibly creating an unhealthy microenvironment for the cells. Here also, increases in culture thickness will serve to delay clearance of the waste from deep within 3D constructs.

Overall, mass transport must surpass specific metabolic thresholds for 3D culture systems to support viable cells, and given the inter-relationship between cell density (determining the effective area for diffusion) and culture thickness, changes in one parameter may compensate for another provided mass transport thresholds remain surpassed. In particular, the mathematical analysis was a useful tool to demonstrate that increasing the cell density in 3D effectively hinders mass transport by decreasing the available area for diffusion, increasing tortuosity, and increasing the rates of nutrient consumption and waste production. This analysis supports our conclusion that mass transport limitations were the predominant reason that 3D neuronal cultures exhibited poor viability at high cell densities ($\geq$6,250 cells/mm^3) [51]. Indeed, it has been demonstrated that enhanced mass transport through forced interstitial convection through the 3D neural cell cultures using custom-built micro-bioreactors was sufficient to support neuronal and neuronal-astrocytic co-cultures at cell densities of 10,000–50,000 cells/mm^3 (depending on total culture volume) [32, 33]. Thus, using interstitial convection to enhance mass transport enabled the construction of thick neural constructs approaching the cell densities of 100,000–1,000,000 cells/mm^3 reported in various cortical regions in vivo [52, 53].

Thus, there are important design considerations and limitations that must be acknowledged for 3D in vitro systems. Although our analysis focused on the effects of cell density and culture thickness in relatively large volume cultures ($\sim$200 μL), it is important to note that these factors alone will not dictate mass transport thresholds. To be complete, the overall construct dimensions, geometry, and physical properties need to be taken into account (e.g., surface area to volume ratio, thickness, pore size, tortuosity, etc.). For instance, reaggregate cultures (or neurospheres) may have a cell density of 1,000,000 cells/mm^3, which at several hundred microns in diameter have a high surface area to volume ratio potentially sufficient for support via passive diffusion (although there are necrotic cores in many cases). In non-optimized 3D cultures, cells may experience diffusional transport limitations, causing essential nutrients to be absent from some cells and possibly leading to accumulation of toxic waste products. Such phenomena may have devastating effects on cell survival, or insidiously may confound experimental results by altering gene expression, detrimentally effecting protein production and fidelity, and leading to appreciable deviations from in vivo behavior [54]. However, properly designed and optimized 3D models, where diffusional limits are not approached, may more faithfully recapitulate elements of native tissue than 2D models.

2.6 Application of 3D Neural Cell Cultures

Neural cell culture models have been reconstituted in 3D constructs to study neural mechanobiological phenomena within 3D microenvironments [14, 17, 31, 32, 45]. These models consisted of either predominantly primary cortical neurons or separately isolated primary cortical neurons and cortical astrocytes, each harvested from rodents, dissociated using standard techniques, and mixed in controlled ratios

[14, 31, 32]. These neural cells were homogeneously distributed throughout the full thickness of bioactive scaffolds (typically 500–600 μm thick, up to 1 mm in some cases), and were maintained using a defined, serum-free, Neurobasal medium. Various ECM- or hydrogel-based scaffolds, including Matrigel (i.e., laminin, collagen IV, entactin, heparan sulfate proteoglycan [55, 56]), collagen IV, and/or agarose have been used. Notably, these systems may be engineered from the bottom-up, with precise control over cellular (e.g., phenotypes, ratios, densities) and scaffold parameters (controlling mechanical properties and extent of bioactive motifs), to assess the influences of specific cellular and environmental factors on mechanobiological responses.

Neural cell survival, neurite growth, and functional maturation within these constructs were characterized. At the initial time of 3D plating, the dissociated neural cells had a spherical morphology (absent neurites) and were entrapped throughout the thickness of the matrix. Over days in culture, there was considerable process outgrowth resulting in the formation of 3D, interconnected neural networks by one week [31, 32]. Plating density was a critical parameter for neurons in 3D, with an optimal cell viability obtained at 3,750 cells/mm^3 (based on the starting surface area to volume ratio) [49]. Astrocytic presence significantly improved long-term culture viability, as cell viability in neuronal cultures was ~90% at 7 days, but <70% at 21 days, whereas viability in co-cultures was >95% up to 21 days. Notably, long-term survival of 3D neuronal-astrocytic co-cultures was observed out to over 60 days in culture (unpublished observation).

Moreover, neuronal maturation was demonstrated through the expression of mature isoforms of neuron-specific cytoskeletal proteins and proper electrophysiological function over weeks in culture [45]. The presence of astrocytes also enhanced expression of neuronal functional markers via an increased rate of synapse formation and increased number of synapses per neuron. Neurons in 3D co-culture were found to have normal resting membrane potentials (average − 56 mV), expressed voltage-sensitive ion channels (Na^+ and K^+ currents), displayed both spontaneous and evoked action potentials (average spike height 70 mV), and exhibited functional synapse formation and network properties. These tissue engineered 3D neural cell cultures provide an innovative platform for neurophysiological and mechanobiological investigations, and serve as an important step in the development of more physiologically-relevant neural tissue models.

2.7 Organotypic Slice Cultures

An alternative to artificial constructs, slices of brain tissue can be cultured for extended periods of time (i.e. in excess of 1 week) given the appropriate environment, thereby allowing the tissue to recover from the trauma of dissection. These cultures are surprisingly stable, maintaining the in vivo anatomy with high fidelity. Organotypic cultures are typically produced from young animals (<P11), and efforts have characterized the in vitro maturation process in detail, suggesting

that they mature in vitro, albeit at a slower pace than in vivo. These culture systems represent an in vitro substrate between the complexity of in vivo models and simplicity of dissociated cell culture, and have been used in models of mechanical injury [57, 58]. Cultured structures include hippocampus [57–59], cortex [60], thalamo-cortical slices [61], whole coronal brain sections [62], and transverse sections of the spinal cord [63, 64].

A critical determinant of organotypic culture health is gas transport which has been addressed through a number of strategies. In one method, tissue was adhered within a plasma clot to a cover-slip placed in rotating tubes with medium [65]. Cultures were sequentially submerged in medium and exposed to the atmosphere ensuring adequate gas exchange and nutrient supply. One limitation was that the cultures thinned to one or two cell layers thick [65–67]. An alternative strategy which produced much thicker cultures many cell layers thick maintained cultures on a silicone membrane with gentle rocking [57, 62]. A third method maintained thick cultures atop a porous membrane with medium delivered through the pores [68–70].

Slice culturing has been observed to begin with an initial period of degeneration which was mostly cleared away by 5 days in vitro (DIV) [69, 71]. Over time, the total number of cells remained constant with only a 12% decrease at 21 DIV [69]. Neuronal numbers remained constant over 21 DIV in the thicker interface cultures [69]. Since total protein content remained constant, the thinning of the interface cultures may be explained by compaction of the ECS which is reminiscent of developmental compaction in vivo [69, 70].

At the single cell and ultrastructural levels, some initial modifications due to the harvesting procedure were noted, but after about a week, cells and synapses developed normally [69, 71–74]. As early as 4 DIV, synaptic density began to recover [65, 69, 71], and by 28 DIV synaptic density was similar to age-matched in vivo levels [69, 72]. The neuronal projections of CA1 and CA3 were fairly resistant to remodeling, although some occurred [67].

Microtubule associated protein-2 (MAP2), a cytoskeletal component, isoform expression (protein and message) mirrored the in vivo developmental program [75, 76]. Tau, another cytoskeletal protein, was initially expressed as a single isoform, but by 14 DIV, four isoforms were expressed, a process not replicated in dissociated cultures [75, 76]. Myelin basic protein (86%) and NCAM (neural cell adhesion molecule) isoforms (71–80%) were expressed at close to adult levels [76]. After an initial drop, MK-801 and AMPA binding, indicative of functional glutamate receptor expression, recovered by 5–10 DIV to in vivo levels [76, 77]. Glutamate receptor subtypes GluR1, Glur2/3, and NMDAR1 were expressed at comparable levels to age-matched controls (50, 40, and 43%, respectively), although GluR4 was aberrantly expressed [76]. The anatomical pattern of excitotoxic death was similar to the in vivo pattern suggesting an anatomically correct distribution of NMDA, AMPA, and kainate receptors [78, 79].

Within the slices, the DG to CA3 to CA1 excitatory pathway was well preserved, although with some sprouting [80]. Over the first weeks in vitro, field population spike and excitatory post-synaptic potential (EPSP) amplitudes increased until approximately 28 DIV and then stabilized [72, 81, 82]. Induction of long term potentiation (LTP), a cellular correlate of learning, was possible after 9–15 DIV, whereas in vivo, LTP can be induced after post-natal day 14 (P14) [72, 81]. Paired pulse facilitation (PPF), a measure of short-term synaptic potentiation, emerged in culture at approximately the same time as in vivo [72, 81]. Cultures grown on cover-slips were more excitable than interface cultures, possibly because of the substantial thinning and loss of cells, potentially including inhibitory interneurons [83, 84]. In contrast, thicker cultures were less excitable and did not burst when followed out to 80 DIV [70, 85].

Progress has been made in the long-term culture of hippocampal tissue from either P20 [86] or as old as P30 [85] rat pups, although long-term culture of fully adult hippocampal tissue is not currently possible [87]. Greater success has been achieved with transverse spinal cord cultures which can be produced from mice as old as P21–42 [63].

3 Neural Mechanical Properties

Existing experimental data and scaling relationships have been used to empirically derive mechanical thresholds to predict physiological outcome in animal experiments [88]. In order to examine cell and tissue level responses, high resolution finite element models (FEMs) require material properties on a corresponding scale. It is imperative that these assigned material properties coincide with those of living tissue to ensure accurate predictions of mechanical response to stress and strain in vivo. In addition to providing increased fidelity to FEMs, the material properties of living tissue can be emulated in culture systems that study mechanical cellular response.

3.1 Mechanical Properties of Brain

Measuring material properties of the brain poses particular challenges, in part due to its high cellularity as detailed above. Without the proper environmental conditions, e.g. adequate gas exchange, nutrient medium, hydration, correct osmolarity, temperature, and pH, cells within the excised tissue will die rapidly altering mechanical properties of the tissue. Determining mechanical properties under the same conditions required for culturing brain tissue avoids many of these pitfalls which may have confounded earlier investigations. A large range of shear moduli have been reported for different species, times postmortem, preconditioning, and frequencies/times, but a consensus is emerging that the shear modulus of brain is around 1 kPa, increasing with rate or frequency [89, 90].

Brain is one of the softest tissues in the body. Therefore to measure reaction forces during testing requires either very sensitive transducers or large samples. Previous studies using relatively large samples (10–30 mm) have used shear [91–95], compression [96, 97], or tension [98, 99]. See in Chapter "Brain Tissue Mechanical Properties" of this volume for detailed discussions of brain properties at macroscopic scales. The use of relatively large tissue samples (at least 10–30 mm) makes the production of a homogeneous tissue sample difficult (depending on the structure of interest and species), thereby complicating interpretation of results. Brain is structurally heterogeneous with gray and white matter at the grossest level of distinction with many finer features (on the order of microns and millimeters) apparent upon histological examination. The use of smaller samples allows measurement of material properties of sub-regions and smaller structures and is aided by the use of novel, highly sensitive microprobes.

One motivation to measure material properties of anatomical structures in the brain is that inclusion of heterogeneous material properties in computational models affects the induced strain fields and hence the pattern of tissue injury [100]. Only by inclusion of particular anatomical structures within a model can the predicted injury pattern be compared and validated against the in vivo histological injury pattern. For example, the CA3 pyramidal layer of the hippocampus is particularly vulnerable to fluid percussion injury (FPI) models of TBI [101]. Predicting this preferential cell loss in CA3 with a computational model would be a substantial step toward its validation.

An alternate testing methodology that is less constrained by sample size is indentation [102–106]. Anatomical regions of interest can be tested in situ, limited by the size of the indenter and the sensitivity of the force transducer. Recently, this approach was used to measure linear viscoelastic mechanical properties of anatomical structures within coronal sections of the rat brain through stress relaxation experiments with a flat circular probe of radius 250 μm [106]. The CA1 subfield of the hippocampus was the stiffest brain region tested, whereas other hippocampal regions (CA3 and dentate gyrus) were significantly softer. The cerebellum was the most compliant region of the brain tested. On average, white matter regions were softer than gray matter, and about a two-fold difference between the stiffest and most compliant brain regions was observed (Fig. 1). Overall, the brain became stiffer with animal age.

To capture the time-dependent relaxation behavior, multiple linear viscoelastic constitutive models were fit to the data, including a Prony series approximation (Eq. 5), a continuous phase lag model (Eq. 6) [107, 108], and a power law model (Eq. 7) [109].

$$G(t) = G_{\infty} + \sum_{j} G_j \cdot \mathrm{e}^{-\frac{t}{\tau_j}} \tag{5}$$

$$G(t) = G_{\text{inst}} \left[\frac{1 + c\left[\int_{\tau_2}^{\tau_1} \frac{e^{\frac{-t}{x}}}{x} dx\right]}{1 + c \ln\left[\frac{\tau_2}{\tau_1}\right]} \right] \tag{6}$$

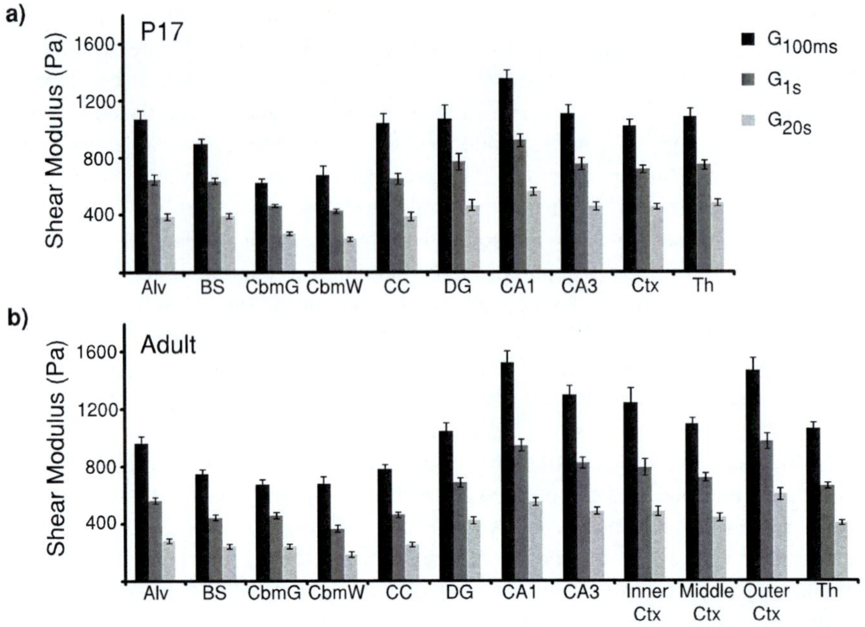

Fig. 1 Time dependent shear modulus determined from stress relaxation indentation experiments is presented for brain tissue of different aged rats: **a** P17 and **b** adult. Modulus is presented at specific time points during the relaxation after indentations of approximately 10% strain (mean ± standard error of the mean). Overall, adult brain was slightly stiffer. White matter was softer than gray; the cerebellum was the most compliant structure tested. *Alv* alveus, *BS* brain stem, *CbmG* cerebellar gray matter, *CbmW* cerebellar white matter, *CC* corpus callosum, *DG* dentate gyrus, *CA1* CA1 region of the hippocampus, *CA3* CA3 region of the hippocampus, *Ctx* cortex, *Th* thalamus. Reproduced with permission from Mary Ann Liebert [106]

$$G(t) = \frac{k}{\Gamma(1-\beta)} t^{-\beta}; k = \frac{A}{\tau^{-\beta}} \tag{7}$$

The Prony series fit the complete data set with the largest R^2 value, although all model fits were greater than 0.92. Although the Prony series fit yields a non-unique set of parameters, its implementation into FEMs is straightforward, since it is usually an included constitutive description.

The main limitation of the indentation methodology is that the strain field beneath the indenter is not uniform. For a flat circular indenter, strains are maximal at the indenter edge consisting of both compressive and tensile strains. An analytical solution for calculating shear modulus exists:

$$G(t) = \frac{1}{4} \cdot \frac{P(t)(1-v)}{R\delta} \tag{8}$$

in which P is the reaction force, R is the indenter radius, δ is the indentation depth, and ν is Poisson's ratio [110, 111]. The analysis assumes the tissue is homogeneous, isotropic, an infinite half-space, and that linear elasticity applies (i.e. small strains). A hereditary integral can be used to determine $G(t)$ if δ is not an ideal step function. Deviations from the analytical solution are influenced by the ratio of indenter radius to specimen thickness [112]. Additionally, although our study utilized a small indenter ($R = 250$ μm), the probe was a similar size to some of the anatomical structures of interest. Actuator vibration and force transducer sensitivity limited the use of smaller probes and faster indentation rates.

To further improve the spatial resolution of testing methods, the atomic force microscope (AFM) has been adapted for indentation of living brain tissue or brain cells [104, 105, 113]. With nano-Newton force resolution and micron spatial resolution, the AFM is well-suited for mechanical testing of brain allowing the use of micron scaled probes. Using a spherical tip of radius 12.5 μm so as to measure tissue properties as opposed to sub-cellular properties, significant differences were found between different anatomical structures within the hippocampus [104, 105]. For example, the CA1 pyramidal layer which contains neuronal cell bodies was the stiffest structure in the adult brain [105]. It was significantly stiffer than the adjacent CA1 stratum radiatum which is comprised of neuronal processes and glia. Another region within the hippocampus, the dentate gyrus was only 25% as stiff as the CA1 pyramidal layer in the adult brain at an indentation depth of 3 μm (approximately 30% indentation strain). These differences may be large enough to affect the induced strain field caused by injurious loading, thereby strongly influences the pattern of traumatic cell death. Future computational models of TBI may need to include this level of spatial detail to accurately predict regional cell death.

A further complication for characterizing the mechanical properties of brain tissue is its non-linearity with respect to strain. For indentation depths greater than 0.5 μm with a spherical probe of radius 12.5 μm in AFM studies, brain tissue became stiffer with indentation depth [105]. Several other studies using larger samples have reported strain stiffening for brain [95, 114, 115], while others have reported strain softening but usually for lower strains than those analyzed in the AFM study [93, 98]. To capture this non-linear behavior, AFM data was fit to several non-linear elastic constitutive models including Neo Hookean (Eq. 9] [116], Mooney Rivlin (Eq. 10) [117, 118], or Ogden (Eq. 11) [119]:

$$W = \frac{E}{2}(I_1 - 3) \tag{9}$$

$$W = \frac{C_1}{2}(I_1 - 3) - \frac{C_2}{2}(I_2 - 3) \tag{10}$$

$$W = \frac{2\mu}{\alpha^2}(\lambda_1^\alpha + \lambda_2^\alpha + \lambda_3^\alpha - 3) \tag{11}$$

in which W was the strain energy density, I_1 and I_2 were the first and second invariants of the strain tensor, λ_1, λ_2, and λ_3, were stretch ratios in the principle directions, and E, C_1, C_2, μ, and α were material parameters. The Ogden material description fit the material non-linearity with the smallest mean square error.

Brain mechanical properties were also shown to be age dependent [105]. The brains of younger rats were softer and more homogeneous than adult rats. The increased stiffness correlated with increased lipid content and decreased water content. The effect of age is an active debate as results from another group indicated that younger brain was stiffer [120, 121]. Nonetheless, these results suggest that the induced strain field for younger animals will differ from that of older animals. Such age-dependence may be relevant for designing age-appropriate safety systems for the prevention of TBI.

Because the AFM methodology is inherently an indentation approach, analysis of force deflection curves suffer from the same limitations as described above for the flat circular punch indentation. An added complication for a spherical probe is that contact radius is not constant during indentation. Analytical solutions for spherical indenters exist but make several simplifying assumptions including sample homogeneity, isotropy, infinite thickness, and small strains. The classic solution makes an additional assumption that, for small indentation, the spherical indenter profile can be approximated as a paraboloid. Contact radius was assumed to be $a = \sqrt{R\delta}$ (R and δ defined above) to provide an expression for shear modulus [110]:

$$G(t) = \frac{3}{8} \cdot \frac{P(t)(1 - v)}{\sqrt{R\delta^3}} \tag{12}$$

A solution for an explicitly spherical indenter also exists, but is slightly more complex as the expression for contact radius is transcendental and cannot be inverted [122, 123]:

$$\delta = \frac{1}{2} \cdot a \cdot \ln\left[\frac{R + a}{R - a}\right] \tag{13}$$

$$G(t) = \frac{p(t)(1 - v)}{\left[(a^2 + R^2) \cdot \ln\left(\frac{R+a}{R-a}\right) - 2aR\right]} \tag{14}$$

Note that Ref. [124] contains an error and is missing a 2 in front of the aR term.

The AFM approach can be extended to the viscoelastic regime by applying a small sinusoidal oscillation to the probe tip and measuring the phase difference between the scanner displacement and the probe deflection. Taking the Taylor series expansion of the analytical solution for indentation with a spherical probe for an oscillatory input yields [125–127]:

$$f^*_{osc} = 4\sqrt{R\delta_\circ}\frac{G}{1 - v}\tilde{\delta}^* \tag{15}$$

in which f^*_{osc}, the AFM probe force, and $\tilde{\delta}^*$, the tissue indentation, are complex quantities with magnitude and phase. Splitting this into real and imaginary components yields the storage (G′) and loss (G″) modulus:

$$
\begin{aligned}
G' &= \frac{(1-\nu)}{4\sqrt{R\delta_{\circ}}} \cdot \frac{a_d k a_r \cos\varphi - k a^2}{a_d^2 - 2a_d a_r \cos\varphi + a_r^2} \\
G'' &= -\frac{(1-\nu)}{4\sqrt{R\delta_{\circ}}} \cdot \left[\frac{a_d k a_r \sin\varphi}{a_d^2 - 2a_d a_r \cos\varphi + a_r^2} - \omega \cdot \gamma\right]
\end{aligned} \tag{16}
$$

in which δ_0 is the static indentation depth, k is the cantilever stiffness, a_d is the magnitude of the scanner displacement, a_r is the magnitude of the cantilever deflection, φ is the phase lag between the two, γ is the drag of the probe in the fluid, and ω is the oscillation frequency. This methodology has been applied to measure viscoelastic properties of brain, producing both frequency and indentation-depth dependent storage and loss moduli (Fig. 2, unpublished observations). One of the recurring challenges is accurate contact point identification.

Mechanical properties of single brain cells have been measured with a similar approach [113]. Hippocampal neurons and astrocytes were indented with a 3 μm sphere with superimposed oscillations of 30, 100, and 200 Hz, although no indentation depth was reported. Neurons were stiffer than astrocytes. Shear modulus increased with frequency and ranged from about 100–300 Pa.

Several sources of non linear behavior can complicate analysis of dynamic indentation data. The first is the effect of large deformation on material response. If data from large-deformation indentation experiments is analyzed with linear elastic theory, a linearly elastic material will appear non-linear, exhibiting strain hardening behavior [128]. Simple shear excitation does not suffer the same complication. The second is due to the intrinsic non-linear material properties of the tissue which arises as the excitation strain is increased above the linear limit. A third source of non-linearity in response to a single frequency excitation is the existence of higher order harmonics in the tissue response. These higher order harmonics violate the basic assumptions of linear frequency analysis. Several strategies have been described to analyze this non-linear material behavior [129–131]. A Fourier transform of the tissue response will quickly identify whether higher order harmonics are contaminating the data.

3.2 Mechanical Properties of 3D Neural Cultures

In light of what is known about brain tissue properties, rheological properties (specifically, complex modulus) of Matrigel (7.5 mg/mL), a commonly used 3D cell scaffold, were measured over days in vitro as a function of frequency for the

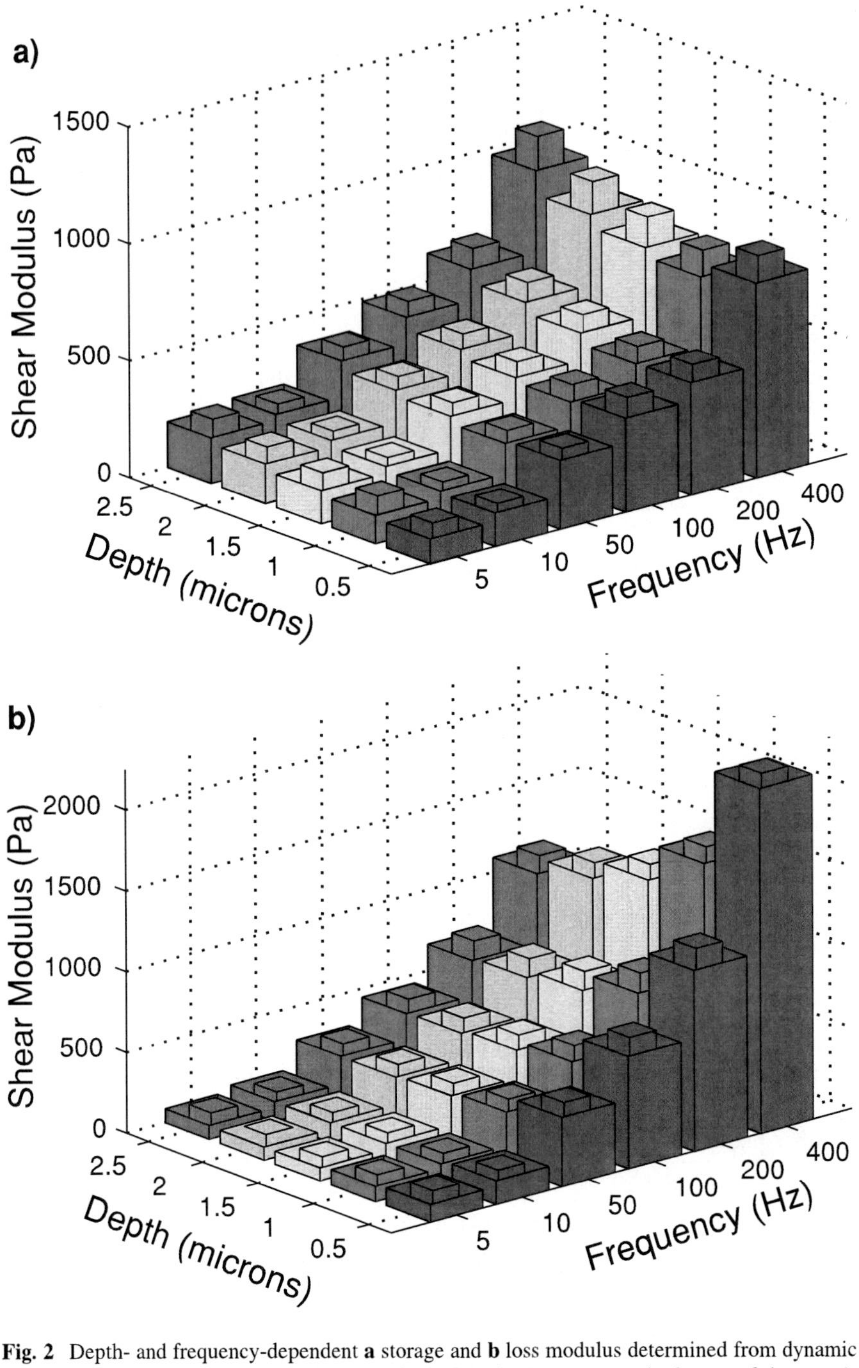

Fig. 2 Depth- and frequency-dependent **a** storage and **b** loss modulus determined from dynamic AFM indentation of the pig alveus with a spherical probe (mean + standard error of the mean). Modulus increased with excitation frequency, consistent with a viscoelastic material. Storage modulus increased with indentation depth, suggesting non-linear behavior

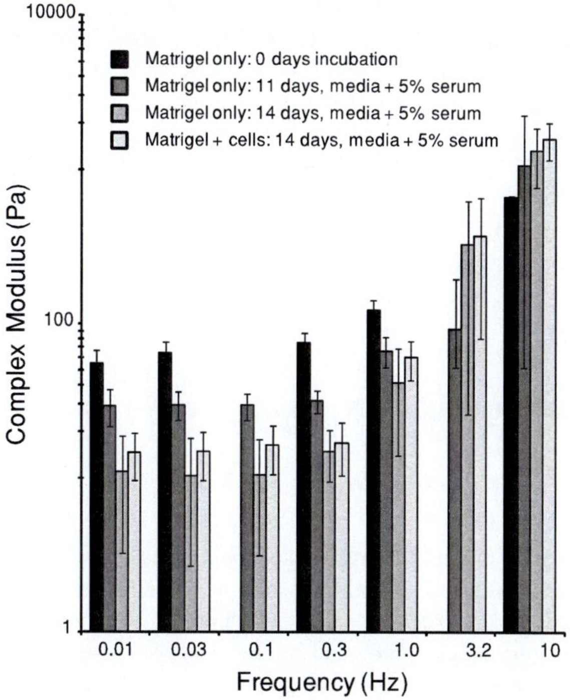

Fig. 3 The complex modulus of thick (500 μm) Matrigel constructs was examined as a function of incubation days, the presence of serum, and the presence of cortical neurons/astrocytes. All dimensions and incubation/rheology conditions were constant for the experimental groups (n = 2–4 samples per condition) for a frequency sweep of 0.01–10 Hz. As expected, complex modulus increased with frequency. The presence of serum during gelation and incubation time periods did not affect the baseline properties (i.e. those at 0 days with no serum). Furthermore, the presence of cells did not significantly affect the overall material properties of the 3D culture after 14 days in culture. Data presented as mean ± standard error of the mean

following conditions and time points: (1) Matrigel alone at 0 (immediately upon gelation), 11, and 14 days following casting in serum free cell culture medium at 37°C and 5% CO_2; (2) Matrigel with neurons and astrocytes at 2,500 cells/mm^3 at 14 days in culture at 37°C and 5% CO_2. All groups except the 0 days group contained 5% serum. All rheology experiments were performed in hydrated samples on a Bohlin CVO Rheometer (East Brunswick, NJ) at 37°C under 0.005 strain with a frequency sweep from 0.01 to 10 Hz (n = 2–4 samples per condition). The trends among the conditions tested were similar across treatment groups with respect to frequency and time under culture conditions, suggesting that 3D cultures of cortical neurons/astrocytes in Matrigel retain the gross mechanical properties over 14 days and that this was independent of the presence of cells (Fig. 3). Furthermore, the mechanical behavior was similar to that of brain [103–106, 121, 132, 133], further supporting the use of 3D cultures and emphasizing the need for mechanical characterization. Of note was that the protein concentration and cell density used were optimized previously for viability [51]. At 0.01 Hz, the order of the complex modulus was: 0 days without serum >11 days with serum >14 days (with or without cells, both with serum). At 10 Hz, the standard deviations were large, and as such, these results can only be taken as a trend, however, the following was the order of the central tendencies of the complex modulus: 0 days without serum <11 days with serum <14 days with 5% serum <14 days with cells and 5% serum. Further studies can help resolve the effects of the culture conditions on rheological properties, such as scaffold degradation, serum presence, and cell density.

4 Mechanical Loading of Neural Cell and Tissue Models

4.1 Mechanotransduction in Neural Tissues

Cells from different organ systems in the body experience different mechanical stimuli (from physiological to pathophysiological) and have different thresholds for activating biochemical and gene-level responses. The process of mechanical stimulation causing biochemical and molecular changes is termed mechanotransduction. Mechanotransduction, in general, is the biological mechanism by which cells interact with their environment and is comprised of a set of complex and dynamic processes. Transmembrane receptors, such as integrins, attach to ECM proteins such as collagen, fibronectin, and laminin. Adhesion receptors have specificity to the ECM ligand molecule and thereby, use different extracellular cues to direct various cell behaviors, such as migration and neurite outgrowth. This process involves an intracellular domain that binds to the cytoskeleton and intracellular signaling molecules such as vinculin, talin, paxillin, focal adhesion kinase (FAK), and proline-rich tyrosine kinase 2 (PYK2). While the function of neural adhesion receptors has been well studied under physiological conditions, the role of these receptors in pathologic mechanotransduction remains elusive.

Cells of the CNS are not normally exposed to large deformations and as such, the mechanisms of signal transduction in response to mechanical stimuli in neurons, in general, are poorly understood. Notable exceptions are in the areas of sensory transduction and neural development. Sensory cells such as inner ear hair cells, for example, which are specialized mechanosensing cells, contain inherent mechanisms for transducing stretch to electrical activity [134]. The physical surrounding of a cell as it interacts with other cells directly or through cell–matrix interactions has been shown in many studies to influence normal function. For example, neurons grown on ECM coated elastomer could be electrically stimulated by mechanical indentation of the elastomer; furthermore this was shown to be dependent on intact ECM-cytoskeletal connections [135].

4.2 Mechanical Response of Neural Cultures

Mechanotransduction has been studied in vitro by controlling the mechanical properties (i.e. rigidity, tortuosity, ligand density) of extracellular matrices and observing cell behavior [30, 136–139]. It has been shown that neurons tend to support longer neuritic outgrowth on compliant substrates compared to rigid support scaffolds [25, 35, 140, 141], whereas fibroblasts, spread and migrate more readily as rigidity was increased [142, 143]. When primary cortical neurons were cultured in a 3D environment with controlled mechanical properties and laminin density, however, it was observed that the more rigid formulation ($G' = 565$ Pa) supported more neurite extension than a softer ($G' = 175$ Pa) gel, suggesting that the 3D environment may provide additional extracellular cues for supporting neurons (unpublished observation). Of note, the stiffer gel mechanical properties were closer to that of brain compared to the softer culture scaffold [103–106, 121, 132, 133], suggesting that cell specific tissue properties should be emulated when creating culture mimics. This concept of matching in vivo mechanical properties to culture conditions has been supported by other studies as well [140, 144, 145]. The response to mechanical stimulation may also be affected by the degree of differentiation, type and level of stress, cytoarchitectural complexity, and physical environment. Controlled biophysical cell and tissue culture conditions are especially critical in which the mechanical environment can be independently varied and the neural response directly measured.

4.3 Mechanical Injury to Neural Tissue

By applying a range of mechanical stimuli to neural tissue models, one can better understand specific cellular responses (e.g., neurotransmitter release, signaling pathways). Here, we focus on the mechanical responses to injurious physical insults observed in 2D and 3D neural cultures and the organotypic slice culture models. Ultimately, the identification of mechanotransduction mechanisms can best be achieved through a multidisciplinary approach, with a variety of cellular and tissue models, animal and human studies, and computer simulations. Traumatic insults to the brain and spinal cord deform cellular elements. Persistent injury can occur from a single mechanical event and subsequent insults may exacerbate the initial response, suggesting increased sensitivity due to an abnormal state [146].

Cellular models of traumatic injury have been developed that mimic the mechanical parameters associated with large, high rate deformation to cellular elements of the brain. These in vitro injury model paradigms include transection, compression, stretch, and shear. Two-dimensional in vitro cultures are valuable tools for studying penetrating and blunt impact injury [147, 148], as well as stretch injury [149–151]. Both 3D cultures derived from dissociated cells and brain slices have been used for neural injury studies [17, 152]. The most appropriate injury models, however, may be those that mimic the 3D in vivo configuration and that capture forces experienced during injury in vivo. In vitro models of TBI allow for well-characterized and repeatable injuries, the parameters of which can be controlled independently and precisely. While various studies have examined cellular and tissue levels thresholds to mechanical loading [31, 153–155], the input parameters inherently vary from system to system, pointing to the continuing need to systematically study a range of mechanical stimuli to different culture systems in a comparative and parallel manner. Some general observations have been made across some laboratories. For example, the rate of application of mechanical loading dictates, to some degree, the severity of the injury [57, 60, 154, 156–160], likely due to the viscoelastic response. Strain level, independent of rate, has also been associated with injury severity [31, 57, 60, 160–162]. It is likely that varying loading regimes, input parameters, cell preparations, time points for outcome measures, and other experimental variables play a role in these different responses. In addition, one must consider the nonlinearities that may be part of the injury response and the range of mechanical parameters with respect to the analogous in vivo loading conditions, many of which are difficult to link to specific pathologies.

Experimental models of superthreshold mechanical stimuli must be defined by: (1) the culture composition (including density, extracellular constituents, cell types, and developmental age) and mechanical properties; and (2) the mechanical parameters (e.g., stress, strain, and modality) that mimic the phenomenon being studied. In vitro systems eliminate systemic responses, while maintaining the ability to manipulate and observe cells during and after physical stimulation [57, 163–166]. Cellular and tissue neural models preserve the ability to apply a more controlled mechanical input than in a correlate in vivo environment. Because complexity, such as additional cell types or extracellular stimuli can be added in a controlled fashion, such models play an important role in discerning the mechanical principles that govern physical and functional response and thresholds.

5 Defining Tolerance Criteria for Neural Tissue

5.1 *Definition of Failure*

Failure is a central theme of our studies on brain tissue and cultures, and our studies have benefited from a definition that is broader than the typical engineering definition of ultimate failure. Mechanical failure may be an adequate definition for load-bearing tissues of the human body like bone and even for soft tissues in the most extreme cases. However, tearing (fracture) of tissue or cells is only one mode of failure, and many others may exist for biological tissues as discussed below.

A fundamental difference between inert engineering materials and biological tissues is that the latter are alive and perform some form of active, physiological function. Therefore, the very concept of failure must be revised. The exact definition may be tissue-specific such as fracture for bone or laceration for skin. However, these definitions may be too insensitive to identify the onset of injury for the brain, i.e. tolerance criteria. For non load-bearing, highly cellularized, living tissues, failure can be defined in myriad ways and may occur far below mechanical failure limits.

5.2 Tolerance Criteria for Living Brain Tissue

Quantification of tolerance criteria for brain has a number of requirements including a living preparation, a means to apply precisely controlled mechanical stimuli, and a mechanism to verify tissue biomechanics. For brain and other highly cellular tissues, cell death may be a widely applicable outcome. This working definition of injury has been applied to determine tissue-level tolerance criteria for cell death in the rat hippocampus and cortex [60, 160]. In these studies, organotypic slice cultures of brain were subjected to controlled mechanical deformation, and the resultant cell death quantified in a region-specific manner over time [57]. An empirical relationship between input mechanical stimuli (tissue strain and strain rate) and cell death was calculated. The goal was to provide these transfer functions for incorporation into FEMs, providing them with the capability to predict biological responses in addition to typical mechanical parameters such as stress and strain.

It was found that cell death was not immediate in response to deformation that was below cellular ultimate failure limits [57, 60, 160]. Instead, cell death increased over 4 days after injury, highlighting a fundamental difference to engineering materials. Both CA3 and CA1 of the hippocampus were the most vulnerable regions to mechanical injury. Dentate gyrus was less vulnerable than other hippocampal regions. Cortex was much less susceptible than the hippocampus as a whole; interestingly, cell death in the cortex increased with increasing strain rate in contrast to hippocampal cell loss which was insensitive to rate. To illustrate, after 0.20 biaxial strain, approximately 35% of cells in CA1 and CA3 died versus 20% in the dentate gyrus and less than 5% in the cortex.

By quantifying both the input mechanical stimulus and the output biological response, it was possible to determine numerical relationships for predicting the time course and magnitude of post-injury cell death (Fig. 4). These three empirical functions (Eq. 17) could be incorporated into FEMs, thereby equipping them with biological predictions to supplement mechanical predictions of stress and strain.

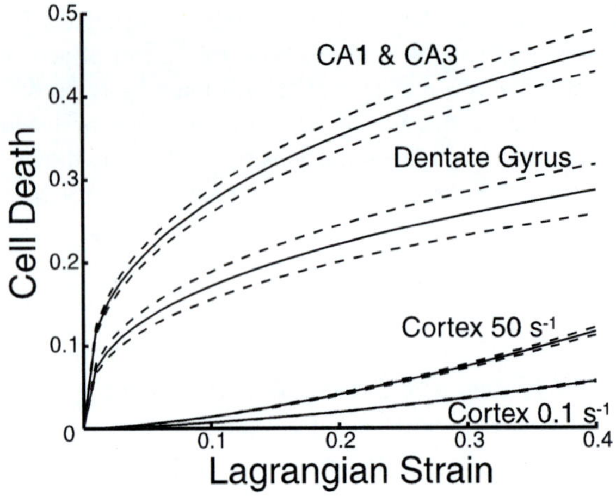

Fig. 4 Mathematical function representing tolerance criteria relating input biaxial strain to resultant cell death 4 days post-injury were derived from experimental data. The *solid line* represents the best fit with 95% confidence intervals indicated with *dashed lines*. The CA1 and CA3 regions of the hippocampus were the most vulnerable, followed by the dentate gyrus. The cortex was much less susceptible. Cortical cell death was dependent on applied strain rate whereas hippocampal cell death was not. Modified with permission from the Stapp Association [60] and with permission from Elsevier [160]

$$\begin{aligned}\mathrm{CA1\&CA3} &= 0.0389(\pm 0.0011)\cdot \mathrm{Strain}^{0.3663(\pm 0.0029)}\cdot \mathrm{Time}^{2.0150(\pm 0.0216)}\\ \mathrm{DG} &= 0.0323(\pm 0.0017)\cdot \mathrm{Strain}^{0.3721(\pm 0.0056)}\cdot \mathrm{Time}^{1.8209(\pm 0.0407)}\\ \mathrm{Cortex} &= 0.094(\pm 0.0021)\cdot \mathrm{Strain}^{1.5293(\pm 0.0125)}\cdot \mathrm{Time}^{0.8337(\pm 0.0120)}\cdot \mathrm{Rate}^{0.1175(\pm 0.0029)}\end{aligned} \tag{17}$$

In a similar model of biaxial stretch applied to dissociated cells, neurons from the hippocampus were significantly more vulnerable to injury than were neurons from the cortex (167). Hippocampal neurons died after 30% stretch suggesting a threshold below this level. In contrast, few cortical neurons died after the same mechanical stimulus, suggesting a higher threshold in corroboration of the slice culture findings [60, 160, 167].

An important consideration in modeling mechanobiological responses in neural cells is the complexity of the loading regime. For instance, all cellular strain fields in the brain, whether physiological or traumatic, will be three-dimensional, an inherent consequence of the three-dimensionality of complex neural morphologies and cell–cell/cell-ECM linkages. Moreover, even for relatively homogeneous bulk loading of the head and brain tissue, local cellular strain distributions may be heterogeneous combinations of tensile, compressive, and shear strains. To investigate the effects of strain field complexity on neuronal responses to high strain rate deformation, both 3D neuronal cultures and 2D neuronal cultures (sandwiched between layers of a 3D matrix) were subjected to bulk mechanical loading

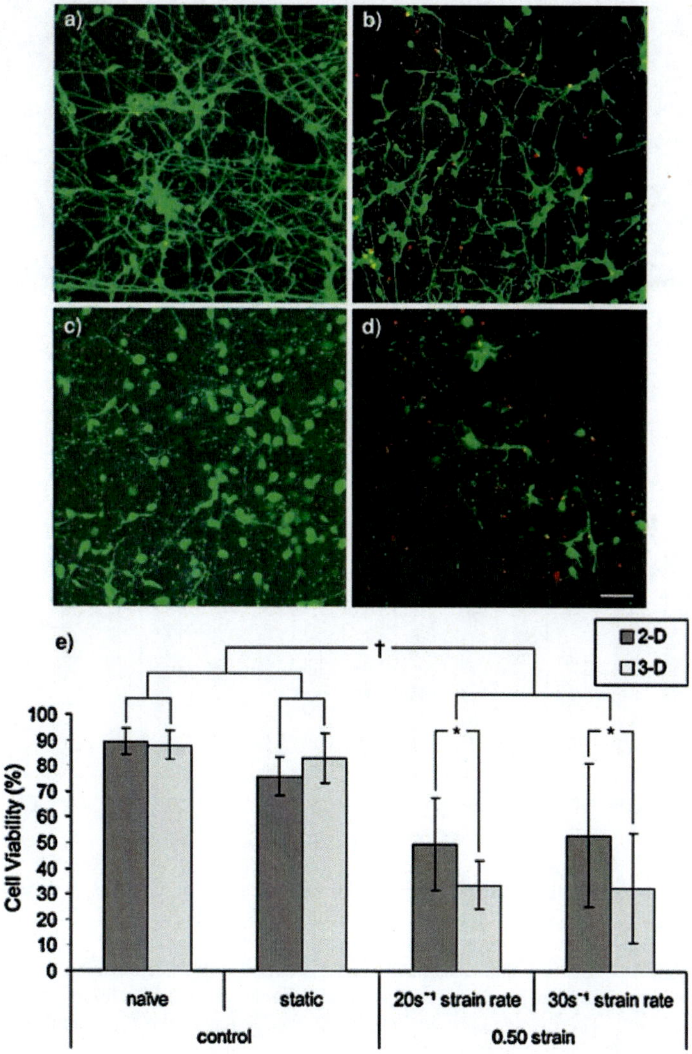

Fig. 5 Cell viability in networks of neurons cultured in 2D and 3D following shear strain. Cells were subjected to high strain rate and magnitude shear deformation using a custom-built device. Viability was assessed 24 h after the application of control or strain conditions (live cells stained in cytosol and appear as larger areas with branching neurite patterns (*green*), nuclei of dead cells stained as punctate smaller areas (*red*)). Confocal reconstructions of neuronal cultures in 2D after **a** static control or **b** 50% strain, $30s^{-1}$ strain rate. Neuronal cultures in 3D after **c** static control or **d** 50% strain, $30s^{-1}$ strain rate (scale bar $= 50\,\mu m$). **e** Graphical representation of cell viability. There was not a significant difference in viability in naive controls compared to static controls for either 2D or 3D culture; however, there was a significant decrease in both 2D and 3D viability after high rate deformation versus uninjured controls (†$p < 0.05$). Moreover, there was a significant decrease in viability in 3D versus 2D under matched loading conditions (*$p < 0.05$), implicating the complexity of the local cellular strain fields in more detrimental outcome following traumatic loading. Error bars represent standard deviation. Reprinted with permission from Mary Ann Liebert [31]

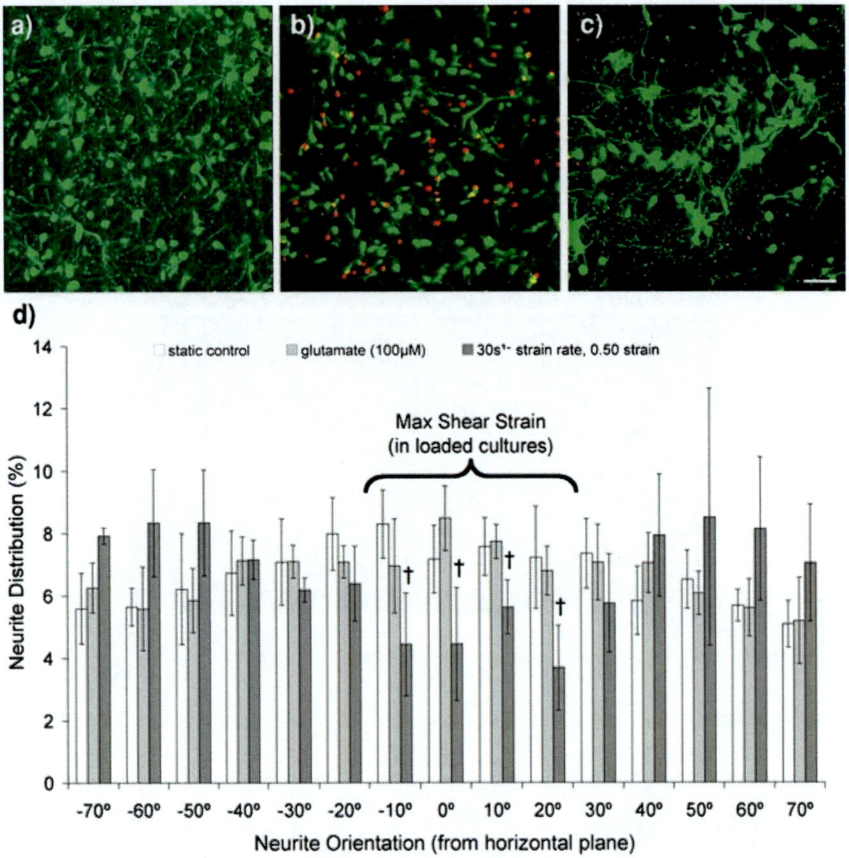

Fig. 6 Neurite loss in 3D neuronal cultures following shear strain. Confocal reconstructions of 3D neuronal cultures at 24 h following **a** sham conditions, **b** excitotoxicity via 100 μM glutamate exposure (positive control), or **c** shear deformation at 50% strain, $30s^{-1}$ strain rate (scale bar = 50 μm). Live neurons and remaining neurites stained with calcein can be seen as rounded structures with typical branching patterns (*green*) and the nuclei of dead cells are seen as punctate bright areas (*red*). There was approximately a 50% reduction in the total number of neurites following either chemical injury (excitotoxicity) or mechanical injury. **d** The orientation of remaining neurites was measured following these conditions. Following excitotoxic injury, the distribution of remaining neurites was consistent across all angles of orientation, indicating indiscriminate neurite loss. However, following high strain rate shear deformation, there was preferential loss of neurites at angles close to the horizontal plane (†$p < 0.05$), correlating with orientations predicted to experience maximum shear strain, suggesting heightened sensitivity to shear strain. Error bars represent standard deviation. Adapted with permission from Elsevier [17]

consistent with tissue-level strains in closed-head TBI [31]. At matched strain magnitude (50% shear strain) and strain rates (20–$30s^{-1}$) to the 3D cell-containing matrices, neuronal death was greater when neurons were cultured in 3D versus monolayer (Fig. 5). These data implicate the complexity of the local cellular strain field, in addition to the magnitude and rate of individual strain components, in thresholds for neuronal death, and also indicate that tolerances were lower in realistic tissue constructs. Moreover, within these 3D constructs, the unique strain fields of individual neurons and neurites (micro-strains) may be calculated based on the cell orientation with respect to the bulk loading (meso-strains) [17]. This analysis revealed that the majority of neurite loss correlated with the direction of maximum shear strain (Fig. 6), demonstrating increased sensitivity to the strain mode in complex loading regimes.

In comparison to neurons, significant astrocyte cell death was observed after 25% shear strain in the same constructs [17]. Although astrocyte death did not significantly increase after 50% shear compared to 25%, astrocyte death significantly increased as the strain rate was increased from 1 to $30s^{-1}$ [14, 17]. Interestingly, high shear strain (50%), quasi-static ($1s^{-1}$) loading did not induce cell death, although this caused significant astrocyte activation measured by proliferation and hypertrophy, suggesting a direct mechanism of mechano-activation in astrocytes. However, at an intermediate strain rate of $10s^{-1}$ inducing significant cell death, astrocyte activation was maximal [14], demonstrating that local cell death may amplify the astrocyte responses to trauma.

Multiple models injure cultured tissue with uniaxial as opposed to biaxial strains, either by fluid shear [151] or substrate deformation [159]. After fluid shear stress of NT-2 cells, a neuronal cell line, significant lactate dehydrogenase (LDH) release was measured immediately after 16% cell strain and at 24 h after 12% strain at strain rates below $7s^{-1}$ [151, 156]. After uniaxial strain applied to hippocampal cultures by substrate deformation, cell death was significantly increased after 50% but not 17% strain, suggesting a threshold above 17%, although in a follow up study, significant cell death was reported after 6% strain [168, 169]. When hippocampal slice cultures were subjected to uniaxial strain, no activated caspase 3 or calpain was detected after 50% strain. Caspase 3 is essential for execution of the intrinsic apoptotic pathway whereas calpain is activated by elevated intracellular calcium and is thought to play a role in necrotic cell death. Calpain was consistently activated after 75 and 100% uniaxial strain, whereas caspase 3 was activated after 75% but not 100% strain [170]. Although cell death was not directly measured, these results suggest that injury may activate different cell-death pathways depending on severity.

The reasons for the discrepancy in cell-death response after uniaxial loading between the fluid shear stress and substrate strain models may be due to experimental considerations. Verification of cell/tissue deformation was absolutely essential for each experiment. Although tissue may appear to be adhered to the substrate after injury, it may be anchored in only one region such that it does not stretch, but does not float away either. In the shear stress model, cell strain during injury was confirmed, whereas in the uniaxial slice model, it was not.

In a comparison of uniaxial and biaxial strain applied to cortical neurons, the post-injury calcium response was significantly greater after 30 and 50% biaxial strain with very few neurons responding to uniaxial strain [171]. Cell membrane permeability was greater after biaxial deformation, as well [171]. Although no cell death was measured 24 h after any injury up to 50% strain, these results suggested that biaxial strain may be a more severe injury than uniaxial strain [171].

The brain carries out essential functions of information processing, motor control, and consciousness which are dependent on neuronal electrophysiological function at the cellular level. One can image many injury mechanisms which may alter neuronal function without requiring cell death. Therefore, an additional tolerance criterion may be more appropriate for nervous system tissue: disruption of electrophysiological function. After either 5 or 10% biaxial strain applied to hippocampal slice cultures, evoked activity was significantly altered [172]. The maximal evoked response (R_{max}) was decreased by up to 75%, and the tissue became less excitable with up to 120% increase in the stimulus required to generate a half maximal response (I_{50}) [172]. These studies were conducted with microelectrode arrays (MEA) which have the advantage of sampling tissue activity from up to 60 locations simultaneously, providing unbiased and comprehensive quantification of activity throughout the neuronal network. MEA recordings are also compatible with dissociated cultures. After uniaxial strain via fluid shear stress, acute changes in activity were noted including an increase in burst interval [173].

6 Conclusions and Future directions

Neural cell and tissue culture models offer a unique platform for studying mechanical response to physiological forces as well as neural trauma, where physical forces may cause structural failure due to excessive stress and strain, or result in sublethal dysfunction. Cells in culture, however, may differ in cell–cell configuration, material properties, cell signaling, and gene expression from in vivo counterparts. Recreating these structural components at the nano-, micro-, meso- and macro-scale, together with physiological and physical properties, allows development of robust culture systems, which in turn, aids in developing both constitutive models and guides for computer simulations. The formulation of analytical models that consider multi-scale tissue characteristics then become very powerful tools for assessment of the biomechanical response to various stimuli.

In vitro models of neural cells have proven to be effective in the systematic identification of specific mechanisms of mechanotransduction and injury causation. The primary attributes of in vitro models include control of cellular composition, control of local environment, accessibility for imaging, experimental manipulability, and elimination of systemic effects. In vitro neural models must mimic the in vivo environment in terms of composition, cell type, extracellular space, cell volume fraction, tortuosity, cell cytoarchitecture, distribution of cell–cell/cell–matrix interactions, and the multicellular composition. Properly designed cellular models may serve as a more accurate guide to cellular growth, interaction, and responses to biochemical and/or mechanical stimuli.

In the continued drive to increase the biofidelity of ex vivo constructs to the in vivo brain, new scaffold materials may provide improved substrates to support higher cell and process density to more closely approximate in vivo brain. Patterning cell types in 3D may allow for replication of the complex connectivity of multiple cell types that typifies the CNS. Our studies have demonstrated the benefits of perfusion for supporting these high cell densities. Microfluidic perfusion systems may hold promise for making these types of cultures more commonplace. Explant culture systems can reproduce a degree of cytoarchitectural complexity in an effort to maintain in vivo characteristics. However, a major limitation is that only post-natal/juvenile tissue can be cultured successfully. Future efforts to culture adult brain tissue would fill a critical need for determining whether tissue-level tolerance criteria are affected by age.

The ex vivo systems provide unfettered access to the cultures at multiple spatial scales. At the tissue level, it has been possible to quantify tissue biomechanics during the applied dynamic event, whether it is injury or mechanical testing. However, dynamic measurement of cellular or subcellular deformation has remained difficult, but could significantly add to our understanding of cell-level biomechanics which may be one source of heterogeneous cellular response. Such insight is essential for multi-scale modeling of TBI with strategies such as representative volume element modeling.

The biological response, as a consequence of mechanotransduction cascades, occurs on a slower time scale than an injurious stimulus. Through the use of fluorescent reporter systems such as optogenetic approaches, continuous monitoring of physiological parameters becomes possible. Correlating the activation of cell death cascades with mechanical injury parameters will further define tolerance criteria and the limits of safe exposures, which are absolutely critical for safety system design.

Analysis of indentation data for the determination of mechanical properties is inherently complex given the heterogeneous strain field beneath the indenter. However, sophisticated numerical analysis methods such as inverse FE approaches could be leveraged to more accurately extract constitutive properties. Such numerical methods explicitly include factors normally ignored in classical indentation solutions. The accuracy of derived constitutive properties is increased by explicitly modeling the complex strain field beneath the indenter, as well. Indentation methods such as these will provide material property data at high spatial resolution, which will be instrumental for the next generation of computation models for simulating TBI and predicting anatomical patterns of cell death. These models, in turn, may be used to design the next generation of safety systems to reduce the socioeconomic costs of TBI.

Acknowledgments The authors thank Varadraj Vernekar, Ph.D. for contributions to the 3D culture mechanical testing study, and Benjamin S. Elkin, Ph.D. for the dynamic AFM indentation studies.

References

1. Kindle, E., Schwartz, J., Jessell, T.: Principles of Neural Science. McGraw-Hill, New York (2000)
2. Sycoma, E., Nicholson, C.: Diffusion in brain extracellular space. Physiol. Rev. **88**, 1277–1340 (2008)
3. Sycoma, E.: Extra synaptic volume transmission and diffusion parameters of the extracellular space. Neuroscience **129**, 861–876 (2004)
4. Sycoma, E.: Diffusion properties of the brain in health and disease. Neurochem. Int. **45**, 453–466 (2004)
5. Fedoroff, S., Richardson, A.: Protocols for Neural Cell Culture, 3rd edn. Humana Press, Totowa (2001)
6. Hayflick, L., Moorhead, P.: The serial cultivation of human diploid cell strains. Exp. Cell Res. **25**, 585–621 (1961)
7. Jacobson, M.: Clonal analysis and cell lineages of the vertebrate central nervous system. Annu. Rev. Neurosci. **8**, 71–102 (1985)
8. Potter, S.M., DeMarse, T.B.: A new approach to neural cell culture for long-term studies. J. Neurosci. Methods **110**, 17–24 (2001)
9. Dichter, M.A.: Rat cortical neurons in cell culture: culture methods, cell morphology, electrophysiology, and synapse formation. Brain. Res. **149**, 279–293 (1978)
10. Corey, J.M., Wheeler, B.C., Brewer, G.J.: Compliance of hippocampal neurons to patterned substrate networks. J. Neurosci. Res. **30**, 300–307 (1991)
11. Choi, H.K., Won, L., Heller, A.: Dopaminergic neurons grown in three-dimensional reaggregate culture for periods of up to one year. J. Neurosci. Methods **46**, 233–244 (1993)
12. Hsiang, J., Heller, A., Hoffmann, P.C., Mobley, W.C., Wainer, B.H.: The effects of nerve growth factor on the development of septal cholinergic neurons in reaggregate cell cultures. Neuroscience **29**, 209–223 (1989)
13. O'Connor, S.M., Andreadis, J.D., Shaffer, K.M., Ma, W., Pancrazio, J.J., Stenger, D.A.: Immobilization of neural cells in three-dimensional matrices for biosensor applications. Biosens. Bioelectron. **14**, 871–881 (2000)
14. Cullen, D.K., Simon, C.M., LaPlaca, M.C.: Strain rate-dependent induction of reactive astrogliosis and cell death in three-dimensional neuronal-astrocytic co-cultures. Brain Res. **1158**, 103–115 (2007)
15. O'Connor, S.M., Stenger, D.A., Shaffer, K.M., Ma, W.: Survival and neurite outgrowth of rat cortical neurons in three-dimensional agarose and collagen gel matrices. Neurosci. Lett. **304**, 189–193 (2001)
16. Woerly, S., Plant, G.W., Harvey, A.R.: Cultured rat neuronal and glial cells entrapped within hydrogel polymer matrices: a potential tool for neural tissue replacement. Neurosci. Lett. **205**, 197–201 (1996)
17. LaPlaca, M.C., Cullen, D.K., McLoughlin, J.J., Cargill 2nd, R.S.: High rate shear strain of three-dimensional neural cell cultures: a new in vitro traumatic brain injury model. J. Biomech. **38**, 1093–1105 (2005)
18. Edelman, D.B., Keefer, E.W.: A cultural renaissance: in vitro cell biology embraces three-dimensional context. Exp. Neurol. **192**, 1–6 (2005)
19. Cukierman, E., Pankov, R., Stevens, D.R., Yamada, K.M.: Taking cell-matrix adhesions to the third dimension. Science **294**, 1708–1712 (2001)
20. Cukierman, E., Pankov, R., Yamada, K.M.: Cell interactions with three-dimensional matrices. Curr. Opin. Cell. Biol. **14**, 633–639 (2002)
21. Schmeichel, K.L., Bissell, M.J.: Modeling tissue-specific signaling and organ function in three dimensions. J. Cell Sci. **116**, 2377–2388 (2003)
22. Yamada, K.M., Pankov, R., Cukierman, E.: Dimensions and dynamics in integrin function. Braz. J. Med. Biol. Res. **36**, 959–966 (2003)

23. Miller, B.E., Miller, F.R., Heppner, G.H.: Factors affecting growth and drug sensitivity of mouse mammary tumor lines in collagen gel cultures. Cancer Res. **45**, 4200–4205 (1985)
24. Grinnell, F.: Fibroblast biology in three-dimensional collagen matrices. Trends Cell Biol. **13**, 264–269 (2003)
25. Balgude, A.P., Yu, X., Szymanski, A., Bellamkonda, R.V.: Agarose gel stiffness determines rate of DRAG neurite extension in 3M cultures. Biomaterials **22**, 1077–1084 (2001)
26. Coates, P.W., Fermini, B., Strahlendorf, J.C., Strahlendorf, H.K.: Utilization of three-dimensional culture for early morphometric and electrophysiological analyses of solitary cerebellar neurons. Dev. Neurosci. **14**, 35–43 (1992)
27. Coates, P.W., Nathan, R.D.: Feasibility of electrical recordings from unconnected vertebrate CNS neurons cultured in a three-dimensional extracellular matrix. J. Neurosci. Methods **20**, 203–210 (1987)
28. O'Shaughnessy, T.J., Lin, H.J., Ma, W.: Functional synapse formation among rat cortical neurons grown on three-dimensional collagen gels. Neurosci. Lett. **340**, 169–172 (2003)
29. Spector, D.H., Boss, B.D., Strecker, R.E.: A model three-dimensional culture system for mammalian dopaminergic precursor cells: application for functional intracerebral transplantation. Exp. Neurol. **124**, 253–264 (1993)
30. Bellamkonda, R., Ranieri, J.P., Aebischer, P.: Laminin oligopeptide derivatized agarose gels allow three-dimensional neurite extension in vitro. J. Neurosci. Res. **41**, 501–509 (1995)
31. Cullen, D.K., LaPlaca, M.C.: Neuronal response to high rate shear deformation depends on heterogeneity of the local strain field. J. Neurotrauma **23**, 1304–1319 (2006)
32. Cullen, D.K., Lessing, M.C., LaPlaca, M.C.: Collagen-dependent neurite outgrowth and response to dynamic deformation in three-dimensional neuronal cultures. Ann Biomed. Eng. **35**, 835–846 (2007)
33. Vukasinovic, J., Cullen, D.K., LaPlaca, M.C., Glezer, A.: A microperfused incubator for tissue mimetic 3D cultures. Biomed. Microdevices **11**, 1155–1165 (2009)
34. Dillon, G.P., Yu, X., Sridharan, A., Ranieri, J.P., Bellamkonda, R.V.: The influence of physical structure and charge on neurite extension in a 3D hydrogel scaffold. J. Biomater. Sci. Polym. Ed. **9**, 1049–1069 (1998)
35. Willits, R.K., Skornia, S.L.: Effect of collagen gel stiffness on neurite extension. J. Biomater. Sci. Polym. Ed. **15**, 1521–1531 (2004)
36. Schense, J.C., Hubbell, J.A.: Three-dimensional migration of neurites is mediated by adhesion site density and affinity. J. Biol. Chem. **275**, 6813–6818 (2000)
37. Yu, T.T., Shoichet, M.S.: Guided cell adhesion and outgrowth in peptide-modified channels for neural tissue engineering. Biomaterials **26**, 1507–1514 (2005)
38. Yu, X., Bellamkonda, R.V.: Dorsal root ganglia neurite extension is inhibited by mechanical and chondroitin sulfate-rich interfaces. J. Neurosci. Res. **66**, 303–310 (2001)
39. Borkenhagen, M., Clemence, J.F., Sigrist, H., Aebischer, P.: Three-dimensional extracellular matrix engineering in the nervous system. J. Biomed. Mater. Res. **40**, 392–400 (1998)
40. Tsacopoulos, M.: Metabolic signaling between neurons and glial cells: a short review. J. Physiol. Paris **96**, 283–288 (2002)
41. Tsacopoulos, M., Magistretti, P.J.: Metabolic coupling between glia and neurons. J. Neurosci. **16**, 877–885 (1996)
42. Aschner, M.: Neuron-astrocyte interactions: implications for cellular energetics and antioxidant levels. Neurotoxicology **21**, 1101–1107 (2000)
43. Pardo, B., Honegger, P.: Differentiation of rat striatal embryonic stem cells in vitro: monolayer culture vs three-dimensional coculture with differentiated brain cells. J. Neurosci. Res. **59**, 504–512 (2000)
44. Pulliam, L., Stubblebine, M., Hyun, W.: Quantification of neurotoxicity and identification of cellular subsets in a three-dimensional brain model. Cytometry **32**, 66–69 (1998)
45. Irons, H.R., Cullen, D.K., Shapiro, N.P., Lambert, N.A., Lee, R.H., LaPlaca, M.C.: Three-dimensional neural constructs: a novel platform for neurophysiological investigation. J. Neural. Eng. **5**, 333–341 (2008)

46. Ahmed, S.M., Rzigalinski, B.A., Willoughby, K.A., Sitterding, H.A., Ellis, E.F.: Stretch-induced injury alters mitochondrial membrane potential and cellular ATP in cultured astrocytes and neurons. J. Neurochem. **74**, 1951–1960 (2000)
47. Steinschneider, R., Delmas, P., Nedelec, J., Gola, M., Bernard, D., Boucraut, J.: Appearance of neurofilament subunit epitopes correlates with electrophysiological maturation in cortical embryonic neurons cocultured with mature astrocytes. Brain. Res. Dev. Brain. Res. **95**, 15–27 (1996)
48. Nakanishi, K., Nakanishi, M., Kukita, F.: Dual intracellular recording of neocortical neurons in a neuron-glia co-culture system. Brain Res. Brain. Res. Protoc. **4**, 105–114 (1999)
49. LaPlaca, M.C., Vernekar, V.N., Shoemaker, J.T., Cullen, D.K.: Three-dimensional neuronal cultures. In: Berthiaume, F., Morgan, J. (eds.) Methods in Bioengineering: 3D Tissue Engineering. Artech House Publishers, London (2010)
50. Nicholson, C., Sykova, E.: Extracellular space structure revealed by diffusion analysis. Trends Neurosci. **21**, 207–215 (1998)
51. Cullen, D.K., Vukasinovic, J., Glezer, A., Laplaca, M.C.: Microfluidic engineered high cell density three-dimensional neural cultures. J. Neural Eng. **4**, 159–172 (2007)
52. Braitenberg, V.: Brain size and number of neurons: an exercise in synthetic neuroanatomy. J. Comput. Neurosci. **10**, 71–77 (2001)
53. Gabbott, P.L., Stewart, M.G.: Distribution of neurons and glia in the visual cortex (area 17) of the adult albino rat: a quantitative description. Neuroscience **21**, 833–845 (1987)
54. Alves, P., Moreira, J., Rodrigues, J., Aunins, J., Carrondo, M.: Two-dimensional versus three-dimensional culture systems: effects on growth and productivity of BHK cells. Biotechnol. Bioeng. **52**, 429–432 (1996)
55. Kleinman, H.K., McGarvey, M.L., Hassell, J.R., Star, V.L., Cannon, F.B., et al.: Basement membrane complexes with biological activity. Biochemistry **25**, 312–318 (1986)
56. Vukicevic, S., Kleinman, H.K., Luyten, F.P., Roberts, A.B., Roche, N.S., Reddi, A.H.: Identification of multiple active growth factors in basement membrane Matrigel suggests caution in interpretation of cellular activity related to extracellular matrix components. Exp. Cell Res. **202**, 1–8 (1992)
57. Morrison III, B., Cater, H.L., Benham, C.D., Sundstrom, L.E.: An in vitro model of traumatic brain injury utilising two-dimensional stretch of organotypic hippocampal slice cultures. J. Neurosci. Meth. **150**, 192–201 (2006)
58. Stoppini, L., Buchs, P.A., Muller, D.: Lesion-induced neurite sprouting and synapse formation in hippocampal organotypic cultures. Neuroscience **57**, 985–994 (1993)
59. Gahwiler, B.H.: Organotypic monolayer cultures of nervous tissue. J. Neurosci. Meth. **4**, 329–342 (1981)
60. Elkin, B.S., Morrison III, B.: Region-specific tolerance criteria for the living brain. Stapp Car Crash J. **51**, 127–138 (2007)
61. Sieg, F., Wahle, P., Pape, H.C.: Cellular reactivity to mechanical axonal injury in an organotypic in vitro model of neurotrauma. J. Neurotrauma **16**, 1197–1213 (1999)
62. Morrison III, B., Eberwine, J.H., Meaney, D.F., McIntosh, T.K.: Traumatic injury induces differential expression of cell death genes in organotypic brain slice cultures determined by complementary DNA array hybridization. Neuroscience **96**, 131–139 (2000)
63. Krassioukov, A.V., Ackery, A., Schwartz, G., Adamchik, Y., Liu, Y., Fehlings, M.G.: An in vitro model of neurotrauma in organotypic spinal cord cultures from adult mice. Brain Res. Brain Res. Protoc. **10**, 60–68 (2002)
64. Stavridis, S.I., Dehghani, F., Korf, H.W., Hailer, N.P.: Characterisation of transverse slice culture preparations of postnatal rat spinal cord: preservation of defined neuronal populations. Histochem. Cell Biol. **123**, 377–392 (2005)
65. Gahwiler, B.H.: Development of the hippocampus in vitro: cell types, synapses, and receptors. Neuroscience **11**, 751–760 (1984)
66. Del. Rio., J.A., Heimrich, B., Soriano, E., Schwegler, H., Frotscher, M.: Proliferation and differentiation of glial fibrillary acidic protein immunoreactive glial cells in organotypic slice cultures of rat hippocampus. Neuroscience **43**, 335–347 (1991)

67. Caeser, M., Aertsen, A.: Morphological organization of rat hippocampal slice cultures. J. Comp. Neurol. **307**, 87–106 (1991)
68. Stoppini, L., Buchs, P.A., Muller, D.: A simple method for organotypic cultutres of nervous tissue. J. Neurosci. Meth. **37**, 173–182 (1991)
69. Buchs, P.A., Stoppini, L., Muller, D.: Structural modifications associated with synaptic development in area CA1 of rat hippocampal organotypic cultures. Brain Res. Dev. Brain Res. **71**, 81–91 (1993)
70. Mielke, J.G., Comas, T., Woulfe, J., Monette, R., Chakravarthy, B., Mealing, G.A.: Cytoskeletal, synaptic, and nuclear protein changes associated with rat interface organotypic hippocampal slice culture development. Brain Res. Dev. Brain Res. **160**, 275–286 (2005)
71. De Simoni, A., Griesinger, C.B., Edwards, F.A.: Development of rat CA1 neurones in acute versus organotypic slices: role of experience in synaptic morphology and activity. J. Physiol. **550**, 135–147 (2003)
72. Collin, C., Miyaguchi, K., Segal, M.: Dendritic spine density and LTP induction in cultured hippocampal slices. J. Neurophysiol. **77**, 1614–1623 (1997)
73. Gahwiler, B.H.: Slice cultures of cerebellar, hippocampal, and hypothalamic tissue. Experientia **40**, 235–243 (1984)
74. Robain, O., Barbin, G., de Billette, V., Jardin, L., Jahchan, T., Ben-Ari, Y.: Development of mossy fiber synapses in hippocampal slice culture. Brain Res. Dev. Brain Res. **80**, 244–250 (1994)
75. Hartel, R., Matus, A.: Cytoskeletal maturation in cultured hippocampal slices. Neuroscience **78**, 1–5 (1997)
76. Bahr, B.A., Kessler, M., Rivera, S., Vanderklish, P.W., Hall, R.A., et al.: Stable maintenance of glutamate receptors and other synaptic components in long-term hippocampal slices. Hippocampus **5**, 425–439 (1995)
77. Martens, U., Wree, A.: Distribution of [3H]MK-801, [3H]AMPA and [3H]kainate binding sites in rat hippocampal long-term slice cultures isolated from external afferents. Anat. Embryol. (Berl) **203**, 491–500 (2001)
78. Morrison III, B., Pringle, A.K., McManus, T., Ellard, J., Bradley, M., et al.: L-Arginyl-3, 4-Spermidine is neuroprotective in several in vitro models of neurodegeneration and in vivo ischaemia without suppressing synaptic transmission. Brit. J. Pharm. **137**, 1255–1268 (2002)
79. Vornov, J.J., Tasker, R.C., Lost, D.: Direct observation of the agonist-specific regional vulnerability to glutamate, NMDA, and kainate neurotoxicity in organotypic hippocampal cultures. Exp. Neurol. **114**, 11–22 (1991)
80. Gutierrez, R., Heinemann, U.: Synaptic reorganization in explanted cultures of rat hippocampus. Brain Res. **815**, 304–316 (1999)
81. Parent, J.M., Yu, T.W., Leibowitz, R.T., Geschwind, D.H., Sloviter, R.S., Lowenstein, D.H.: Dentate granule cell neurogenesis is increased by seizures and contributes to aberrant network reorganization in the adult rat hippocampus. J. Neurosc. Off. J. Soc. Neurosci. **17**, 3727–3738 (1997)
82. Muller, D., Buchs, P.A., Stoppini, L.: Time course of synaptic development in hippocampal organotypic cultures. Brain Res. Mol. Brain Res. **71**, 93–100 (1993)
83. McBain, C.J., Boden, P., Hill, R.G.: Rat hippocampal slices 'in vitro' display spontaneous epileptiform activity following long-term organotypic culture. J. Neurosci. Meth. **27**, 35–49 (1989)
84. Fowler, J., Bornstein, M.B., Crain, S.M.: Sustained hyperexcitability elicited by repetitive electric stimulation of organotypic hippocampal explants. Brain Res. **378**, 398–404 (1986)
85. Xiang, Z., Hrabetova, S., Moskowitz, S.I., Casaccia-Bonnefil, P., Young, S.R., et al.: Long-term maintenance of mature hippocampal slices in vitro. J. Neurosci. Meth. **98**, 145–154 (2000)
86. Finley, M., Fairman, D., Liu, D., Li, P., Wood, A., Cho, S.: Functional validation of adult hippocampal organotypic cultures as an in vitro model of brain injury. Brain Res. **1001**, 125–132 (2004)

87. Wilhelmi, E., Schoder, U.H., Benabdallah, A., Sieg, F., Breder, J., Reymann, K.G.: Organotypic brain-slice cultures from adult rats: approaches for a prolonged culture time. Altern. Lab Anim. **30**, 275–283 (2002)
88. Takhounts, E.G., Eppinger, R.H., Campbell, J.Q., Tannous, R.E., Power, E.D.: On the development of the SIMon finite element head model. Stapp Car Crash J. **47**, 107–133 (2003)
89. Cheng, S., Clarke, E.C., Bilston, L.E.: Rheological properties of the tissues of the central nervous system: a review. Med. Eng. Phys. **30**, 1318–1337 (2008)
90. Hrapko, M., van Dommelen, J.A., Peters, G.W., Wismans, J.S.: The influence of test conditions on characterization of the mechanical properties of brain tissue. J. Biomech. Eng. **130**, 031003 (2008)
91. Fallenstein, G.T., Hulce, V.D.: Dynamic mechanical properties of human brain tissue. J. Biomech. **2**, 217–226 (1969)
92. Arbogast, K.B., Margulies, S.S.: Material characterization of the brainstem from oscillatory shear tests. J. Biomech. **31**, 801–807 (1998)
93. Nicolle, S., Lounis, M., Willinger, R., Palierne, J.F.: Shear linear behavior of brain tissue over a large frequency range. Biorheology **42**, 209–223 (2005)
94. Bilston, L.E., Liu, Z., Phan-Thien, N.: Large strain behaviour of brain tissue in shear: some experimental data and differential constitutive model. Biorheology **38**, 335–345 (2001)
95. Takhounts E., Crandall, J.R., Darvish, K.K.: On the importance of nonlinearity of brain tissue under large deformations. Stapp Car Crash J. **47**, 79–92 (2003)
96. Tamura, A., Hayashi, S., Watanabe, I., Nagayama, K., Matsumoto, T.: Mechanical characterization of brain tissue in high-rate compression. J. Biomech. Sci. Eng. **2**, 115–126 (2007)
97. Cheng, S., Bilston, L.E.: Unconfined compression of white matter. J. Biomech. **40**, 117–124 (2007)
98. Miller, K., Chinzei, K.: Mechanical properties of brain tissue in tension. J. Biomech. **35**, 483–490 (2002)
99. Franceschini, G., Bigoni, D., Regitnig, P., Holzapfel, G.A.: Brain tissue deforms similarly to filled elastomers and follows consolidation theory. J. Mech. Phys. Solids **54**, 2592–2620 (2006)
100. Mao, H., Jin, X., Zhang, L., Yang, K.H., Igarashi, T., et al.: Finite element analysis of controlled cortical impact-induced cell loss. J. Neurotrauma **27**, 877–888 (2010)
101. Hicks, R., Soares, H., Smith, D., McIntosh, T.: Temporal and spatial characterization of neuronal injury following lateral fluid-percussion brain injury in the rat. Acta Neuropathol. **91**, 236–246 (1996)
102. van Dommelen, J.A., van der Sande, T.P., Hrapko, M., Peters, G.W.: Mechanical properties of brain tissue by indentation: interregional variation. J. Mech. Behav. Biomed. Mater. **3**, 158–166 (2010)
103. Gefen, A., Margulies, S.S.: Are in vivo and in situ brain tissues mechanically similar? J. Biomech. **37**, 1339–1352 (2004)
104. Elkin, B.S., Azeloglu, E.U., Costa, K.D., Morrison III, B.: Mechanical heterogeneity of the rat hippocampus measured by AFM indentation. J. Neurotrauma **24**, 812–822 (2007)
105. Elkin, B.S., Ilankovan, A., Morrison, III B.: Age-dependent regional mechanical properties of the rat hippocampus and cortex. J. Biomech. Eng. **132**, 011010 (2010)
106. Elkin, B.S., Ilankovan, A., Morrison, III B.: A detailed viscoelastic characterization of the P17 and adult rat brain. J. Neurotrauma (2011, in press)
107. Tripathy, S., Berger, E.J.: Measuring viscoelasticity of soft samples using atomic force microscopy. J. Biomech. Eng. **131**, 094507 (2009)
108. Neubert, H.K.P.: A simple model representing internal damping in solid materials. Aeronaut. Quart. **14**, 187–210 (1963)
109. Schiessel, H., Metzler, R., Blumen, A., Nonnenmacher, T.F.: Generalized viscoelastic models: their fractional equations with solutions. J. Phys. A Math. Gen. **28**, 6567–6584 (1995)

110. Harding, J.W., Sneddon, I.N.: The elastic stresses produced by the indentation of the plane surface of a semi-infinite elastic solid by a rigid punch. Proc. Camb. Philol. Soc. **41**, 16–26 (1945)
111. Cheng, L., Xia, X., Yu, W., Scriven, L.E., Gerberich, W.W.: Flat-punch indentaiton of viscoelastic material. J. Polym. Sci. B **38**, 10–22 (2000)
112. Hayes, W.C., Keer, L.M., Herrmann, G., Mockros, L.F.: A mathematical analysis for indentation tests of articular cartilage. J. Biomech. **5**, 541–551 (1972)
113. Lu, Y.B., Franze, K., Seifert, G., Steinhauser, C., Kirchhoff, F., et al.: Viscoelastic properties of individual glial cells and neurons in the CNS. Proc. Natl. Acad. Sci. U.S.A **103**, 17759–17764 (2006)
114. Darvish, K.K., Crandall, J.R.: Nonlinear viscoelastic effects in oscillatory shear deformation of brain tissue. Med. Eng. Phys. **23**, 633–645 (2001)
115. Donnelly, B.R., Medige, J.: Shear properties of human brain tissue. J. Biomech. Eng. **119**, 423–432 (1997)
116. Rivlin, R.S.: Large elastic deformations of isotropic material I fundamental concepts. Philol. Trans. R. Soc. Lond. A **240**, 459–490 (1948)
117. Mooney, M.: A theory of large elastic deformation. J. Appl. Phys. **11**, 582–592 (1940)
118. Adkins, J.E., Rivlin, R.S.: Large elastic deformations of isotropic materials IX. The deformation of thin shells. Philol. Trans. R. Soc. Lond. A **244**, 505–531 (1952)
119. Ogden, R.W.: Large deformation isotropic elasticity—correlation of theory and experiment for compressible rubberlike solids. Proc. R. Soc. Lond. A **328**, 567–583 (1972)
120. Gefen, A., Gefen, N., Zhu, Q., Raghupathi, R., Margulies, S.S.: Age-dependent changes in material properties of the brain and braincase of the rat. J. Neurotrauma **20**, 1163–1177 (2003)
121. Prange, M.T., Margulies, S.S.: Regional, directional, and age-dependent properties of the brain undergoing large deformation. J. Biomech. Eng. **124**, 244–252 (2002)
122. Ting, T.C.T.: Contact stresses between a rigid indenter and a viscoelastic half-space. J. Appl. Mech. **33**, 845–854 (1966)
123. Segedin, C.M.: The relationship between load and penetration for a spherical punch. Mathematika **4**, 156–161 (1957)
124. Sneddon, I.N.: The relation between load and penetration in the axisymmetric Boussinesq problem for a punch of arbitrary profile. Int. J. Eng. Sci. **3**, 47–57 (1965)
125. Mahaffy, R.E., Park, S., Gerde, E., Kas, J., Shih, C.K.: Quantitative analysis of the viscoelastic properties of thin regions of fibroblasts using atomic force microscopy. Biophys. J. **86**, 1777–1793 (2004)
126. Mahaffy, R.E., Shih, C.K., MacKintosh, F.C., Kas, J.: Scanning probe-based frequency-dependent microrheology of polymer gels and biological cells. Phys. Rev. Lett. **85**, 880–883 (2000)
127. Alcaraz, J., Buscemi, L., Puig-de-Morales, M., Colchero, J., Baro, A., Navajas, D.: Correction of microrheological measurements of soft samples with atomic force microscopy for the hydrodynamic drag on the cantilever. Langmuir **18**, 716–721 (2002)
128. Lodge AS. 1964. *Elastic Liquids*. London: Academic Press
129. Wilhelm, M.: Fourier-transform rheology. Macromol. Mater. Eng. **287**, 83–105 (2002)
130. Cho, K.S., Hyun, K., Ahn, K.H., Lee, S.J.: A geometrical interpretation of large amplitude oscillatory shear response. J. Rheol. **49**, 747–758 (2005)
131. Ewoldt, R.H., Hosoi, A.E., McKinley, G.H.: New measures for characterizing nonlinear viscoelasticity in large amplitude oscillatory shear. J. Rheol. **52**, 1427–1458 (2008)
132. Coats, B., Margulies, S.S.: Material properties of porcine parietal cortex. J. Biomech. **39**, 2521–2525 (2006)
133. Thibault, K.L., Margulies, S.S.: Age-dependent material properties of the porcine cerebrum: effect on pediatric inertial head injury criteria. J. Biomech. **31**, 1119–1126 (1998)
134. Gillespie, P.G., Muller, U.: Mechanotransduction by hair cells: models, molecules, and mechanisms. Cell **139**, 33–44 (2009)

135. Lin, Y.W., Cheng, C.M., Leduc, P.R., Chen, C.C.: Understanding sensory nerve mechanotransduction through localized elastomeric matrix control. PLoS One **4**, e4293 (2009)
136. Stabenfeldt, S.E., Garcia, A.J., LaPlaca, M.C.: Thermoreversible laminin-functionalized hydrogel for neural tissue engineering. J. Biomed. Mater. Res. A **77A**, 718–725 (2006)
137. Dai, W.G., Belt, J., Saltzman, W.M.: Cell-binding peptides conjugated to poly(ethylene glycol) promote neural cell-aggregation. Bio-Technology **12**, 797–801 (1994)
138. Hern, D.L., Hubbell, J.A.: Incorporation of adhesion peptides into nonadhesive hydrogels useful for tissue resurfacing. J. Biomed. Mater. Res. **39**, 266–276 (1998)
139. Mahoney, M.J., Anseth, K.S.: Three-dimensional growth and function of neural tissue in degradable polyethylene glycol hydrogels. Biomaterials **27**, 2265–2274 (2006)
140. Georges, P.C., Miller, W.J., Meaney, D.F., Sawyer, E.S., Janmey, P.A.: Matrices with compliance comparable to that of brain tissue select neuronal over glial growth in mixed cortical cultures. Biophys. J. **90**, 3012–3018 (2006)
141. Flanagan, L.A., Ju, Y.E., Marg, B., Osterfield, M., Janmey, P.A.: Neurite branching on deformable substrates. Neuroreport **13**, 2411–2415 (2002)
142. Guo, W.H., Frey, M.T., Burnham, N.A., Wang, Y.L.: Substrate rigidity regulates the formation and maintenance of tissues. Biophys. J. **90**, 2213–2220 (2006)
143. Pelham Jr., R.J., Wang, Y.: Cell locomotion and focal adhesions are regulated by substrate flexibility. Proc. Natl. Acad. Sci. U.S.A. **94**, 13661–13665 (1997)
144. Engler, A.J., Griffin, M.A., Sen, S., Bonnemann, C.G., Sweeney, H.L., Discher, D.E.: Myotubes differentiate optimally on substrates with tissue-like stiffness: pathological implications for soft or stiff microenvironments. J. Cell. Biol. **166**, 877–887 (2004)
145. Georges, P.C., Janmey, P.A.: Cell type-specific response to growth on soft materials. J. Appl. Physiol. **98**, 1547–1553 (2005)
146. Slemmer, J.E., Matser, E.J., de Zeeuw, C.I., Weber, J.T.: Repeated mild injury causes cumulative damage to hippocampal cells. Brain **125**, 2699–2709 (2002)
147. Balentine, J.D., Greene, W.B., Bornstein, M.: In vitro spinal cord trauma. Lab. Invest. **58**, 93–99 (1988)
148. Regan, R.F., Choi, D.W.: The effect of NMDA, AMPA/kainate, and calcium channel antagonists on traumatic cortical neuronal injury in culture. Brain Res. **633**, 236–242 (1994)
149. Cargill, R.S., Thibault, L.E.: Acute alterations in [Ca2 +]i in NG108–15 cells subjected to high strain rate deformation and chemical hypoxia: An in vitro model for neural trauma. J. Neurotrauma **13**, 395–407 (1996)
150. Ellis, E.F., McKinney, J.S., Willoughby, K.A., Liang, S., Povlishock, J.T.: A new model for rapid stretch-induced injury of cells in culture: Characterization of the model using astrocytes. J. Neurotrauma **12**, 325–339 (1995)
151. LaPlaca, M.C., Thibault, L.E.: An in vitro traumatic injury model to examine the response of neurons to a hydrodynamicaly induced deformation. Ann. Biomed. Eng. **25**, 665–677 (1997)
152. Morrison III, B., Meaney, D.F., McIntosh, T.K.: Mechanical characterization of an in vitro device designed to quantitatively injure living brain tissue. Ann. Biomed. Eng. **26**, 381–390 (1998)
153. Smith, D.H., Wolf, J.A., Lusardi, T.A., Lee, V.M., Meaney, D.F.: High tolerance and delayed elastic response of cultured axons to dynamic stretch injury. J. Neurosci. **19**, 4263–4269 (1999)
154. Morrison III, B., Cater, H.L., Wang, C.C., Thomas, F.C., Hung, C.T., et al.: A tissue level tolerance criterion for living brain developed with an in vitro model of traumatic mechanical loading. Stapp Car Crash J. **47**, 93–105 (2003)
155. Elkin, B.S., Morrison 3rd, B.: Region-specific tolerance criteria for the living brain. Stapp Car Crash J. **51**, 127–138 (2007)
156. LaPlaca, M.C., Lee, V.M., Thibault, L.E.: An in vitro model of traumatic neuronal injury: loading rate-dependent changes in acute cytosolic calcium and lactate dehydrogenase release. J. Neurotrauma **14**, 355–368 (1997)

157. Geddes, D.M., Cargill 2nd, R.S., LaPlaca, M.C.: Mechanical stretch to neurons results in a strain rate and magnitude-dependent increase in plasma membrane permeability. J. Neurotrauma **20**, 1039–1049 (2003)
158. Pfister, B.J., Weihs, T.P., Betenbaugh, M., Bao, G.: An in vitro uniaxial stretch model for axonal injury. Ann. Biomed. Eng. **31**, 589–598 (2003)
159. Lusardi, T.A., Rangan, J., Sun, D., Smith, D.H., Meaney, D.F.: A device to study the initiation and propagation of calcium transients in cultured neurons after mechanical stretch. Ann. Biomed. Eng. **32**, 1546–1558 (2004)
160. Cater, H.L., Sundstrom, L.E., Morrison III, B.: Temporal development of hippocampal cell death is dependent on tissue strain but not strain rate. J. Biomech. **39**, 2810–2818 (2006)
161. Bain, A.C., Meaney, D.F.: Tissue-level thresholds for axonal damage in an experimental model of central nervous system white matter injury. J. Biomech. Eng. **122**, 615–622 (2000)
162. Galle, B., Ouyang, H., Shi, R., Nauman, E.: Correlations between tissue-level stresses and strains and cellular damage within the guinea pig spinal cord white matter. J. Biomech. **40**, 3029–3033 (2007)
163. Barbee, K.A.: Mechanical cell injury. Ann. N. Y. Acad. Sci. **1066**, 67–84 (2005)
164. Kumaria, A., Tolias, C.M.: In vitro models of neurotrauma. Br. J. Neurosurg **22**, 200–206 (2008)
165. LaPlaca, M.C., Simon, C.M., Prado, G.R., Cullen, D.K.: CNS injury biomechanics and experimental models. Prog. Brain Res. **161**, 13–26 (2007)
166. Morrison III, B., Saatman, K.E., Meaney, D.F., McIntosh, T.K.: In vitro central nervous system models of mechanically induced trauma: a review. J. Neurotrauma **15**, 911–928 (1998)
167. Geddes, D.M., LaPlaca, M.C., Cargill, R.S.: Susceptibility of hippocampal neurons to mechanically induced injury. Exp. Neurol. **184**, 420–427 (2003)
168. Lusardi, T.A., Wolf, J.A., Putt, M.E., Smith, D.H., Meaney, D.F.: Effect of acute calcium influx after mechanical stretch injury in vitro on the viability of hippocampal neurons. J. Neurotrauma **21**, 61–72 (2004)
169. Lusardi, T.A., Smith, D.H., Wolf, J.A., Meaney, D.F.: The separate roles of calcium and mechanical forces in mediating cell death in mechanically injured neurons. Biorheology **40**, 401–409 (2003)
170. DeRidder, M.N., Simon, M.J., Siman, R., Auberson, Y.P., Raghupathi, R., Meaney, D.F.: Traumatic mechanical injury to the hippocampus in vitro causes regional caspase-3 and calpain activation that is influenced by NMDA receptor subunit composition. Neurobiol. Dis. **22**, 165–176 (2006)
171. Geddes-Klein, D.M., Schiffman, K.B., Meaney, D.F.: Mechanisms and consequences of neuronal stretch injury in vitro differ with the model of trauma. J. Neurotrauma **23**, 193–204 (2006)
172. Yu, Z., Morrison III, B.: Experimental mild traumatic brain injury induces functional alteration of the developing hippocampus. J. Neurophysiol. **103**, 499–510 (2010)
173. Prado, G.R., Ross, J.D., Deweerth, S.P., LaPlaca, M.C.: Mechanical trauma induces immediate changes in neuronal network activity. J. Neural. Eng. **2**, 148–158 (2005)

Author Index

Stud Mechanobiol Tissue Eng Biomater (2011) 3: 287–288
DOI: 10.1007/978-3-642-13890-4